Building Business-Ready Generative AI Systems

Build Human-Centered AI Systems with Context Engineering, Agents, Memory, and LLMs for Enterprise

Denis Rothman

Building Business-Ready Generative AI Systems

Portfolio Director: Gebin George
Relationship Lead: Ali Abidi
Project Manager: Prajakta Naik
Content Engineer: Tanya D'cruz
Technical Editor: Rahul Limbachiya
Copy Editor: Safis Editing
Indexer: Hemangini Bari
Proofreader: Tanya D'cruz
Production Designer: Shantanu Zagade
Growth Lead: Nimisha Dua
Marketing Owner: Dipali Malwatkar

First published: July 2025

Production reference: 2180925

Published by Packt Publishing Ltd.
Grosvenor House
11 St Paul's Square
Birmingham
B3 1RB, UK.

ISBN 978-1-83702-069-0
www.packtpub.com

I would like to dedicate this book to my family and friends who are my source of happiness.

– Denis Rothman

Contributors

About the author

Denis Rothman is a graduate of Sorbonne University and Paris-Diderot University. He pioneered one of the first patented word2matrix embedding algorithms and AI-powered conversational agents. Early in his career, Denis developed a cognitive NLP chatbot adopted by Moët & Chandon and other global brands as an automated language trainer. He later created an AI resource optimizer for Airbus (formerly Aérospatiale), which was implemented by IBM and leading companies in the apparel industry. His Advanced Planning and Scheduling (APS) solution, used worldwide, has shaped supply chain intelligence across sectors.

Through his books, Denis shares his experience of innovation with a global community of thinkers, builders, and learners committed to shaping AI with purpose.

About the reviewers

Martin Yanev is an experienced software engineer who has spent nearly a decade tackling complex challenges in fields like aerospace and medical technology. As an instructor and professor of computer science at Fitchburg State University, Martin is deeply familiar with OpenAI's suite of APIs and excels at building, training, and refining practical AI systems. He is also a recognized author who enjoys making AI development accessible to others.

Leonid Ganeline is a machine learning engineer with extensive experience in natural language processing. He has worked at several startups, creating AI applications, models, and production systems. He is an active contributor to LangChain, with a particular interest in model quantization and AI agent architectures.

Lavanya Gupta is a Carnegie Mellon University (CMU) alumna from the Language Technologies Institute (LTI). She currently works as a Senior AI/ML Applied Associate at JPMorgan Chase, within their specialized Machine Learning Center of Excellence (MLCOE). She has delivered several talks at events and conferences, including WiDS, PyData, SciPy, Illuminate AI, and the TensorFlow User Group. She also actively supports early-career professionals through mentorship programs with Anita Borg and Women in Coding and Data Science (WiCDS).

Subscribe for a Free eBook

New frameworks, evolving architectures, research drops, production breakdowns—*AI_Distilled* filters the noise into a weekly briefing for engineers and researchers working hands-on with LLMs and GenAI systems. Subscribe now and receive a free eBook, along with weekly insights that help you stay focused and informed.

Subscribe at `https://packt.link/TRO5B` or scan the QR code below.

Table of Contents

Chapter 5: Adding Multimodal, Multifunctional Reasoning with Chain of Thought 151

Chapter 6: Reasoning E-Marketing AI Agents 189

Preface

In the rapidly evolving AI landscape, standalone **large language models** (**LLMs**) alone don't deliver business value. This comprehensive guide enables you to become a generative AI architect, building powerful ChatGPT-grade systems from scratch that are adaptable to any platform. You'll go beyond basic chatbots, developing systems capable of semantic and sentiment analysis—with context-aware AI controllers at their core.

You'll walk through the complete journey of designing an AI controller architecture with multi-user memory retention, enabling the system to adapt to diverse user and system inputs. Building on this foundation, you'll architect a dynamic **retrieval-augmented generation** (**RAG**) system using Pinecone, designed to intelligently combine instruction scenarios.

With powerful multimodal capabilities—including image generation, voice interactions, and machine-driven reasoning—you'll expand your system's intelligence using chain-of-thought orchestration to tackle the cross-domain automation challenges that real business environments demand. You'll also be fully equipped to integrate cutting-edge models such as OpenAI's LLMs and DeepSeek-R1 into your AI system—confident that it will remain stable, scalable, and production-ready even in the turbulent pace of today's AI ecosystem.

By the end, your **generative AI system** (**GenAISys**) will be capable of trajectory analysis and human mobility prediction, even when working with incomplete data. It will bring neuroscience-inspired insight to your marketing messages, integrate seamlessly into human workflows, visualize complex delivery and movement scenarios, and connect to live external data—all wrapped in a polished, investor-ready interface.

Who this book is for

This book is for AI and machine learning engineers, software architects, and enterprise developers seeking to architect and build a comprehensive GenAISys from scratch for enterprise applications. It will particularly benefit those interested in building AI agents, creating advanced orchestration systems, and leveraging AI for automation in marketing, production, and logistics. Software architects and enterprise developers looking to build scalable AI-driven systems will also find immense value in this guide. No prior superintelligence experience is necessary, but familiarity with AI concepts is recommended.

What this book covers

Chapter 1, Defining a Business-Ready Generative AI System, explains how to design robust, ChatGPT-level AI systems tailored for business use, incorporating AI controllers, agents, contextual awareness, advanced memory retention, and strategic human collaboration. You will explore how to go beyond basic model integrations to build AI solutions that deliver real-world business results across industries and become an architect who can adapt to any environment.

Chapter 2, Building the Generative AI Controller, dives into hands-on development of adaptive AI controllers, combining conversational agents and orchestrators. You'll become proficient in building a GenAISys that dynamically responds to user input, leverages advanced memory management, and orchestrates tasks such as sentiment and semantic analysis, which are essential for real-world, data-intensive applications.

Chapter 3, Integrating Dynamic RAG into the GenAISys, teaches you how to architect a scalable, dynamic RAG system using Pinecone, effectively combining instruction scenarios and classical data. You will discover how to build adaptable generative AI solutions capable of responding rapidly to real-world business disruptions and domain-specific challenges.

Chapter 4, Building the AI Controller Orchestration Interface, shows how to build a collaborative, event-driven generative AI interface designed for rapid, real-world business responses. We'll create a practical, multi-user conversational AI that integrates seamlessly into human workflows, enhancing decision-making during critical scenarios.

Chapter 5, Adding Multimodal, Multifunctional Reasoning with Chain of Thought, enhances our GenAISys with OpenAI's powerful multimodal capabilities, including image generation, voice interactions, and machine-driven reasoning. We'll expand our system's intelligence using chain-of-thought orchestration—enabling smart automation tailored to complex, cross-domain business scenarios.

Chapter 6, Reasoning E-Marketing AI Agents, harnesses the power of AI-driven memory analysis to create marketing messages customers remember. We'll build a consumer memory agent that intelligently analyzes reviews and emotions, producing personalized marketing content enhanced by multimodal reasoning—bringing neuroscience insights directly into our GenAISys.

Chapter 7, Enhancing the GenAISys with DeepSeek, navigates the rapid evolution of AI without compromising stability. You will learn how to strategically integrate cutting-edge models such as DeepSeek-R1 into our GenAISys, using a flexible handler selection mechanism—enabling our system to remain adaptable, scalable, and production-ready in a fast-changing market.

Chapter 8, GenAISys for Trajectory Simulation and Prediction, brings advanced AI-driven trajectory analysis into our GenAISys to precisely predict human mobility and deliveries, even with incomplete data. We'll build an intuitive pipeline that leverages LLMs to simulate, analyze, and visualize complex delivery and movement scenarios, transforming spatial data into actionable insights.

Chapter 9, Upgrading the GenAISys with Data Security and Moderation for Customer Service, connects our GenAISys securely to real-world data by integrating live weather information, travel information, robust moderation tools, and advanced security features. You will discover how to safely extend our AI's capabilities beyond internal data, enabling personalized marketing, dynamic activity recommendations, and real-time customer interactions.

Chapter 10, Presenting Your Business-Ready Generative AI System, will show you how to turn your GenAISys proof of concept into a compelling, investor-ready showcase. We'll learn how to effectively present our AI system with clear messaging, practical demos, and a polished user interface, capturing audience attention, demonstrating value, and proving readiness to scale in a crowded AI marketplace.

To get the most out of this book

You don't need to be an AI expert to benefit from this guide, just some familiarity with the basics of artificial intelligence and programming. If you've worked with Python or explored LLMs like ChatGPT, you'll feel right at home.

This book introduces concepts like AI agent orchestration, memory retention, and RAG in clear, hands-on steps. Whether you're a developer, engineer, or curious technologist, you'll find tools, code, and explanations designed to support real-world applications.

No deep math or theoretical background is required, only a desire to build meaningful AI systems that solve business problems and scale in production environments.

The code is available in Google Colab notebooks with automated installations or you can download the code locally.

Download the example code files

The code bundle for the book is hosted on GitHub at `https://github.com/Denis2054/Building-Business-Ready-Generative-AI-Systems`. We also have other code bundles from our rich catalog of books and videos available at `https://github.com/PacktPublishing`. Check them out!

Download the color images

We also provide a PDF file that has color images of the screenshots/diagrams used in this book. You can download it here: `https://packt.link/gbp/9781837020690`.

Conventions used

There are a number of text conventions used throughout this book.

`CodeInText`: Indicates code words in text, database table names, folder names, filenames, file extensions, pathnames, dummy URLs, user input, and Twitter/X handles. For example: "Next, we incorporate `instruct_selector` into the existing interface layout (`VBox`)."

A block of code is set as follows:

```
# Ensure 'Instructions' exists in the memory_selector options
instruct_selector = Dropdown(
    options=["None","Analysis", "Generation"],
    value="None",  # Ensure default active_memory is in the options
    description='Reasoning:',
    layout=Layout(width='50%')
)
```

Any command-line input or output is written as follows:

```
Response: The dialog begins by explaining the formation of Hawaii's
volcanic islands as the Pacific Plate moves over a stationary hotspot,
leading to active volcanoes like Kilauea….
```

Bold: Indicates a new term, an important word, or words that you see on the screen. For instance, words in menus or dialog boxes appear in the text like this. For example: "The user then reenters the sentence, but this time with the **Generation** option and the **Files** option checked, so that the image generated with the text will be displayed."

Warnings or important notes appear like this.

Tips and tricks appear like this.

Get in touch

Feedback from our readers is always welcome.

General feedback: If you have questions about any aspect of this book or have any general feedback, please email us at `customercare@packt.com` and mention the book's title in the subject of your message.

Errata: Although we have taken every care to ensure the accuracy of our content, mistakes do happen. If you have found a mistake in this book, we would be grateful if you reported this to us. Please visit `http://www.packt.com/submit-errata`, click **Submit Errata**, and fill in the form.

Piracy: If you come across any illegal copies of our works in any form on the internet, we would be grateful if you would provide us with the location address or website name. Please contact us at `copyright@packt.com` with a link to the material.

If you are interested in becoming an author: If there is a topic that you have expertise in and you are interested in either writing or contributing to a book, please visit `http://authors.packt.com/`.

Share your thoughts

Once you've read *Building Business-Ready Generative AI Systems*, we'd love to hear your thoughts! Scan the QR code below to go straight to the Amazon review page for this book and share your feedback.

`https://packt.link/r/1837020698`

Your review is important to us and the tech community and will help us make sure we're delivering excellent quality content.

Join our Discord and Reddit space

You're not the only one navigating fragmented tools, constant updates, and unclear best practices. Join a growing community of professionals exchanging insights that don't make it into documentation.

Stay informed with updates, discussions, and behind-the-scenes insights from our authors. Join our Discord at `https://packt.link/z8ivB` or scan the QR code below: 	Connect with peers, share ideas, and discuss real-world GenAI challenges. Follow us on Reddit at `https://packt.link/0rExL` or scan the QR code below: 

Your Book Comes with Exclusive Perks – Here's How to Unlock Them

Unlock this book's exclusive benefits now

Scan this QR code or go to `https://packtpub.com/unlock`, then search this book by name. Ensure it's the correct edition.

Note: Keep your purchase invoice ready before you start.

Enhanced reading experience with our Next-gen Reader:

Multi-device progress sync: Learn from any device with seamless progress sync.

Highlighting and notetaking: Turn your reading into lasting knowledge.

Bookmarking: Revisit your most important learnings anytime.

Dark mode: Focus with minimal eye strain by switching to dark or sepia mode.

Learn smarter using our AI assistant (Beta):

Summarize it: Summarize key sections or an entire chapter.

✦ **AI code explainers:** In the next-gen Packt Reader, click the **Explain** button above each code block for AI-powered code explanations.

Note: The AI assistant is part of next-gen Packt Reader and is still in beta.

Learn anytime, anywhere:

Access your content offline with DRM-free PDF and ePub versions—compatible with your favorite e-readers.

Unlock Your Book's Exclusive Benefits

Your copy of this book comes with the following exclusive benefits:

Next-gen Packt Reader

✦ AI assistant (beta)

DRM-free PDF/ePub downloads

Use the following guide to unlock them if you haven't already. The process takes just a few minutes and needs to be done only once.

How to unlock these benefits in three easy steps

Step 1

Keep your purchase invoice for this book ready, as you'll need it in *Step 3*. If you received a physical invoice, scan it on your phone and have it ready as either a PDF, JPG, or PNG.

For more help on finding your invoice, visit `https://www.packtpub.com/unlock-benefits/help`.

Note: Did you buy this book directly from Packt? You don't need an invoice. After completing Step 2, you can jump straight to your exclusive content.

Step 2

Scan this QR code or go to `https://packtpub.com/unlock`.

On the page that opens (which will look similar to *Figure 0.1* if you're on desktop), search for this book by name. Make sure you select the correct edition.

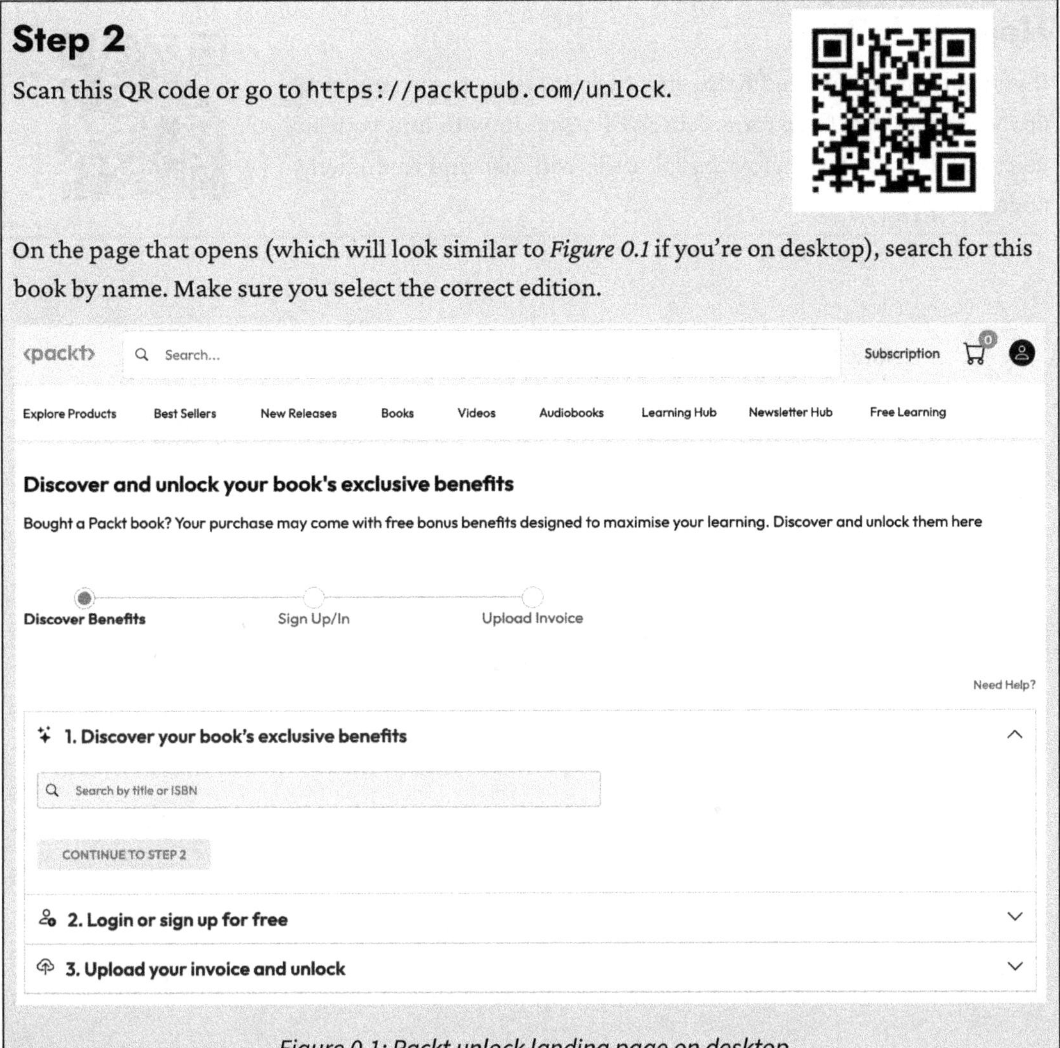

Figure 0.1: Packt unlock landing page on desktop

Step 3

Sign in to your Packt account or create a new one for free. Once you're logged in, upload your invoice. It can be in PDF, PNG, or JPG format and must be no larger than 10 MB. Follow the rest of the instructions on the screen to complete the process.

Need help?

If you get stuck and need help, visit `https://www.packtpub.com/unlock-benefits/help` for a detailed FAQ on how to find your invoices and more. The following QR code will take you to the help page directly:

Note: If you are still facing issues, reach out to `customercare@packt.com`.

1

Defining a Business-Ready Generative AI System

Implementing a **generative AI system (GenAISys)** in an organization doesn't stop at simply integrating a standalone model such as GPT, Grok, Llama, or Gemini via an API. While this is often a starting point, we often mistake it as the finish line. The rising demand for AI, as it expands across all domains, calls for the implementation of advanced AI systems that go beyond simply integrating a prebuilt model.

A business-ready GenAISys should provide ChatGPT-grade functionality in an organization, but also go well beyond it. Its capabilities and features must include **natural language understanding (NLU)**, contextual awareness through memory retention across dialogues in a chat session, and agentic functions such as autonomous image, audio, and document analysis and generation. This requires thoughtful **context engineering**, where we strategically manage the information given to the model to guide its responses. Think of a generative AI model as an entity with a wide range of functions, including AI agents as agentic co-workers.

We will begin the chapter by defining what a business-ready GenAISys is. From there, we'll focus on the central role of a generative AI model, such as GPT-4o, that can both orchestrate and execute tasks. Building on that, we will lay the groundwork for contextual awareness and memory retention, discussing four types of generative AI memory: memoryless, short-term, long-term, and multiple sessions. We will also define a new approach to **retrieval-augmented generation (RAG)** that introduces an additional dimension to data retrieval: instruction and agentic reasoning scenarios. Adding instructions stored in a vector store takes RAG to another level by retrieving instructions that we can add to a prompt. In parallel, we will examine a critical component of a

GenAISys: human roles. We will see how, throughout its life cycle, an AI system requires human expertise. Additionally, we will define several levels of implementation to adapt the scope and scale of a GenAISys, not only to business requirements but also to available budgets and resources.

Finally, we'll illustrate how contextual awareness and memory retention can be implemented using OpenAI's LLM and multimodal API. A GenAISys cannot work without solid memory retention functionality—without memory, there's no context, and without context, there's no sustainable generation. Throughout this book, we will create modules for memoryless, short-term, long-term, and multisession types depending on the task at hand. By the end of this chapter, you will have acquired a clear conceptual framework for what makes an AI system business-ready and practical experience in building the first bricks of an AI controller.

In a nutshell, this chapter covers the following topics:

- Components of a business-ready GenAISys
- AI controllers and agentic functionality (model-agnostic)
- Hybrid human roles and collaboration with AI
- Business opportunities and scope
- Contextual awareness through memory retention

Let's begin by defining what a business-ready GenAISys is.

Components of a business-ready GenAISys

A business-ready GenAISys is a modular orchestrator that seamlessly integrates standard AI models with multifunctional frameworks to deliver hybrid intelligence. By combining generative AI with agentic functionality, RAG, **machine learning** (**ML**), web search, non-AI operations, and multiple-session memory systems, we are able to deliver scalable and adaptive solutions for diverse and complex tasks. Take ChatGPT, for example; people use the name "ChatGPT" interchangeably for the generative AI model as well as for the application itself. However, behind the chat interface, tools such as ChatGPT and Gemini are part of larger systems—online copilots—that are fully integrated and managed by intelligent AI controllers to provide a smooth user experience.

It was Tomczak (2024) who took us from thinking of generative AI models as a collective entity to considering complex GenAISys architectures. His paper uses the term "GenAISys" to describe these more complex platforms. Our approach in this book will be to expand the horizon of a GenAISys to include advanced AI controller functionality and human roles in a business-ready ecosystem. There is no single silver-bullet architecture for a GenAISys. However, in this section, we'll define the main components necessary to attain ChatGPT-level functionality. These include a generative

AI model, memory retention functions, modular RAG, and multifunctional capabilities. How each component contributes to the GenAISys framework is illustrated in *Figure 1.1*:

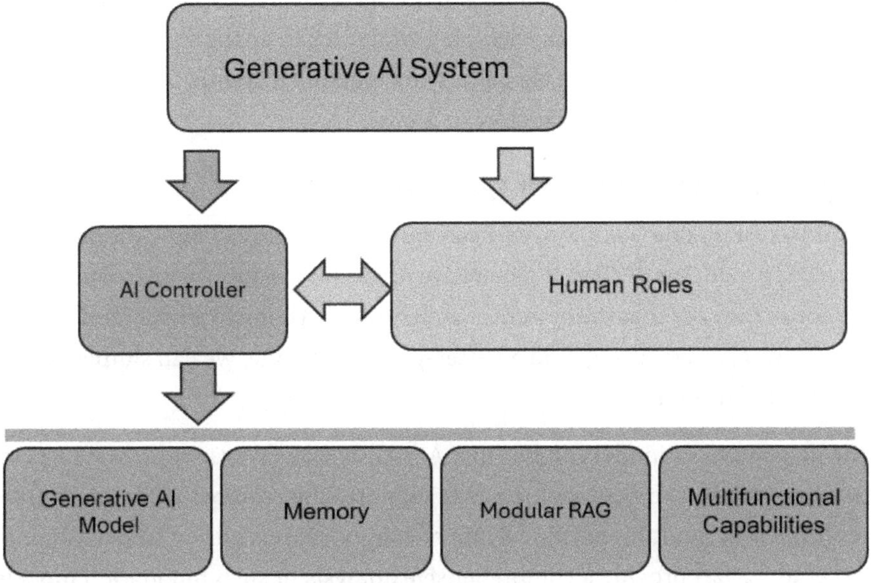

Figure 1.1: GenAISys, the AI controller, and human roles

Let's now define the architecture of the AI controllers and human roles that make up a GenAISys.

AI controllers

At the heart of a business-ready GenAISys is an **AI controller** that activates custom ChatGPT-level features based on the context of the input. Unlike traditional pipelines with predetermined task sequences, the AI controller operates without a fixed order, dynamically adapting tasks—such as web search, image analysis, and text generation—based on the specific context of each input. This agentic context-driven approach enables the AI controller to orchestrate various components seamlessly, ensuring effective and coherent performance of the generative AI model.

A lot of work is required to achieve effective results with a custom ChatGPT-grade AI controller. However, the payoff is a new class of AI systems that can withstand real-world pressure and produce tangible business results. A solid AI controller ecosystem can support use cases across multiple domains: customer support automation, sales lead generation, production optimization (services and manufacturing), healthcare response support, supply chain optimization, and any other domain the market will take you! A GenAISys, thus, requires an AI controller to orchestrate multiple pipelines, such as contextual awareness to understand the intent of the prompt and memory retention to support continuity across sessions.

The GenAISys must also define human roles, which determine which functions and data can be accessed. Before we move on to human roles, however, let's first break down the key components that power the AI controller. As shown in *Figure 1.1*, the generative AI model, memory, modular RAG, and multifunctional capabilities each play vital roles in enabling flexible, context-driven orchestration. Let's explore how these elements work together to build a business-ready GenAISys. We will first define the role of the generative AI model.

Model-agnostic approach to generative AI

When we build a sustainable GenAISys, we need model *interchangeability*—the flexibility to swap out the underlying model as needed. A generative AI model should serve as a component within the system, not as the core that the system is built around. That way, if our model is deprecated or requires updating, or we simply find a better-performing one, we can simply replace it with another that better fits our project.

As such, the generative AI model can be OpenAI's GPT, Google's Gemini, Meta's Llama, xAI's Grok, or any Hugging Face model, as long as it supports the required tasks. Ideally, we should choose a multipurpose, multimodal model that encompasses text, vision, and reasoning abilities. Bommasani et al. (2021) provide a comprehensive analysis of such foundation models, whose scope reaches beyond LLMs.

A generative AI model has two main functions, as shown in *Figure 1.2*:

- **Orchestrates** by determining which tasks need to be triggered based on the input. This input can be a user prompt or a system request from another function in the pipeline. The orchestration function agent can trigger web search, document parsing, image generation, RAG, ML functions, non-AI functions, and any other function integrated into the GenAISys.
- **Executes** the tasks requested by the orchestration layer or executes a task directly based on the input. For example, a simple query such as requesting the capital of the US will not necessarily require complex functionality. However, a request for document analysis might require several functions (chunking, embedding, storing, and retrieving).

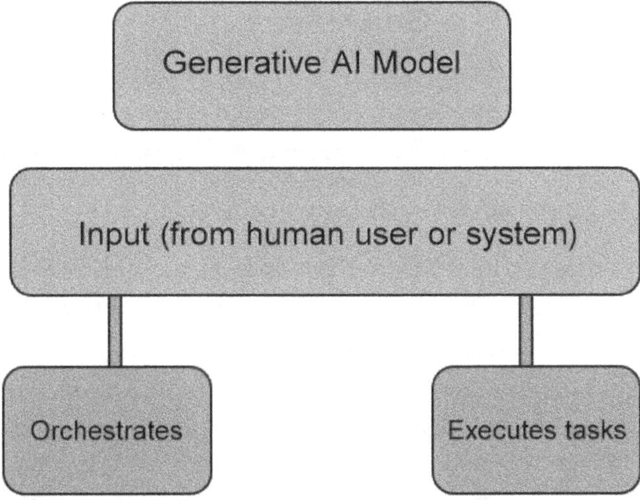

Figure 1.2: A generative AI model to orchestrate or execute tasks

Notice that *Figure 1.2* has a unique feature. There are no arrows directing the input, orchestration, and execution components. Unlike traditional hardcoded linear pipelines, a flexible GenAISys has its components unordered. We build the components and then let automated scenarios selected by the orchestration function order the tasks dynamically.

This flexibility ensures the system's adaptability to a wide range of tasks. We will not be able to build a system that solves every task, but we can build one that satisfies a wide range of tasks within a company. Here are two example workflows that illustrate how a GenAISys can dynamically sequence tasks based on the roles involved:

- Human roles can be configured so that, in some cases, the user input executes a simple API call to provide a straightforward response, such as requesting the capital of a country. In this case, the generative AI model executes a request directly.
- System roles can be configured dynamically to orchestrate a set of instructions, such as searching the web first and then summarizing the web page. In this case, the system goes through an orchestration process to produce an output.

The possibilities are unlimited; however, all the scenarios will rely on the memory to ensure consistent, context-aware behavior. Let's look at memory next.

Building the memory of a GenAISys

Advanced generative AI models such as OpenAI's GPT, Meta's Llama, xAI's Grok, Google's Gemini, and many Hugging Face variants are *context-driven* regardless of their specific version or performance level. You will choose the model based on your project, but the basic rule remains simple:

No-context => No meaningful generation

When we use ChatGPT or any other copilot, we have nothing to worry about as contextual memory is handled for us. We just start a dialogue, and things run smoothly as we adapt our prompt to the level of responses we are obtaining. However, when we develop a system with a generative AI API from scratch, we have to explicitly build contextual awareness and memory retention. This process of building and managing memory is a fundamental form of context engineering.

Four approaches stand out among the wide range of possible memory retention strategies with an API:

- **Stateless and memoryless session:** A request is sent to the API, and a response is returned with no memory retention functionality.
- **Short-term memory session:** The exchanges between the requests and responses are stored in memory during the session but not beyond.
- **Long-term memory of multiple sessions:** The exchanges between the requests and responses are stored in memory and memorized even after the session ends.
- **Long-term memory of multiple cross-topic sessions:** This feature links the long-term memory of multiple sessions to other sessions. Each session is assigned a role: a system or multiple users. This feature is not standard in platforms such as ChatGPT but is essential for workflow management within organizations.

Figure 1.3 sums up these four memory architectures. We'll demonstrate each configuration in Python using GPT-4o in the upcoming section, *Contextual awareness and memory retention.*

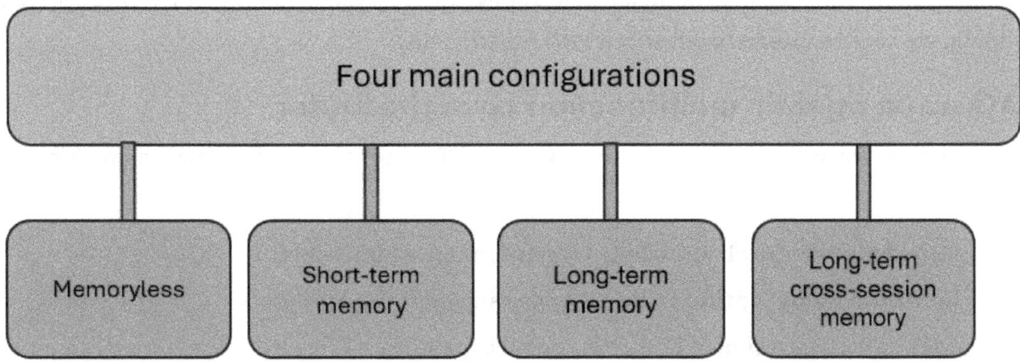

Figure 1.3: Four different GenAISys memory configurations

These four memory types serve as a starting point that can be expanded as necessary when developing a GenAISys. However, practical implementations often require additional functionality, including the following:

- **Human roles** to define users or groups of users that can access session history or sets of sessions on multiple topics. This will take us beyond ChatGPT-level platforms. We will introduce this aspect in *Chapter 2, Building the Generative AI Controller*.
- **Storage strategies** to define what we need to store and what we need to discard. We will introduce storage strategies and take this concept further with a Pinecone vector store in *Chapter 3, Integrating Dynamic RAG into the GenAISys.*

There are native distinctions between two key categories of memorization in generative models:

- **Semantic memory**, which contains facts such as hard science
- **Episodic memory**, which contains personal timestamped memories such as personal events in time and business meetings

We can see that building a GenAISys's memory requires careful design and deliberate development to implement ChatGPT-grade memory and additional memory configurations, such as long-term, cross-topic sessions. The ultimate goal, however, of this advanced memory system is to enhance the model's contextual awareness. While generative AI models such as GPT-4o have inbuilt contextual awareness, to expand the scope of a context-driven system such as the GenAISys we're building, we need to integrate advanced RAG functionality.

RAG as an agentic multifunction co-orchestrator

In this section, we explain the motivations for using RAG for three core functions within a GenAISys:

- **Knowledge retrieval:** Retrieving targeted, nuanced information
- **Context window optimization:** Engineering optimized prompts
- **Agentic orchestration of multifunctional capabilities:** Triggering functions dynamically

Let's begin with knowledge retrieval.

1. Knowledge retrieval

Generative AI models excel when it comes to revealing parametric knowledge that they have learned, which is embedded in their weights. This knowledge is learned during training and embedded in models such as GPT, Llama, Grok, and Gemini. However, that knowledge stops at the cutoff date when no additional data is fed to the model. At that point, to update or supplement it, we have two options:

- **Implicit knowledge**: Fine-tune the model so that more trained knowledge is added to its weights (parametric). This process can be challenging if you are working with dynamic data that changes daily, such as weather forecasts, newsfeeds, or social media messages. It also comes with costs and risks if the fine-tuning process doesn't work that well for your data.
- **Explicit knowledge**: Store the data in files or embed data in vector stores. The knowledge will then be structured, accessible, traceable, and updated. We can then retrieve the information with advanced queries.

It's important to note here that static implicit knowledge cannot scale effectively without dynamic explicit knowledge. More on that in the upcoming chapters.

2. Context window optimization

Generative AI models are expanding the boundaries of context windows. For example, at the time of writing, the following are the supported context lengths:

- Llama 4 Scout: 10 million tokens
- Gemini 2.0 Pro Experimental: 2 million tokens
- Claude 3.7 Sonnet: 200,000 tokens
- GPT-4o: 128,000 tokens

While impressive, these large context windows can be expensive in terms of token costs and compute. Furthermore, the main issue is that their precision diminishes when the context becomes too large. Also, we don't need the largest context window but only the one that best fits our project. This can justify implementing RAG if necessary to optimize a project.

The chunking process of RAG splits large content into more nuanced groups of tokens. When we embed these chunks, they become vectors that can be stored and efficiently retrieved from vector stores. Through effective context engineering, this approach ensures we use only the most relevant information per task, minimizing token usage and maximizing response quality. Thus, we can rely on generative AI capabilities for parametric implicit knowledge and RAG for large volumes of explicit non-parametric data in vector stores. We can take RAG further and use the method as an orchestrator.

3. Agentic orchestrator of multifunctional capabilities

The AI controller bridges with RAG through the generative AI model. RAG is used to augment the model's input with a flexible range of instructions. Now, using RAG to retrieve instructions might seem counterintuitive at first—but think about it. If we store instructions as vectors and retrieve the best set for a task, we get a fast, adaptable way to enable agentic functionality, generate effective results, and avoid the need to fine-tune the model every time we change our instruction strategies for how we want it to behave.

These instructions act as optimized prompts, tailored to the task at hand. In this sense, RAG becomes part of the orchestration layer of the AI system. A vector store such as Pinecone can store and return this functional information, as illustrated in *Figure 1.4*:

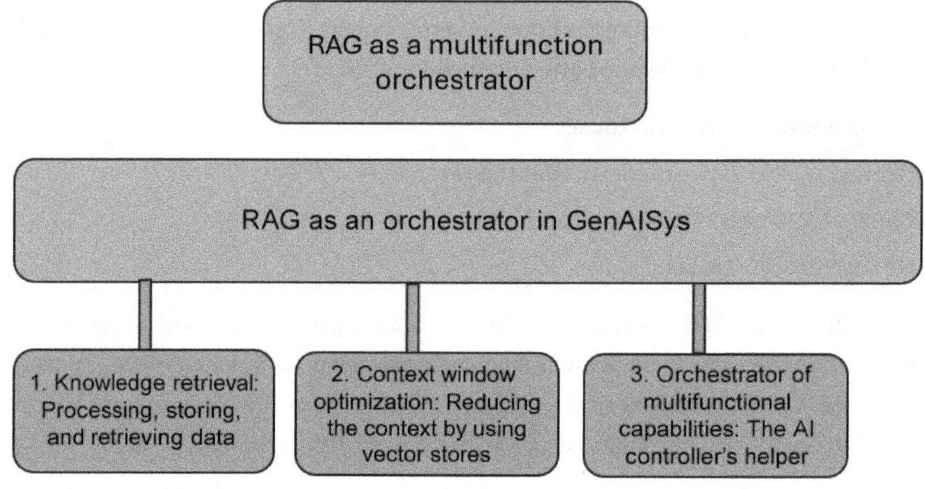

Figure 1.4: RAG orchestration functionality

The orchestration of these scenarios is performed through the following:

- **Scenario retrieval**: The AI controller will receive structure instructions (scenarios) from a vector database, such as Pinecone, adapted to the user's query
- **Dynamic task activation**: Each scenario specifies a series of tasks, such as web search, ML algorithms, standard SQL queries, or any function we need

Adding classical functions and ML functionality to the GenAISys enhances its capabilities dramatically. The modular architecture of a GenAISys makes this multifunctional approach effective, as in the following use cases:

- **Web search** to perform real-time searches to augment inputs
- **Document analysis** to process documents and populate the vector store
- **Document search** to retrieve parts of the processed documents from the vector store
- **ML** such as **K-means clustering** (**KMC**) to group data and **k-nearest neighbors** (**KNN**) for similarity searches
- **SQL queries** to execute rule-based retrieval on structured datasets
- Any other function required for your project or workflow

RAG remains a critical component of a GenAISys, which we will build into our GenAISys in *Chapter 3, Integrating Dynamic RAG into the GenAISys*. In *Chapter 3, Integrating Dynamic RAG into the GenAISys*, we will also enhance the system with multifunctional features.

We'll now move on to the human roles, which form the backbone of any GenAISys.

Human roles

Contrary to popular belief, the successful deployment and operation of a GenAISys—such as the ChatGPT platform—relies heavily on human involvement throughout its entire life cycle. While these tools may seem to handle complex tasks effortlessly, behind the scenes are multiple layers of human expertise, oversight, and coordination that make their smooth operation possible.

Software professionals must first design the architecture, process massive datasets, and fine-tune the system on million-dollar servers equipped with cutting-edge compute resources. After deployment, large teams are required to monitor, validate, and interpret system outputs—continuously adapting them in response to errors, emerging technologies, and regulatory changes. On top of that, when it comes to deploying these systems within organizations—whether inside corporate intranets, public-facing websites, research environments, or learning management systems—it takes cross-functional coordination efforts across multiple domains.

These tasks require high levels of expertise and qualified teams. Humans are, therefore, not just irreplaceable; they are critical! They are architects, supervisors, curators, and guardians of the AI systems they create and maintain.

GenAISys implementation and governance teams

Implementing a GenAISys requires technical skills and teamwork to gain the support of end users. It's a collaborative challenge between AI controller design, user roles, and expectations. To anyone who thinks that deploying a real-world AI system is just about getting access to a model—such as the latest GPT, Llama, or Gemini—a close look at the resources required will reveal the true challenges. A massive number of human resources might be involved in the development, deployment, and maintenance of an AI system. Of course, not every organization will need all of these roles, but we must recognize the range of skills involved, such as the following:

- **Project manager** (**PM**)
- Product manager
- Program manager
- **ML engineer** (**MLE**)/data scientist

- Software developer/**backend engineer** (**BE**)
- **Cloud engineer** (**CE**)
- **Data engineer** (**DE**) and privacy manager
- UI/UX designer
- Compliance and regulatory officer
- Legal counsel
- **Security engineer** (**SE**) and security officer
- Subject-matter experts for each domain-specific deployment
- **Quality assurance engineer** (**QAE**) and tester
- Technical documentation writer
- System maintenance and support technician
- User support
- Trainer

These are just examples—just enough to show how many different roles are involved in building and operating a full-scale GenAISys. *Figure 1.5* shows that designing and implementing a GenAISys is a continual process, where human resources are needed at every stage.

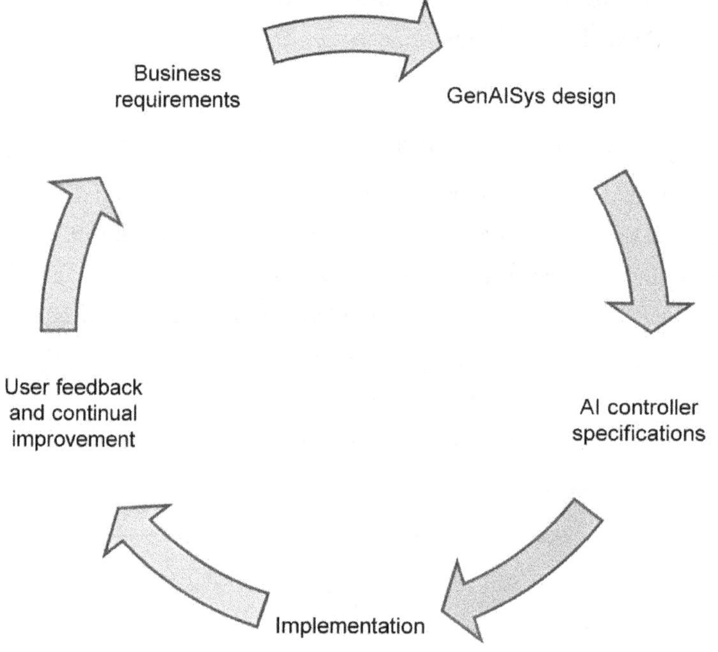

Figure 1.5: A GenAISys life cycle

We can see that a GenAISys life cycle is a never-ending process:

- **Business requirements** will continually evolve with market constraints
- **GenAISys** design will have to adapt with each business shift
- **AI controller** specifications must adapt to technological progress
- **Implementation** must adapt to ever-changing business specifications
- **User feedback** will drive continual improvement

Real-world AI relies heavily on human abilities—the kind of contextual and technical understanding that AI alone cannot replicate. AI can automate a wide range of tasks effectively. But it's humans who bring the deep insight needed to align those systems with real business goals.

Let's take this further and look at a RACI heatmap to show why humans are a critical component of a GenAISys.

GenAISys RACI

Organizing a GenAISys project requires human resources that go far beyond what AI automation alone can provide. **RACI** is a responsibility assignment matrix that helps define roles and responsibilities for each task or decision by identifying who is **Responsible**, **Accountable**, **Consulted**, and **Informed**. RACI is ideal for managing the complexity of building a GenAISys. It adds structure to the growing list of human roles required during the system's life cycle and provides a pragmatic framework for coordinating their involvement.

As in any complex project, teams working on a GenAISys need to collaborate across disciplines, and RACI helps define who does what. Each letter in RACI stands for a specific type of role:

- **R (Responsible):** The person(s) who works actively on the task. They are responsible for the proper completion of the work. For example, an MLE may be responsible for processing datasets with ML algorithms.
- **A (Accountable):** The person(s) answerable for the success or failure of a task. They oversee the task that somebody else is responsible for carrying out. For example, the **product owner** (**PO**) will have to make sure that the MLE's task is done on time and in compliance with the specifications. If not, the PO will be accountable for the failure.
- **C (Consulted):** The person(s) providing input, advice, and feedback to help the others in a team. They are not responsible for executing the work. For example, a subject-matter expert in retail may help the MLE understand the goal of an ML algorithm.

- **I (Informed)**: The person(s) kept in the loop about the progress or outcome of a task. They don't participate in the task but want to be simply informed or need to make decisions. For example, a **data privacy officer** (**DPO**) would like to be informed about a system's security functionality.

A RACI heatmap typically contains legends for each human role in a project. Let's build a heatmap with the following roles:

- The **MLE** develops and integrates AI models
- The **DE** designs data management pipelines
- The **BE** builds API interactions
- The **frontend engineer** (**FE**) develops end user features
- The **UI/UX designer** designs user interfaces
- The **CE/DevOps engineer** manages cloud infrastructure
- The **prompt engineer** (**PE**) designs optimal prompts
- The **SE** handles secure data and access
- The **DPO** manages data governance and regulation compliance
- The **legal/compliance officer** (**LC**) reviews the legal scope of a project
- The **QAE** tests the GenAISys
- The **PO** defines the scope and scale of a product
- The **PM** coordinates resources and timelines
- The **technical writer** (**TW**) produces documentation
- The **vendor manager** (**VM**) communicates with external vendors and service providers

Not every GenAISys project will include all of these roles, but depending on the scope and scale of the project, many of them will be critical. Now, let's list the key responsibilities of the roles defined above in a typical generative AI project:

- **Model**: AI model development
- **Controller**: Orchestration of APIs and multimodal components
- **Pipelines**: Data processing and integration workflows
- **UI/UX**: User interface and experience design
- **Security**: Data protection and access control

- **DevOps:** Infrastructure, scaling, and monitoring
- **Prompts:** Designing and optimizing model interactions
- **QA:** Testing and quality assurance

We've defined the roles and the tasks. Now, we can show how they can be mapped to a real-world scenario. *Figure 1.6* illustrates an example RACI heatmap for a GenAISys.

Role	Model	Controller	Pipelines	UI/UX	Security	DevOps	Prompts	QA
MLE	R/A	C	C	I	I	I	R/A	C
DE	C	C	R/A	I	C	I	I	C
BE	I	R/A	R	C	I	R	I	C
FE	I	I	I	R/A	I	I	I	C
UX	I	I	I	A/R	I	I	I	C
CE	C	C	C	I	C	R/A	I	C
SE	I	I	C	I	R/A	I	I	C
DPO	I	I	C	I	R	I	I	I
QAE	C	C	C	C	C	C	C	R/A

Figure 1.6: Example of a RACI heatmap

For example, in this heatmap, the MLE has the following responsibilities:

- **(R)**esponsible and **(A)**ccountable for the model, which could be GPT-4o.
- **(R)**esponsible and **(A)**ccountable for the prompts for the model
- **(C)**onsulted as an expert for the controller, the pipeline, and testing (QA)
- **(I)**nformed about the UI/UX, security, and DevOps

We can sum it up with one simple rule for a GenAISys:

No humans -> no system!

We can see that *we* are necessary during the whole life cycle of a GenAISys, from design to maintenance and support, including continual evolutions to keep up with user feedback. Humans have been and will be here for a long time! Next, let's explore the business opportunities that a GenAISys can unlock.

Business opportunities and scope

More often than not, we will not have access to the incredible billion-dollar resources of OpenAI, Meta, xAI, or Microsoft Azure to build ChatGPT-like platforms. The previous section showed that beneath a ChatGPT-like, seemingly simple, seamless interface, there is a complex layer of expensive infrastructure, rare talent, and continuous improvement and evolution that absorb resources only large corporations can afford. Therefore, a smarter path from the start is to determine which project category we are in and leverage the power of existing modules and libraries to build our GenAISys. Whatever the use case, such as marketing, finance, production, or support, we need to find the right scope and scale to implement a realistic GenAISys.

The first step of any GenAISys is to define the project's goal (opportunity), including its scope and scale, as we mentioned. During this step, you will assess the risks, such as costs, confidentiality, and resource availability (risk management).

We can classify GenAISys projects into three main business implementation types depending on our resources, our objectives, the complexity of our use case, and our budget. These are illustrated in *Figure 1.7*:

- **Hybrid approach**: Leveraging existing AI platforms
- **Small scope and scale**: A focused GenAISys
- **Full-scale generative multi-agent AI system**: A complete ChatGPT-level generative AI platform

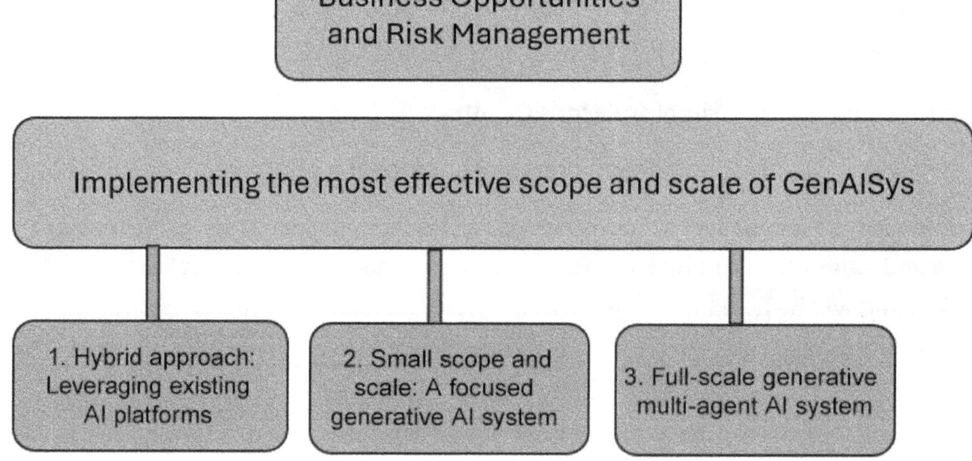

Figure 1.7: The three main GenAISys business implementations

Let's begin with a hybrid approach, a practical way to deliver business results without overbuilding.

Hybrid approach

A hybrid framework enables you to minimize development costs and time by combining ready-to-use SaaS platforms with custom-built components developed only when necessary, such as web search and data cleansing. This way, you can leverage the power of generative AI without developing everything from scratch. Let's go through the key characteristics and a few example use cases.

Key characteristics

- Relying on proven web services such as OpenAI's GPT API, AWS, Google AI, or Microsoft Azure. These platforms provide the core generative functionality.
- Customizing your project by integrating domain-specific vector stores and your organization's proprietary datasets.
- Focusing development on targeted functionality, such as customer support automation or marketing campaign generation.

Use case examples

- Implementing a domain-specific vector store to handle legal, medical, or product-related customer queries
- Building customer support on a social media platform with real-time capabilities

This category offers the ability to do more with less—in terms of both cost and development effort. A hybrid system can be a standalone GenAISys or a subsystem within a larger generative AI platform where full-scale development isn't necessary. Let's now look at how a small-scope, small-scale GenAISys can take us even further.

Small scope and scale

A small-scale GenAISys might include an intelligent, GenAI-driven AI controller connected to a vector store. This setup allows the system to retrieve data, trigger instructions, and call additional functionality such as web search or ML—without needing full-scale infrastructure.

Key characteristics

- A clearly defined profitable system designed to achieve reasonable objectives with optimal development time and cost
- The AI controller orchestrates instruction scenarios that, in turn, trigger RAG, web search, image analysis, and additional custom tasks that fit your needs
- The focus is on high-priority, productive features

Use case examples

- A GenAISys for document retrieval and summarization for any type of document with nuanced analysis through chunked and embedded content
- Augmenting a model such as GPT or Llama with real-time web search to bypass its data cutoff date—ideal for applications such as weather forecasting or news monitoring that don't need continual fine-tuning

This category takes us a step beyond the hybrid approach, while still staying realistic and manageable for small to mid-sized businesses or even individual departments within large organizations.

Full-scale GenAISys

If you're working in a team of experts within an organization that has a large budget and advanced infrastructure, this category is for you. Your team can build a full-scale GenAISys that begins to approach the capabilities of ChatGPT-grade platforms.

Key characteristics

- A full-blown AI controller that manages and orchestrates complex automated workflows, including RAG, instruction scenarios, multimodal functionality, and real-time data
- Requires significant computing resources and highly skilled development teams

Think of the GenAISys we're building in this book as an alpha version—a template that can be cloned, configured, and deployed anywhere in the organization as often as needed.

Use case examples

- GenAISys is already present in healthcare to assist with patient diagnosis and disease prevention. The Institut Curie in Paris, for example, has a very advanced AI research team: `https://institut-curie.org/`.
- Many large organizations have begun implementing GenAISys for fraud detection, weather predictions, and legal expertise.

You can join one of these large organizations that have the resources to build a sustainable GenAISys, whether it be on a cloud platform, local servers, or both.

The three categories—hybrid, small scale, and full scale—offer distinct paths for building a GenAISys, depending on your organization's goals, budget, and technical capabilities. In this book, we'll explore the critical components that make up a GenAISys. By the end, you'll be equipped to contribute to any of these categories and offer realistic, technically grounded recommendations for the projects you work on.

Let's now lift the hood and begin building contextual awareness and memory retention in code.

Contextual awareness and memory retention

In this section, we'll begin implementing simulations of contextual awareness and memory retention in Python to illustrate the concepts introduced in the *Building the memory of a GenAISys* section. The goal is to demonstrate practical ways to manage context and memory—two features that are becoming increasingly critical as generative AI platforms evolve.

Open the `Contextual_Awareness_and_Memory_Retention.ipynb` file located in the `chapter01` folder of the GitHub repository (`https://github.com/Denis2054/Building-Business-Ready-Generative-AI-Systems/tree/main`). You'll see that the notebook is divided into five main sections:

- **Setting up the environment**, building reusable functions, and storing them in the `commons` directory of the repository, so we can reuse them when necessary throughout the book
- **Stateless and memoryless session** with semantic and episodic memory
- **Short-term memory session** for context awareness during a session
- **Long-term memory across multiple sessions** for context retention across different sessions
- **Long-term memory of multiple cross-topic sessions**, expanding long-term memory over formerly separate sessions

The goal is to illustrate each type of memory in an explicit process. These examples are intentionally kept manual for now, but they will be automated and managed by the AI controller we will begin to build in the next chapter.

Due to the probabilistic nature of generative models, you may observe different outputs for the same prompt across runs. Make sure to run the entire notebook in a single session, as memory retention in this notebook is explicit in different cells. In *Chapter 2*, this functionality will become persistent and fully managed by the AI controller

The first step is to install the environment.

Setting up the environment

We will need a `commons` directory for our GenAISys project. This directory will contain the main modules and libraries needed across all notebooks in this book's GitHub repository. The motivation is to focus on designing the system for maintenance and support. As such, by grouping the main modules and libraries in one directory, we can zero in on a resource that requires our attention instead of repeating the setup steps in every notebook. Furthermore, this section will serve as a reference point for all the notebooks in this book's GitHub repository. We'll only describe the downloading of each resource once and then reuse them throughout the book to build our educational GenAISys.

Thus, we can download the notebook resources from the `commons` directory and install the requirements.

The first step is to download `grequests.py`, a utility script we will use throughout the book. It contains a function to download the files we need directly from GitHub:

```
!curl -L https://raw.githubusercontent.com/Denis2054/Building-Business-
Ready-Generative-AI-Systems/master/commons/grequests.py --output
grequests.py
```

Quick tip: Enhance your coding experience with the **AI Code Explainer** and **Quick Copy** features. Open this book in the next-gen Packt Reader. Click the **Copy** button

(1) to quickly copy code into your coding environment, or click the **Explain** button

(2) to get the AI assistant to explain a block of code to you.

```
function calculate(a, b) {
  return {sum: a + b};
};
```

Copy 1 Explain 2

The next-gen Packt Reader is included for free with the purchase of this book. Scan the QR code OR visit `https://packtpub.com/unlock`, then use the search bar to find this book by name. Double-check the edition shown to make sure you get the right one.

The goal of this script is to download a file from any directory of the repository by calling the `download` function from `grequests`:

```
import sys
import subprocess
from grequests import download
download([directory],[file])
```

This function uses a `curl` command to download files from a specified directory and filename. It also includes basic error handling in case of command execution failures.

The code begins by importing `subprocess` to handle paths and commands. The `download` function contains two parameters:

```
def download(directory, filename):
```

- `directory`: The subdirectory of the GitHub repository where the file is stored
- `filename`: The name of the file to download

The base URL for the GitHub repository is then defined, pointing to the raw files we will need:

```
base_url = 'https://raw.githubusercontent.com/Denis2054/Building-Business-
Ready-Generative-AI-Systems/main/'
```

We now need to define the file's full URL with the `directory` and `filename` parameters:

```
file_url = f"{base_url}{directory}/{filename}"
```

The function now defines the `curl` command:

```
curl_command = f'curl -o {filename} {file_url}'
```

Finally, the `download` command is executed:

```
subprocess.run(curl_command, check=True, shell=True)
```

- `check=True` activates an exception if the `curl` command fails
- `shell=True` runs the command through the shell

The `try-except` block is used to handle errors:

```
try:
    # Prepare the curl command with the Authorization header
    curl_command = f'curl -o {filename} {file_url}'

    # Execute the curl command
    subprocess.run(curl_command, check=True, shell=True)
    print(f"Downloaded '{filename}' successfully.")

except subprocess.CalledProcessError:
    print(f"Failed to download '{filename}'. Check the URL and your
internet connection")
```

We now have a standalone download script that we'll use throughout the book. Let's go ahead and download the resources we need for this program.

Downloading OpenAI resources

We need three resources for this notebook:

- `requirements01.py` to install the precise OpenAI version we want
- `openai_setup.py` to initialize the OpenAI API key
- `openai_api_py` contains a reusable function for calling the GPT-4o model, so you don't need to rewrite the same code across multiple cells or notebooks

We will be reusing the same functions throughout the book for standard OpenAI API calls. You can come back to this section any time you want to revisit the installation process. Other scenarios will be added to the `commons` directory when necessary.

We can download these files with the `download()` function:

```
from grequests import download
download("commons","requirements01.py")
download("commons","openai_setup.py")
download("commons","openai_api.py")
```

The first resource is `requirements01.py`.

Installing OpenAI

`requirements01.py` makes sure that a specific version of the OpenAI library is installed to avoid conflicts with other installed libraries. The code thus uninstalls existing versions, force-installs the specified version requested, and verifies the result. The function executes the installation with error handling:

```
def run_command(command):
    try:
        subprocess.check_call(command)

    except subprocess.CalledProcessError as e:
        print(f"Command failed: {' '.join(command)}\nError: {e}")
        sys.exit(1)
```

The first step for the function is to uninstall the current OpenAI library, if there is one:

```
print("Installing 'openai' version 1.57.1...")
run_command([sys.executable, "-m", "pip", "install", "--force-reinstall",
"openai==1.57.1"])
```

The function then installs a specific version of OpenAI:

```
run_command(
    [
        sys.executable, "-m", "pip", "install",
        "--force-reinstall", "openai==1.57.1"
    ]
)
```

Finally, the function verifies that OpenAI is properly installed:

```
try:
    import openai
    print(f"'openai' version {openai.__version__} is installed.")

except ImportError:
    print("Failed to import the 'openai' library after installation.")
    sys.exit(1)
```

The output at the end of the function should be as follows:

```
'openai' version 1.57.1 is installed.
```

We can now initialize the OpenAI API key.

OpenAI API key initialization

There are two methods to initialize the OpenAI API key in the notebook:

1. **Using Google Colab secrets:** Click on the key icon in the left pane in Google Colab, as shown in *Figure 1.8*, then click on **Add new secret** and add your key with the name of the key variable you will use in the notebook:

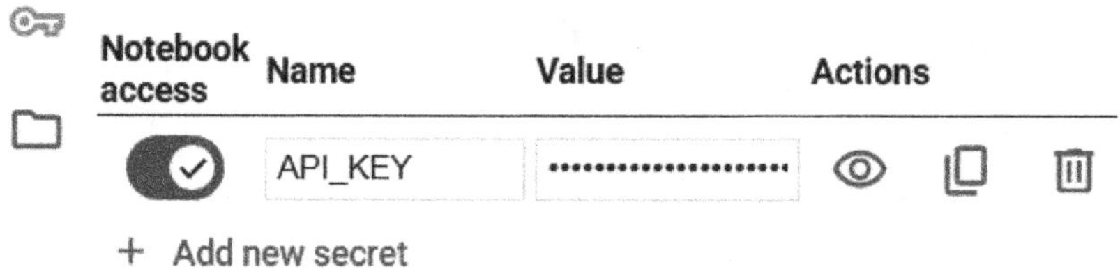

Figure 1.8: Add a new Google secret key

Then, we can use Google's function to initialize the key by calling it in our openai_setup function in openai_setup.py:

```
# Import libraries
import openai
import os
from google.colab import userdata
# Function to initialize the OpenAI API key
def initialize_openai_api():
# Function to initialize the OpenAI API key
def initialize_openai_api():
    # Access the secret by its name
    API_KEY = userdata.get('API_KEY')

    if not API_KEY:
        raise ValueError("API_KEY is not set in userdata!")

    # Set the API key in the environment and OpenAI
    os.environ['OPENAI_API_KEY'] = API_KEY
    openai.api_key = os.getenv("OPENAI_API_KEY")
    print("OpenAI API key initialized successfully.")
```

This method is activated if google_secrets is set to True:

```
google_secrets=True
if google_secrets==True:
    import openai_setup
    openai_setup.initialize_openai_api()
```

- **Custom secure method:** You can also choose a custom method or enter the key in the code by setting google_secrets to False, uncommenting the following code, and entering your API key directly, or any method of your choice:

```
if google_secrets==False: # Uncomment the code and choose any method
you wish to initialize the API_KEY
    import os
    #API_KEY=[YOUR API_KEY]
    #os.environ['OPENAI_API_KEY'] = API_KEY
    #openai.api_key = os.getenv("OPENAI_API_KEY")
    #print("OpenAI API key initialized successfully.")
```

In both cases, the code will create an environment variable:

```
os.environ['OPENAI_API_KEY'] = API_KEY
openai.api_key = os.getenv("OPENAI_API_KEY")
```

The OpenAI API key is initialized. We will now import a custom OpenAI API call.

OpenAI API call

The goal next is to create an OpenAI API call function in `openai_api.py` that we can import in two lines:

```
#Import the function from the custom OpenAI API file
import openai_api
from openai_api import make_openai_api_call
```

The function is thus built to receive four variables when making the call and display them seamlessly:

```
# API function call
response = openai_api.make_openai_api_call(
    uinput,mrole,mcontent,user_role)
print(response)
```

The parameters in this function are the following:

- `input`: Contains the input (user or system), for example, `Where is Hawaii?`
- `mrole`: Defines the system's role, for example, `You are a geology expert.` or simply `System.`
- `mcontent`: Is what we expect the system to be, for example, `You are a geology expert.`
- `user_role`: Defines the role of the user, for example, `user`

The first part of the code in the function defines the model we will be using in this notebook and creates a message object for the API call with the parameters we sent:

```
def make_openai_api_call(input, mrole,mcontent,user_role):
    # Define parameters
    gmodel = "gpt-4o"

    # Create the messages object
    messages_obj = [
        {
```

```
            "role": mrole,
            "content": mcontent
        },
        {
            "role": user_role,
            "content": input
        }
    ]
```

We then define the API call parameters in a dictionary for this notebook:

```
# Define all parameters in a dictionary named params:
    params = {
        "temperature": 0,
        "max_tokens": 256,
        "top_p": 1,
        "frequency_penalty": 0,
        "presence_penalty": 0
    }
```

The dictionary parameters are the following:

- `temperature`: Controls the randomness of a response. `0` will produce deterministic responses. Higher values (e.g., `0.7`) will produce more creative responses.
- `max_tokens`: Limits the maximum number of tokens of a response.
- `top_p`: Produces nucleus sampling. It controls the diversity of a response by sampling from the top tokens with a cumulative probability of 1.
- `frequency_penalty`: Reduces the repetition of tokens to avoid redundancies. `0` will apply no penalty, and 2 a strong penalty. In this case, `0` is sufficient because of the high performance of the OpenAI model.
- `presence_penalty`: Encourages new content by penalizing existing content to avoid redundancies. It applies to the same values as for the frequency penalty. In this case, due to the high performance of the OpenAI model, it doesn't require this control.

We then initialize the OpenAI client to create an instance for the API calls:

```
    client = OpenAI()
```

Finally, we make the API call by sending the model, the message object, and the unpacked parameters:

```
# Make the API call
response = client.chat.completions.create(
    model=gmodel,
    messages=messages_obj,
    **params  # Unpack the parameters dictionary
)
```

The function ends by returning the content of the API's response that we need:

```
#Return the response
return response.choices[0].message.content
```

This function will help us focus on the GenAISys architecture without having to overload the notebook with repetitive libraries and functions.

In the notebook, we have the following:

- The program provides the input, roles, and message content to the function
- `messages_obj` contains the conversation history
- The parameters for the API's behavior are defined in the `params` dictionary
- An API call is made to the OpenAI model using the OpenAI client
- The function returns only the AI's response content

A GenAISys will contain many components—including a generative model. You can choose the one that fits your project. In this book, the models are used for educational purposes only, not as endorsements or recommendations.

Let's now build and run a stateless and memoryless session.

1. Stateless and memoryless session

A stateless and memoryless session is useful if we only want a single and temporary exchange with no stored information between requests. The examples in this section are both stateless and memoryless:

- *Stateless* indicates that each request will be processed independently
- *Memoryless* means that there is no mechanism to remember past exchanges

Let's begin with a semantic query.

Semantic query

This request expects a purely semantic, factual response:

```
uinput = "Hawai is on a geological volcano system. Explain:"
mrole = "system"
mcontent = "You are an expert in geology."
user_role = "user"
```

Now, we call the OpenAI API function:

```
# Function call
response = openai_api.make_openai_api_call(
    uinput,mrole,mcontent,user_role)
print(response)
```

As you can see, the response is purely semantic:

```
Hawaii is located on a volcanic hotspot in the central Pacific Ocean,
which is responsible for the formation of the Hawaiian Islands. This
hotspot is a region where magma from deep within the Earth's mantle rises
to the surface, creating volcanic activity…
```

The next query is episodic.

Episodic query with a semantic undertone

The query in this example is episodic and draws on personal experience. However, there is a semantic undertone because of the description of Hawaii. Here's the message, which is rather poetic:

```
# API message
uinput = "I vividly remember my family's move to Hawaii in the 1970s,
how they embraced the warmth of its gentle breezes, the joy of finding a
steady job, and the serene beauty that surrounded them. Sum this up in one
nice sentence from a personal perspective:"
mrole = "system"
mcontent = "You are an expert in geology."
user_role = "user"
```

`mcontent` is reused from the semantic query example ("You are an expert in geology"), but in this case, it doesn't significantly influence the response. Since the user input is highly personal and narrative-driven, the system prompt plays a minimal role.

We could insert external information before the function call if necessary. For example, we could add some information from another source, such as a text message received that day from a family member:

```
text_message='I agree, we had a wonderful time there.'
uninput=text_message+uinput
text_message="Hi, I agree, we had a wonderful time there."
```

Now, we call the function:

```
# Call the function
response = openai_api.make_openai_api_call(
    uinput,mrole,mcontent,user_role)
print(response)
```

We see that the response is mostly episodic with some semantic information:

```
Moving to Hawaii in the 1970s was a transformative experience for my
family, as they found joy in the island's gentle breezes, the security of
steady employment, and the serene beauty that enveloped their new home.
```

Stateless and memoryless verification

We added no memory retention functionality earlier, making the dialogue stateless. Let's check:

```
# API message
uinput = "What question did I just ask you?"
mrole = "system"
mcontent = "You already have this information"
user_role = "user"
```

When we call the function, our dialogue will be forgotten:

```
# API function call
response = openai_api.make_openai_api_call(
    uinput,mrole,mcontent,user_role
)
print(response)
```

The output confirms that the session is memoryless:

```
I'm sorry, but I can't recall previous interactions or questions. Could
you please repeat your question?
```

The API call is stateless because the OpenAI API does not retain memory between requests. If we were using ChatGPT directly, the exchanges would be memorized within that session. This has a critical impact on implementation. It means we have to build our own memory mechanisms to give GenAISys stateful behavior. Let's start with the first layer: short-term memory.

2. Short-term memory session

The goal of this section is to emulate a short-term memory session using a two-step process:

1. First, we initiate a session that goes from user input to a response:

   User input => Generative model API call => Response

   To achieve this first step, we run the session up to the response:

   ```
   uinput = "Hawai is on a geological volcano system. Explain:"
   mrole = "system"
   mcontent = "You are an expert in geology."
   user_role = "user"
   response = openai_api.make_openai_api_call(
       uinput,mrole,mcontent,user_role)
   print(response)
   ```

   The response's output is stored in `response`:

   ```
   "Hawaii is part of a volcanic system known as a hotspot, which is
   a region of the Earth's mantle where heat rises as a thermal plume
   from deep within the Earth. This hotspot is responsible for the
   formation of the Hawaiian Islands. Here's how the process works:…"
   ```

2. The next step is to feed the previous interaction into the next prompt, along with a follow-up question:

   - Explain the situation: `The current dialog session is:`
   - Add the user's initial input: `Hawai is on a geological volcano system. Explain:`
   - Add the response we obtained in the previous call
   - Add the user's new input: `Sum up your previous response in a short sentence in a maximum of 20 words.`

The goal here is to compress the session log. We won't always need to compress dialogues, but in longer sessions, large context windows can pile up quickly. This technique helps keep the token count low, which matters for both cost and performance. In this particular case, we're only managing one response, so we could keep the entire interaction in memory if we wanted to. Still, this example introduces a useful habit for scaling up.

Once the prompt is assembled:

- Call the API function
- Display the response

The scenario is illustrated in the code:

```
ninput = "Sum up your previous response in a short sentence in a
maximum of 20 words."

uinput = (
    "The current dialog session is: " +
    uinput +
    response +
    ninput
)
response = openai_api.make_openai_api_call(
    uinput, mrole, mcontent, user_role
)

print("New response:", "\n\n", uinput, "\n", response)
```

The output provides a nice, short summary of the dialogue:

```
New response: Hawaii's islands form from volcanic activity over a
stationary hotspot beneath the moving Pacific Plate.
```

This functionality wasn't strictly necessary here, but it sets us up for the longer dialogues we'll encounter later in the book. Next, let's build a long-term simulation of multiple sessions.

Keep in mind: Since the session is still in-memory only, the conversation would be lost if the notebook disconnects. Nothing is stored on disk or in a database yet.

3. Long-term memory of multiple sessions

In this section, we're simulating long-term memory by continuing a conversation from an earlier session. The difference here is that we're not just *remembering* a dialogue from a single session—we're *reusing* content from a past session to extend the conversation. At this point, the term "session" takes on a broader meaning. In a traditional copilot scenario, one user interacts with one model in one self-contained session. Here, we're blending sessions and supporting multiple sub-sessions. Multiple users can interact with the model in a shared environment, effectively creating a single global session with branching memory threads. Think of the model as a guest in an ongoing Zoom or Teams meeting. You can ask the AI guest to participate or stay quiet—and when it joins, it may need a recap.

To avoid repeating the first steps of the past conversation, we're reusing the content from the short-term memory session we just ran. Let's assume the previous session is over, but we still want to continue from where we left off:

```
session01=response
print(session01)
```

The output contains the final response from our short-term memory session:

```
Hawaii's islands form from volcanic activity over a stationary hotspot
beneath the moving Pacific Plate.
```

The process in this section will build on the previous session, similar to how you'd revisit a conversation with an online copilot after some time away:

Save previous session => Load previous session => Add it to the new session's scenario

Let's first test whether the API remembers anything on its own:

```
uinput="Is it safe to go there on vacation"
response = openai_api.make_openai_api_call(
    uinput,mrole,mcontent,user_role
)
print(response)
```

The output shows that it forgot the conversation we were in:

```
I'm sorry, but I need more information to provide a helpful response.
Could you specify the location you're considering for your vacation? …
```

The API forgot the previous call because stateless APIs don't retain past dialogue. It's up to us to decide what to include in the prompt. We have a few choices:

- Do we want to remember everything with a large consumption of tokens?
- Do we want to summarize parts or all of the previous conversations?

In a real GenAISys, when an input triggers a request, the AI controller decides which is the best strategy to apply to a task. The code now associates the previous session's context and memory with a new request:

```
ninput = "Let's continue our dialog."
uinput=ninput + session01 + "Would it be safe to go there on vacation?"
response = openai_api.make_openai_api_call(
    uinput,mrole,mcontent,user_role
)
print("Dialog:", uinput,"\n")
print("Response:", response)
```

The response shows that the system now remembers the past session and has enough information to provide an acceptable output:

```
Response: Hawaii is generally considered a safe destination for vacation,
despite its volcanic activity. The Hawaiian Islands are formed by a
hotspot beneath the Pacific Plate, which creates volcanoes as the plate
moves over it. While volcanic activity is a natural and ongoing process
in Hawaii, it is closely monitored by the United States Geological Survey
(USGS) and other agencies…
```

Let's now build a long-term simulation of multiple sessions across different topics.

4. Long-term memory of multiple cross-topic sessions

This section illustrates how to merge two separate sessions into one. This isn't something standard ChatGPT-like platforms offer. Typically, when we start a new topic, the copilot only remembers what's happened in the current session. But in a corporate environment, we may need more flexibility—especially when multiple users are collaborating. In such cases, the AI controller can be configured to allow groups of users to view and merge sessions generated by others in the same group.

Let's say we want to sum up two separate conversations—one about Hawaii's volcanic systems, and another about organizing a geological field trip to Arizona. We begin by saving the previous long-term memory session:

```
session02=uinput + response
print(session02)
```

Then we can start a separate multi-user sub-session from another location, Arizona:

```
ninput ="I would like to organize a geological visit in Arizona."
uinput=ninput+"Where should I start?"
response = openai_api.make_openai_api_call(
    uinput,mrole,mcontent,user_role
)
#print("Dialog:", uinput,"\n")
```

We now expect a response on Arizona, leaving Hawaii out:

```
Response: Organizing a geological visit in Arizona is a fantastic idea, as
the state is rich in diverse geological features. Here's a step-by-step
guide to help you plan your trip:…
```

The response is acceptable. Now, let's simulate long-term memory across multiple topics by combining both sessions and prompting the system to summarize them:

```
session02=response
ninput="Sum up this dialog in a short paragraph:"
uinput=ninput+ session01 + session02
response = openai_api.make_openai_api_call(
    uinput,mrole,mcontent,user_role
)
#print("Dialog:", uinput,"\n")#optional
print("Response:", response)
```

The system's output shows that the long-term memory of the system is effective. We see that the first part is about Hawaii:

```
Response: The dialog begins by explaining the formation of Hawaii's
volcanic islands as the Pacific Plate moves over a stationary hotspot,
leading to active volcanoes like Kilauea.…
```

Then the response continues to the part about Arizona:

```
It then transitions to planning a geological visit to Arizona, emphasizing
the state's diverse geological features. The guide recommends researching
key sites such as the Grand Canyon…
```

We've now covered the core memory modes of GenAISys—from stateless and short-term memory to multi-user, multi-topic long-term memory. Let's now summarize the chapter's journey and move to the next level!

Summary

A business-ready GenAISys offers functionality on par with ChatGPT-like platforms. It brings together generative AI models, agentic features, RAG, memory retention, and a range of ML and non-AI functions—all coordinated by an AI controller. Unlike traditional pipelines, the controller doesn't follow a fixed sequence of steps. Instead, it orchestrates tasks dynamically, adapting to the context.

A GenAISys typically runs on a model such as GPT-4o—or whichever model best fits your use case. But as we've seen, just having access to an API isn't enough. Contextual awareness and memory retention are essential. While ChatGPT-like tools offer these features by default, we have to build them ourselves when creating custom systems.

We explored four types of memory: memoryless, short-term, long-term, and cross-topic. We also distinguished semantic memory (facts) from episodic memory (personal, time-stamped information). Context awareness depends heavily on memory—but context windows have limits. Even if we increase the window size, models can still miss the nuance in complex tasks. That's where advanced RAG comes in—breaking down content into smaller chunks, embedding them, and storing them in vector stores such as Pinecone. This expands what the system can "remember" and use for reasoning.

We also saw that no matter how advanced GenAISys becomes, it can't function without human expertise. From design to deployment, maintenance, and iteration, people remain critical throughout the system's life cycle. We then outlined three real-world implementation models based on available resources and goals: hybrid systems that leverage existing AI platforms, small-scale systems for targeted business needs, and full-scale systems built for ChatGPT-grade performance.

Finally, we got hands-on—building a series of memory simulation modules in Python using GPT-4o. These examples laid the groundwork for what comes next: the AI controller that will manage memory, context, and orchestration across your GenAISys. We are now ready to build a GenAISys AI controller!

Questions

1. Is an API generative AI model such as GPT an AI controller? (Yes or No)
2. Does a memoryless session remember the last exchange(s)? (Yes or No)
3. Is RAG used to optimize context windows? (Yes or No)
4. Are human roles important for the entire life cycle of a GenAISys? (Yes or No)
5. Can an AI controller run tasks dynamically? (Yes or No)
6. Is a small-scale GenAISys built with a limited number of key features? (Yes or No)
7. Does a full-scale ChatGPT-like system require huge resources? (Yes or No)
8. Is long-term memory necessary across multiple sessions? (Yes or No)
9. Do vector stores such as Pinecone support knowledge and AI controller functions? (Yes or No)
10. Can a GenAISys function without contextual awareness? (Yes or No)

References

- Tomczak, J. M. (2024). *Generative AI Systems: A Systems-based Perspective on Generative AI.* `https://arxiv.org/pdf/2407.11001`
- Zewe, A. (2023, November 9). *Explained: Generative AI.* MIT News. Retrieved from `https://news.mit.edu/2023/explained-generative-ai-1109`
- OpenAI models: `https://platform.openai.com/docs/models`
- Bommasani, R., Hudson, D. A., Adeli, E., Altman, R., Arora, S., von Arx, S., ... & Liang, P. (2021). *On the Opportunities and Risks of Foundation Models.* arXiv preprint arXiv:2108.07258. Retrieved from `https://arxiv.org/abs/2108.07258`

Further reading

- Feuerriegel, S., Hartmann, J., Janiesch, C., & Zschech, P. (2023). *Generative AI. Business & Information Systems Engineering.* `https://doi.org/10.1007/s12599-023-00834-7`
- Eloundou et al. (2023). *GPTs are GPTs: An Early Look at the Labor Market Impact Potential of Large Language Models.* `https://arxiv.org/abs/2303.10130`

2

Building the Generative AI Controller

A **generative AI system (GenAISys)**'s controller requires two key components: a **conversational agent** and an **orchestrator**. The conversational agent—powered by a generative AI model—interacts with human users and system processes. The orchestrator, on the other hand, is a set of generative AI and non-AI functions, such as managing user roles, content generation, activating machine learning algorithms, and running classical queries. We need both to build a functional GenAISys.

If we examine this architecture closely, we'll see that software orchestrators and user interfaces date back to the first computers. Any operating system, with even basic functionality, has orchestrators that trigger disk space alerts, memory usage, and hundreds of other functions. Today's user interfaces are intuitive and have event-driven functionality, but at a high level, the underlying architecture of a GenAISys still echoes decades of software design principles. So, what sets a classical software controller apart from a GenAISys controller?

We can sum up the difference in one word: *adaptability*. In a classical software controller, a sequence of tasks is more or less hardcoded. But in a GenAISys, the user interface is a conversational AI agent that is flexible, and the generative AI model behind it is pre-trained to respond to a wide range of requests with no additional coding. This adaptability is achieved by engineering the conversational context to dynamically guide the model's behavior. Furthermore, the orchestrator isn't locked into static flows either; it can modify the tasks it triggers based on the user (human or system) prompts.

In this chapter, we'll take a hands-on approach to building a custom GenAISys based on the architecture of a GenAISys defined in the previous chapter. We'll begin by defining the structure of our AI controller in Python, breaking it into two parts—the conversational agent and the orchestrator—and exploring how the two interact. Then, we'll build the conversational agent using GPT-4o. We'll automate the contextual awareness and memory retention features from *Chapter 1*. Our system will support both short-term and long-term memory, as well as multi-user and cross-session capabilities—pushing it beyond what standard copilots typically offer.

Finally, we will build the structure of an AI controller to interpret user input and trigger a response scenario. The response will be a sentiment analysis or a semantic (hard science) analysis, depending on the context of what the AI controller will analyze and manage. Our custom GenAISys will lay the groundwork for domain-specific RAG, something a standard ChatGPT-grade system can't offer when you're working with large volumes of data, especially in cases of daily dataset updates, such as the daily sales of a product or service. By the end of this chapter, you'll know how to build the foundations of a GenAISys AI controller that we will enhance throughout the book.

To sum up, this chapter covers the following topics:

- Architecture of the AI controller
- Architecture of an AI conversational agent and its workflow
- Implementing the storage of short- and long-term memory sessions in code
- Architecture of an AI orchestrator and the intent functionality
- Creating a GenAI scenario library containing instruction scenarios
- Processing an input with vector search to orchestrate instructions
- Processing an input with a GPT-4o analysis to orchestrate instructions
- Selecting and executing tasks based on the input with the multipurpose orchestrator

Let's begin by defining the architecture of the AI controller.

Architecture of the AI controller

We'll continue to implement the architecture of GenAISys as we've defined in *Figure 1.1* from *Chapter 1*. *Figure 2.1*, on the other hand, takes us further into the underlying functions of a GenAISys.

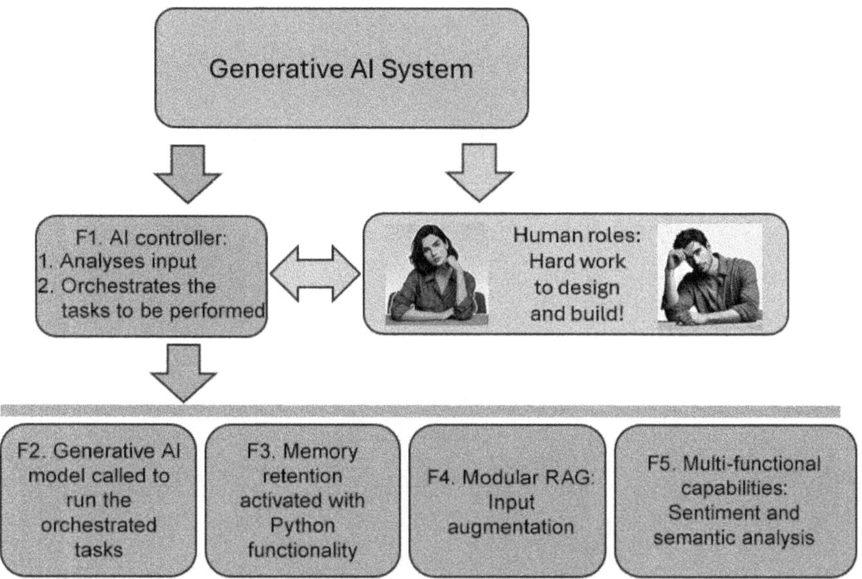

Figure 2.1: Defining the functions to build

We established in the previous chapter that human roles are essential, and the preceding figure acknowledges that fact. We are the core of a GenAISys, no matter how advanced the building blocks (models or frameworks) are. Our first task is designing using our human creativity to find effective ways to implement a GenAISys controller. GenAISys needs human creativity, judgment, and technical decision-making. Under the hood of seamless copilots such as ChatGPT, Gemini, and Microsoft Copilot lie intricate layers of AI and non-AI logic. If we want to build our own ChatGPT-like system, we humans need to do the heavy lifting!

We will build two separate programs:

- A **conversational agent** implemented with GPT-4o, which supports both short- and long-term memory. This will help us enforce contextual awareness across multiple exchanges. It aligns with function **F3** in *Figure 2.1*.
- An **AI controller orchestrator** that will also use GPT-4o to analyze the user input, search a library of instructions, augment the input with the appropriate instructions, and run the function(s) in the instructions.

In this chapter, we'll focus on two scenarios: sentiment analysis and semantic (hard science) analysis, which correspond to functions **F1** and **F2** in our architecture. Functions **F4** and **F5** will be added in *Chapter 3*.

Although these examples are built for OpenAI's API, the logic is model-agnostic. Once you understand how it works, you can adapt the code to use any LLM—such as Meta's Llama, xAI's Grok, Google's Gemini, or Cohere.

Once we've built the conversational agent and controller orchestrator programs separately, we will merge them into a unified intelligence AI controller, as shown in *Figure 2.2*.

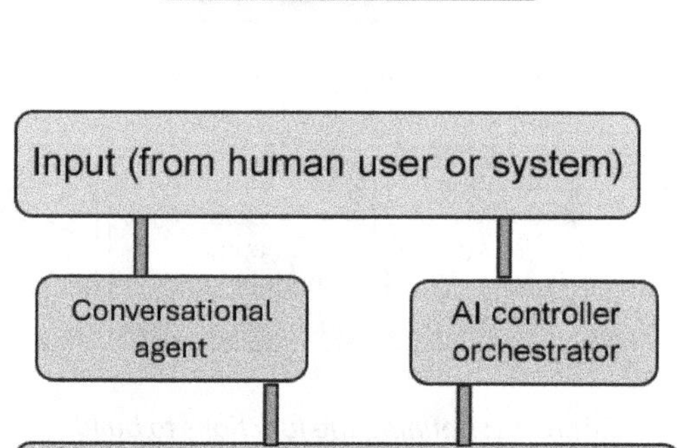

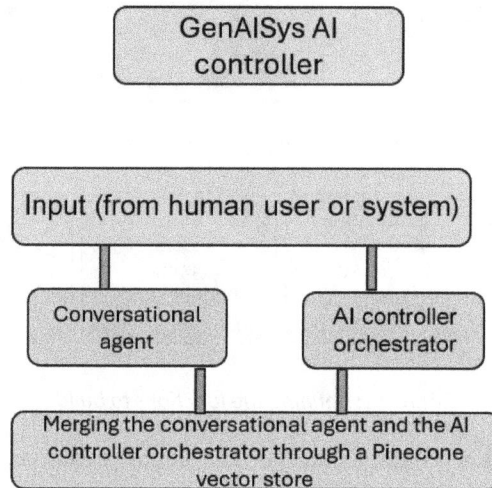

Figure 2.2: Next steps—integrating the AI controller functions through a Pinecone vector store

For now, we need to focus on building each component individually so we can fully understand their behavior. Once that foundation is in place, in *Chapter 3*, we will merge them through a Pinecone vector store. Let's now dive straight down into code and begin developing the conversational agent.

Conversational AI agent

Our two primary goals for this section are to build a conversational AI agent with the following:

- **Short-term memory retention** for a full ChatGPT-like conversational loop. The user and agent can have as many exchanges as they wish; there is no limit to the number of interactions between them.
- **Long-term memory retention** across multiple users and sessions. We'll store in-memory sessions and persist them to a memory storage (in this case, a text file). This will enable multi-user contextual awareness for users such as John, Myriam, and Bob. Our conversational agent will move beyond classic one-to-one ChatGPT-style dialogues toward a custom GenAISys capable of handling multi-session, multi-user interactions.

To get started, open `Conversational_AI_Agent.ipynb` in this chapter's GitHub directory (`https://github.com/Denis2054/Building-Business-Ready-Generative-AI-Systems/tree/main`). This notebook will guide you through the environment setup.

Setting up the environment

We'll reuse the setup process from the previous chapter. If you need a refresher, feel free to revisit that section. Start by installing OpenAI and downloading the required files:

```
!curl -L https://raw.githubusercontent.com/Denis2054/Building-Business-
Ready-Generative-AI-Systems/master/commons/grequests.py --output
grequests.py
from grequests import download
download("commons","requirements01.py")
download("commons","openai_setup.py")
download("commons","openai_api.py")
```

We'll also need to download two additional functions to build our conversational agent:

- `download("commons","conversational_agent.py")`: This contains the functions to manage a full-turn conversation loop and memorize the dialogue.
- `download("commons", "processing_conversations.py")`: This contains tools to load, display, and cleanse past conversations to increase the memory span of the conversational agent across several sessions and users. This custom multisession, multi-user feature goes beyond the scope of standard ChatGPT-like copilots.

Let's now move on to implementing the functions in `conversational_agent.py`, which we'll call throughout our sessions with the conversational AI agent.

Conversational AI agent workflow

The conversation AI agent contains two main parts:

- Starting the initial conversation to initiate a dialogue with the AI agent
- Running the full-turn conversation loop to continue as many in-memory exchanges as a user wishes with the AI agent. At the end of each session, the dialog is saved so it can be resumed later—by the same user or another.

Starting the initial conversation

The initial conversation marks the entry point for a new session. It's handled by the AI controller and illustrated in *Figure 2.3*.

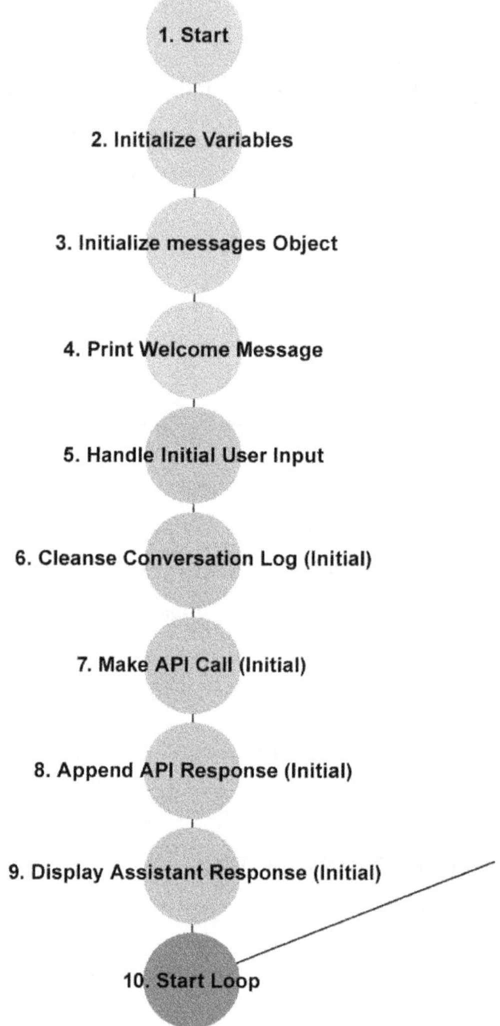

Figure 2.3: The initial conversation controller

We will go through each step of the initial conversation with the generative AI model to understand in detail how a small-scale ChatGPT-like conversational agent works. The 10-step process begins with *Start*.

1. Starting the conversation

The program begins at this entry point through the `run_conversational_agent` function in `openai_api.py`, which will be called in the notebook by `conversational_agent` and its parameters:

```
# Start the conversational agent
def run_conversational_agent(
    uinput, mrole, mcontent, user_role, user_name
):
    conversational_agent(uinput, mrole, mcontent, user_role, user_name)
```

The parameters the conversational agent will process in this case are the following:

- `uinput`: Contains the input (user or system), for example, `Where is Hawaii?`.
- `mrole`: Defines the role of the message. It can be `user` or `system`. You can also assign other roles that the API will interpret, such as defining the AI's persona, for example, `You are a geology expert.`
- `mcontent`: Is what we expect the system to be, for example, `You are a geology expert.`
- `user_role`: Defines the role of the user, for example, `user`.
- `user_name`: The name of the user, for example, `John`.

2–3. Initializing API variables and the messages object

`messages_obj` is initialized with the parameters of the conversation described in the previous step, *Starting the conversation*:

```
messages_obj = [{"role": mrole, "content": mcontent}]
```

`messages_obj` is focusing on the memory of the system. This object will be appended as long as the session lasts with the exchanges with the GPT-4o model. It will be used to log conversations between sessions. The first message contains the role and content for setting up the agent's context.

4. Printing a welcome message

The system is now ready to interact with users. The agent first displays a welcome message and explains how to exit the system once the conversation is over:

```
print("Welcome to the conversational agent! Type 'q' or 'quit' to end the
conversation.")
```

5. Handling the initial user input

The user's initial input is added to messages_obj to provide the agent with memory and provide the direction the agent is expected to follow. The initial user input will be sent from the conversational agent:

```
if initial_user_input:
    print(f"{user_name}: {initial_user_input}")
    messages_obj.append(
        {"role": user_role, "content": initial_user_input}
    )
```

6. Cleansing the initial conversation log

As a practical step in context engineering, messages_obj holds the conversation's history in a structured format. For certain operations within our application, such as generating a simplified display, creating a consolidated log entry, or preparing input for a text-based function, we need to convert this structured log into a single, continuous string. This makes sure that the data is in the correct format for these specific tasks and helps resolve any potential punctuation or formatting quirks that might arise when combining the different message parts:

```
conversation_string = cleanse_conversation_log(messages_obj)
```

The cleansing function cleans the conversation and returns a string:

```
def cleanse_conversation_log(messages_obj):
  conversation_str = " ".join(
        [f"{entry['role']}: {entry['content']}" for entry in messages_obj]
    )
    # Remove problematic punctuations
    return re.sub(r"[^\w\s,.?!:]", "", conversation_str)
```

7. Making the initial API call

The cleansed conversation string is sent to the API for processing. The API provides a response based on the last input and the conversation history. The system now has a memory:

```
agent_response = make_openai_api_call(
    input=conversation_string,
    mrole=mrole,
    mcontent=mcontent,
    user_role=user_role
)
```

8. Appending the initial API response

The assistant's response from the API is processed and appended to `messages_obj`. We are continuing to increase the system's memory and, thus, its contextual awareness:

```
messages_obj.append({"role": "assistant", "content": agent_response})
```

9. Displaying the initial assistant's response

The system's response is displayed for the user to analyze and decide whether to continue or exit the session:

```
print(f"Agent: {agent_response}")
```

10. Starting the conversation loop

The system now enters the conversation loop, where multiple dialogue turns can take place until the user decides to exit the session:

```
while True:
    user_input = input(f"{user_name}: ")
    if user_input.lower() in ["q", "quit"]:
        print("Exiting the conversation. Goodbye!")
        break
```

We are now ready to begin a full-turn conversation loop.

The full-turn conversation loop

The initial conversation is now initialized. We will enter the full-turn conversation loop starting from *step 11* onward, as illustrated in *Figure 2.4*.

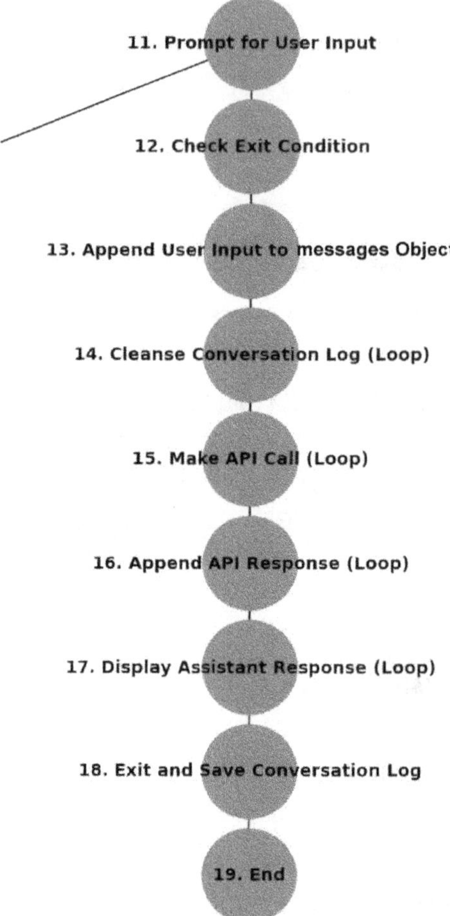

Figure 2.4: The conversation loop starting from step 11

11. Prompting for the user input

The conversation continues the initial dialogue and is memorized through the `messages` object. The user prompt triggers a full-turn conversation loop. The first step is to enter the user's name. This custom takes us beyond the standard ChatGPT-like conversational agents that are limited to one user per session. We are initializing a multi-user conversation:

```
user_input = input(f"{user_name}: ")
```

12. Checking the Exit condition

If q or quit is entered, the session is ended:

```
if user_input.lower() in ["q", "quit"]:
    print("Exiting the conversation. Goodbye!")
    break
```

13. Appending the user input to the messages object

The system is now equipped with a memory of a full-turn conversation loop. It uses the generic API format we defined. The user's input is appended to messages_obj:

```
messages_obj.append({"role": user_role, "content": user_input})
```

14. Cleansing the conversation log (loop)

The updated messages_obj is cleansed to make sure it complies with the API calls, as in *step 6, Cleansing the initial conversation log*:

```
conversation_string = cleanse_conversation_log(messages_obj)
```

15. Making the API call in the conversation loop

In this full-turn conversation loop, the whole conversation is sent to the API. The API will thus return a response based on the context of the whole conversation and the new input:

```
agent_response = make_openai_api_call(
    input=conversation_string,
    mrole=mrole,
    mcontent=mcontent,
    user_role=user_role
)
```

16. Appending the API response in the conversation loop

The API's response is appended to messages_obj at each conversation turn:

```
messages_obj.append({"role": "assistant", "content": agent_response})
```

17. Displaying the assistant's response

The API response is displayed at each conversation turn in the loop:

```
print(f"Agent: {agent_response}")
```

18. Exiting and saving the conversation log

When a user exits the loop, the conversation is saved. This feature will replicate a ChatGPT-like platform that can save dialogue between two sessions with the same user. However, as we will see in our implementation of a conversational agent in the *Running the conversational agent* section, our program will be able to save a multi-user session in a conversation between team members:

```
with open("conversation_log.txt", "w") as log_file:
    log_file.write("\n".join([f"{(user_name if entry['role'] == 'user'
else entry['role'])}: {entry['content']}" for entry in messages_obj]))
```

19. End

The conversational agent terminates the session after memorizing the conversation:

```
print("Conversation saved to 'conversation_log.txt'.")
```

We have explored the conversational agent's functionality.

Now, let's move on to the AI conversational agent program that represents an AI controller.

Running the conversational AI agent

The main program, `Conversational_AI_Agent.ipynb`, calls the necessary functions from `conversational_agent.py` to handle AI interactions. We will be running a conversation through three user sessions with this scenario:

1. John begins with a short-term memory session with the conversational AI agent.
2. John's conversation will be saved in a log file when the session is over.
3. Myriam resumes the session using that same log file.
4. Myriam's conversation will be saved in the same log file as John's when the session is over.
5. Bob will pick up where John and Myriam left off.
6. Bob's conversation will be saved in the same log file as John's and Myriam's when the session is over.

All three users interact in successive sessions. In *Chapter 3*, we'll go further by grouping users through a Pinecone vector store so that multiple users can participate together in a session in real time. For the moment, let's walk through this multi-user setup step by step and see how the conversational AI agent handles these sessions. Let's begin with the first step: John's short-term memory session.

Short-term memory session

The session begins with the parameters described in *step 1, Starting the conversation,* of the conversational agent:

```
uinput = "Hawai is on a geological volcano system. Explain:"
mrole = "system"
mcontent = "You are an expert in geology."
user_role = "user"
```

We are also adding the name of the user like in a ChatGPT-like session:

```
user_name = "John"
```

This simple addition—user_name—is what takes our GenAISys beyond standard ChatGPT-like platforms. It allows us to associate memory with specific users and expand into multi-user conversations within a single system.

We will now import the first function, the OpenAI API functionality, to make a request to OpenAI's API, as described in *Chapter 1*:

```
from openai_api import make_openai_api_call
```

The program now imports the second function, the conversational agent, and runs it as described earlier in this section:

```
from conversational_agent import run_conversational_agent
run_conversational_agent(uinput, mrole, mcontent, user_role,user_name)
```

Let's go through each step of the dialog implemented with our two functions. The agent first welcomes us:

```
Welcome to the conversational agent! Type 'q' or 'quit' to end the
conversation.
```

John, the first user, asks for a geological explanation about Hawaii:

```
John: Hawai is on a geological volcano system. Explain:
```

The agent provides a satisfactory answer:

```
Agent: Hawaii is part of a geological volcanic system known as a
"hotspot"...
```

John now asks about surfing “there”:

```
John: Can we surf there?
```

Thanks to the memory we built into the agent, it now has contextual awareness through memory retention. The agent correctly responds about surfing in Hawaii:

```
Agent: Yes, you can definitely surf in Hawaii! The Hawaiian Islands are
renowned …
```

John now asks about the best places to stay without mentioning Hawaii:

```
John: Where are the best places to stay?
```

The agent answers correctly using contextual awareness:

```
Agent: Hawaii offers a wide range of accommodations …
```

John then quits the session:

```
John: quit
```

The agent exits the conversation and saves the dialogue in a conversation log:

```
Agent:Exiting the conversation. Goodbye!
Conversation saved to 'conversation_log.txt'.
```

The short-term session ends, but thanks to memory retention via `conversation_log.txt`, we can easily pick up from where John left off. We can thus continue the dialogue immediately or at a later time, leveraging memory retention through the `conversation_log.txt` file that was automatically generated.

Long-term memory session

The short-term session is saved. We have three options:

- Stop the program now. In this case, `conversation_log.txt` will only contain John’s session, which can be continued or not.
- Decide to initialize a separate `conversation_log.txt` for the next user, Myriam.
- Continue with a multi-user session by loading John’s conversation into Myriam’s initial dialog context.

The program in this chapter chooses to continue a multi-session, multi-user scenario.

The first step to continue the conversation with John is to load and display the conversation log using the function in `processing_conversations.py` that we downloaded in the *Setting up the environment* section. We now import and run the function that we need to load and display the conversation log:

```
from processing_conversations import load_and_display_conversation_log
conversation_log = load_and_display_conversation_log()
```

The function is a standard `IPython` process using HTML functionality that reads and displays the conversation:

```
from IPython.core.display import display, HTML
import re
# Step 1: Load and Display Conversation Log
def load_and_display_conversation_log():
    try:
        with open("conversation_log.txt", "r") as log_file:
            conversation_log = log_file.readlines()
        # Prepare HTML for display
        html_content = "<h3>Loaded Conversation Log</h3><table
border='1'>"
        for line in conversation_log:
            html_content += f"<tr><td>{line.strip()}</td></tr>"
        html_content += "</table>"
        # Display the HTML
        display(HTML(html_content))
        return conversation_log
    except FileNotFoundError:
        print("Error: conversation_log.txt not found. Ensure it exists in
the current directory.")
        return []
```

The output displays each participant in the conversation, beginning with the system's information, followed by John's request, and then the GPT-4o assistant's response at each turn:

```
system: You are an expert in geology.
John: Hawai is on a geological volcano system. Explain:
assistant: Hawaii is part of a geological volcanic system…
```

Before adding the conversation to the context of the next input, we will clean and prepare it. To achieve this, we successively import `cleanse_conversation_log` and import `initialize_uinput` from `processing_conversations.py`:

```
from processing_conversations import cleanse_conversation_log
from processing_conversations import initialize_uinput
```

Then, we will call the two Python functions that we defined to cleanse and then prepare the new input:

```
cleansed_log = cleanse_conversation_log(conversation_log)
nuinput = initialize_uinput(cleansed_log)
```

The `cleanse` function removes punctuation and potentially problematic characters:

```
# Step 2: Clean the conversation log by removing punctuations and special
characters
def cleanse_conversation_log(conversation_log):
    cleansed_log = []
    for line in conversation_log:
        # Remove problematic punctuations and special characters
        cleansed_line = re.sub(r"[^\w\s,.?!:]", "", line)
        cleansed_log.append(cleansed_line.strip())
    return " ".join(cleansed_log)  # Combine all lines into a single
string
```

Finally, we initialize the new input:

```
# Step 3: Initialize `uinput` with the cleansed conversation log to
continue the conversation
def initialize_uinput(cleansed_log):
    if cleansed_log:
        print("\nCleansed conversation log for continuation:")
        print(cleansed_log)
        return cleansed_log  # Use the cleansed log as the new input
    else:
        print("Error: No data available to initialize `uinput`.")
        return ""
```

The output confirms that the conversation log has been cleansed:

```
Cleansed conversation log for continuation:
system: You are an expert in geology…
```

Then, the output confirms that `nuinput` contains the conversation log for continuation:

```
# `nuinput` now contains the cleansed version of the conversation log and
can be used
print("\nInitialized `nuinput` for continuation:", nuinput)
```

Continuing the previous session

We can now continue the conversation that John began with `nuinput` as the memory retention variable for contextual awareness. We will add the context, `nuinput`, to Myriam's request using the message variables as before:

```
ninput = nuinput+ "What about surfing in Long Beach"
mrole = "system"
mcontent = "You are an expert in geology."
user_role = "user"
user_name = "Myriam"
```

The message call contains two key features:

- `ninput = nuinput+ [user input]`, which shows that the AI controller now has a long-term memory that goes beyond a single session
- `user_name = "Myriam"`, which shows the multi-user feature, proving that our custom small-scale ChatGPT-like AI controller has more flexibility than a standard copilot

The overall process is the same as with John. Myriam asks a question:

```
Myriam: What about surfing in Long Beach
```

The agent responds:

```
Agent:Long Beach, California, offers a different surfing experience
compared to Hawai…
```

Myriam quits:

```
Myriam: quit
```

The agent confirms that the conversation has ended and is saved to the conversation log:

```
Agent:Exiting the conversation. Goodbye!
Conversation saved to 'conversation_log.txt'.
```

The AI controller now has a log of John's session and Myriam's continuation of the session. The controller can take this further and add yet another user to the conversation.

Continuing the long-term multi-user memory

Let's add Bob to the mix to continue the conversation. First, display the conversation log again:

```
# Run the process
conversation_log = load_and_display_conversation_log()
```

You'll see entries for both John and Myriam:

```
system: You are an expert in geology.
Myriam: system: You are an expert …
```

The log is then cleansed and prepared for the next turn of the conversation as previously. `nuinput` now contains John and Myriam's sessions:

```
uinput =nuinput+ "Read the whole dialog then choose the best for geology
research"
mrole = "system"
mcontent = "You are an expert in geology."
user_role = "user"
user_name = "Bob"
```

Bob is focused on the geological mission, not leisure:

```
Bob:"Read the whole dialog then choose the best for geology research"
```

The AI agent provides an accurate response:

```
Agent: For geology research, the most relevant part of the dialogue is the
explanation of Hawaii's geological volcanic system. This section provides
detailed insights into the Hawaiian hotspot, mantle plumes, volcanic
activity,…
```

Bob then quits the session:

```
Bob: quit
```

The agent exits the conversation and saves it in the conversation log:

```
Agent:Exiting the conversation. Goodbye!
Conversation saved to 'conversation_log.txt'.
```

With these three scenarios, we have implemented a conversational agent managed by the AI controller in a multi-user full-turn conversational loop. Let's examine the next steps for this conversational agent.

Next steps

At this point, we have the basic structure of a conversational agent. We need to integrate it into an AI controller orchestrator. Let's sum up the work we did for the conversational agent before beginning to build the AI controller orchestrator.

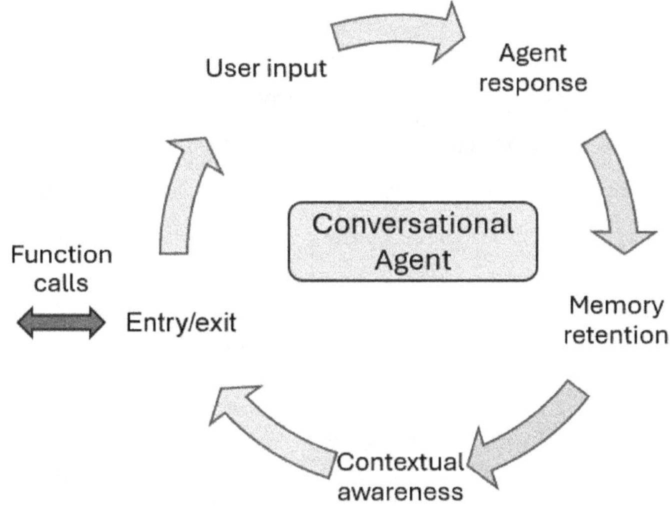

Figure 2.5: The cycle of a conversational agent loop

As illustrated in the preceding figure, the AI conversation agent does the following:

1. The agent processes the input (system or human user).
2. The agent responds.
3. The memory retention function is activated.
4. The conversation is added to the following input as context.
5. The user can quit.

However, the entry/exit point is incomplete. We can enter and exit the conversation but cannot call functions to orchestrate tasks such as activating sentiment analysis and semantic analysis. To complete the architecture of the AI controller, we need to begin building the AI controller orchestrator.

AI controller orchestrator

In this section, we will build the first component of our AI controller orchestrator: the ability to select the right task to perform. We develop this component as a standalone component that we will integrate starting from *Chapter 3*, where we will bridge the conversational agent with the AI controller orchestrator through a Pinecone vector store.

Figure 2.6 illustrates the workflow of the AI controller orchestrator we'll be developing:

- **C1. AI controller entry point input** triggers the process.
- **C2. Analyzes input,** which could be a system or human user prompt.
- **C3. Embeds user input** through GPT-4o's native functionality.
- **C4. Embeds task scenario repository** through GPT-4o's native functionality.
- **C5. Selects a scenario** to execute a task that best matches the input.
- **C6. Executes the scenario** selected by the AI controller orchestrator.

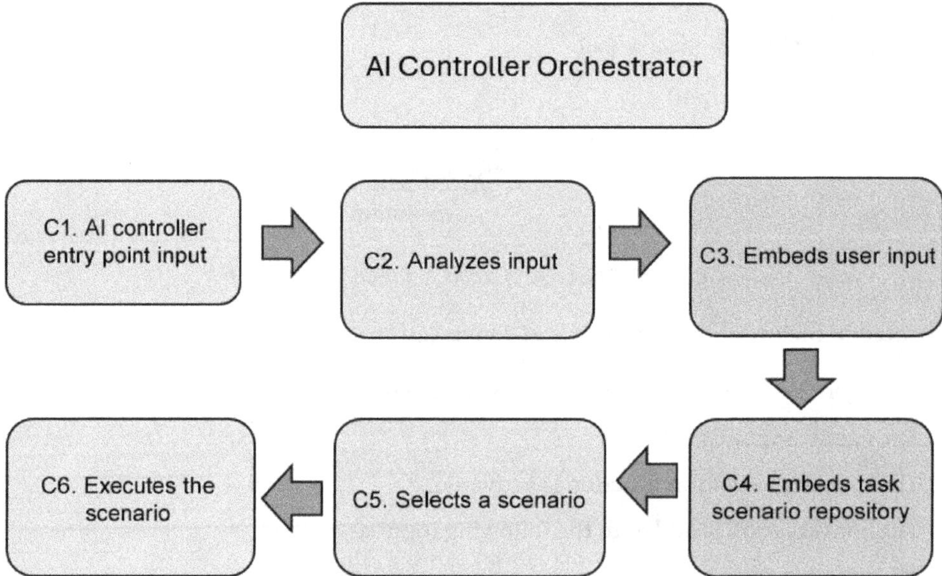

Figure 2.6: Workflow of the AI controller orchestrator

We'll develop this first component of the AI controller orchestrator with OpenAI's GPT-4o API and Python. Additionally, since the idea is to leverage the full power of the generative AI model to perform several tasks requested by the AI controller orchestrator, we will thus avoid overloading the orchestrator with additional libraries to focus on the architecture of the GenAISys.

In this notebook, GPT-4o will perform three key functions in the program, as shown in *Figure 2.7*:

- **Embedding**: GPT-4o systematically embeds all the data it receives through a prompt. The input is embedded before going through the layers of the model. In *Chapter 3*, we will take this further by embedding and upserting reusable data such as instruction scenarios into a Pinecone vector store.
- **Similarity search**: GPT-4o can perform a similarity search with reliable results. GPT-4o doesn't have a deterministic fixed cosine similarity function. It learns to understand relationships through its complex neural network, mimicking similarity judgments in a much more nuanced, less deterministic way.
- **Task execution**: Once a scenario is chosen, GPT-4o can execute a number of standard tasks, such as sentiment and semantic analysis.

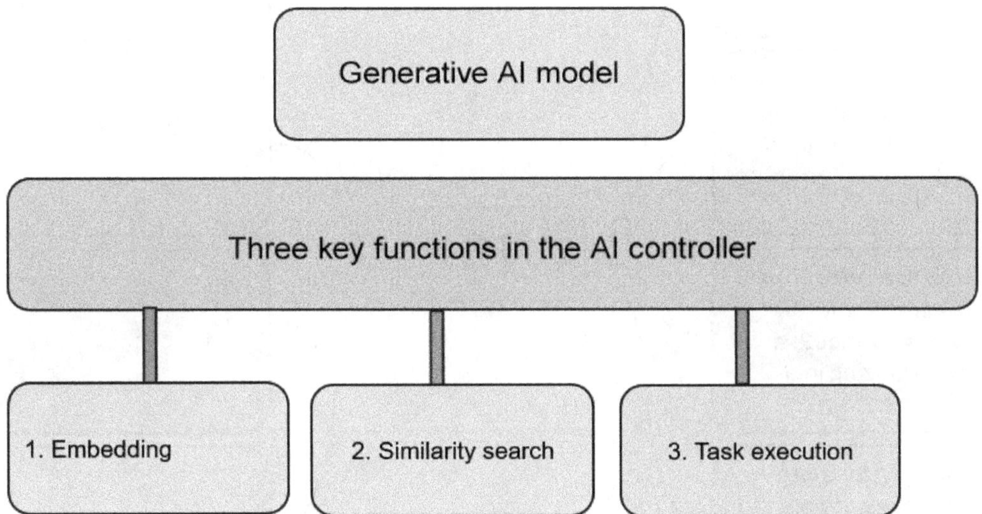

Figure 2.7: Triggering tasks with similarity searches in a list of instructions

We have defined the workflow of the orchestrator and the generative AI model's usage. However, we must examine how a model identifies the task it is expected to perform.

Understanding the intent functionality

No matter how powerful a generative AI model such as GPT-4o is, it cannot guess what a user wants without a prompt that explicitly expresses *intent*. We cannot just say, "The Grand Canyon is a great place to visit in Arizona" and expect the model to guess that we want a sentiment analysis done on our statement. We have to explicitly formulate our intent by entering: "Provide a sentiment analysis of the following text: The Grand Canyon is a great place to visit in Arizona."

To resolve the issue of intent for an AI controller, we have to find a framework for it to orchestrate its tasks. A good place to start is to study the **Text-to-Text Transfer Transformer** (**T5**), which is a text-to-text model (Raffel et al., 2020). A T5 model uses *task tags* or *task-specific prefixes* to provide the intent of a prompt to the transformer model. A task tag contains instructions such as summarization, translation, and classification. The model will detect the tag and know what to do, as shown in *Figure 2.8*.

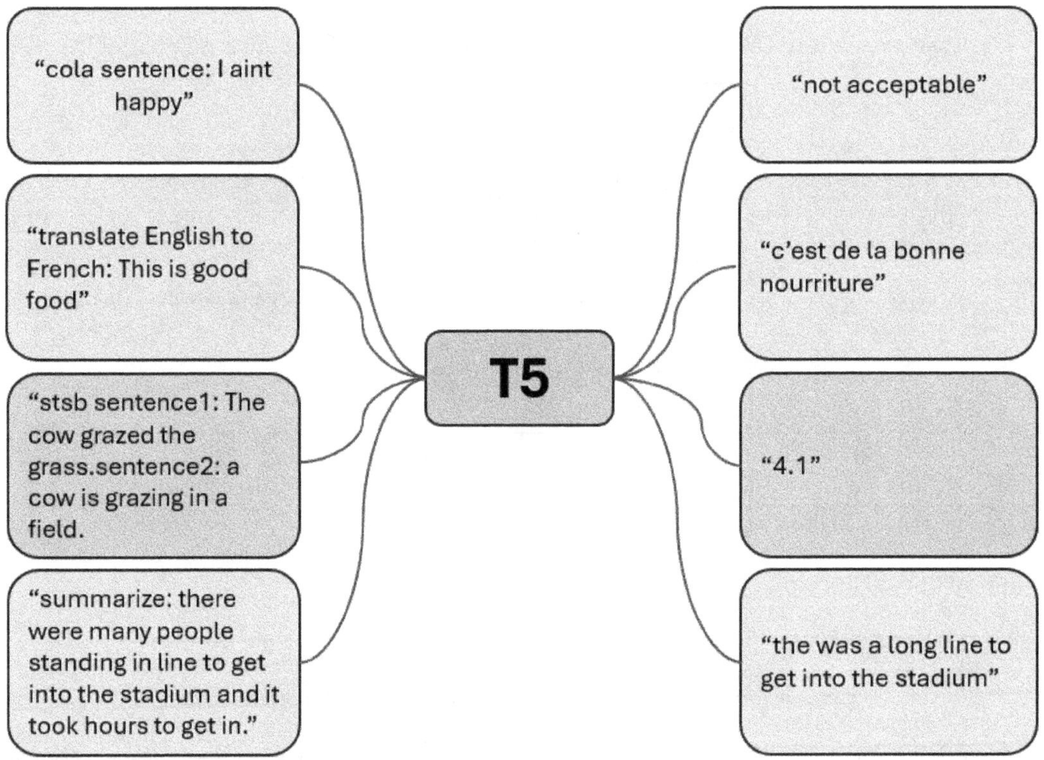

Figure 2.8: T5 with task tags

Training a T5 model involves *explicitly* adding a task tag when creating an input and then providing the response. However, OpenAI GPT models learn which task to perform by analyzing billions of sequences of language, not explicit structures, that contain instructions and responses. A generative AI model using GPT-like architectures will thus learn which task to perform *implicitly* through the context of the prompt. For example, a well-parsed prompt such as "Provide a sentiment analysis of the following text: The Grand Canyon is a great place to visit in Arizona." contains enough context for GPT-4o to infer the desired operation—without requiring an explicit tag.

Let's illustrate how a GPT model works by running T5-style examples with GPT-4o's implicit analysis of which task needs to be performed.

From T5 to GPT models

In this section, we'll write a program to show how GPT-4o interprets instructions—a capability we'll leverage in our orchestrator. The aim is to demonstrate that, although GPT-style models infer intent implicitly, they still need clear instructions.

We'll begin by opening `T52GPT.ipynb` in the `Chapter02` directory on GitHub. Set up the environment exactly as in the *Setting up the environment* subsection of the *Conversational AI agent* section, installing only the OpenAI environment:

```
download("commons","requirements01.py")
download("commons","openai_setup.py")
download("commons","openai_api.py")
```

No additional installations are required. Let's now begin with a CoLA task.

Corpus of Linguistic Acceptability (CoLA)

The **Corpus of Linguistic Acceptability** (**CoLA**) is a public dataset of short English sentences, each tagged as acceptable (grammatical) or unacceptable (ungrammatical). By testing GPT-4o on these examples, we can show that advanced generative models can tackle new tasks purely by understanding language, without any task-specific fine-tuning. This means that we can apply advanced generative AI models to a wide range of tasks we didn't train them for.

Let's first submit the following input to the GPT-4o model to see whether it is acceptable without an explicit task tag:

```
input="This aint the right way to talk."
```

We will provide minimal information to the system:

```
mrole = "system"
user_role = "user"
mcontent = "Follow the instructions in the input"
```

We'll also make an OpenAI API call with the function we have been using throughout this chapter:

```
# API function call
task_response = openai_api.make_openai_api_call(
    input,mrole,mcontent,user_role
)
print(task_response)
```

The output shows that even one of the most powerful generative AI models doesn't have a clue about what to do without a task tag:

```
I apologize if my previous response didn't meet your expectations. Please
let me know how I can assist you better!
```

Now, let's write an instruction with a task tag and the same message:

```
input="Is the following sentence gramatically correct:This aint the right
way to talk."
mrole = "system"
user_role = "user"
mcontent = "Follow the instructions in the input"
# API function call
task_response = openai_api.make_openai_api_call(
    input,mrole,mcontent,user_role
)
print(task_response)
```

The input now contains an indication of what is expected of the generative AI model. The output is now accurate:

```
The sentence "This aint the right way to talk." is not grammatically
correct. The response corrects the sentence:
"This isn't the right way to talk."
Alternatively, if you want to maintain the informal tone, you could write:
"This ain't the right way to talk."
```

```
Note that "ain't" is considered informal and nonstandard in formal
writing.
```

Let's now perform a translation task.

Translation task

The task begins with a task tag that is expressed in natural language:

```
input="Translate this sentence into French: Paris is quite a city to
visit."
mrole = "system"
user_role = "user"
mcontent = "Follow the instructions in the input"
# API function call
task_response = openai_api.make_openai_api_call(
    input,mrole,mcontent,user_role
)
print(task_response)
```

The output we get is accurate:

```
Paris est vraiment une ville à visiter.
```

Let's now perform a **Semantic Textual Similarity Benchmark (STSB)** task.

Semantic Textual Similarity Benchmark (STSB)

STSB-style scoring is an important feature for a GenAISys AI controller, which depends on similarity searches to pick the right instruction scenarios, documents, and other resources. The orchestrator will rely on this very capability. In the test that follows, we submit two sentences to the model and ask it to judge their semantic similarity:

```
input="stsb:Sentence 1: This is a big dog. Sentence 2: This dog is very
big."
mrole = "system"
user_role = "user"
mcontent = "Follow the instructions in the input"
# API function call
task_response = openai_api.make_openai_api_call(
    input,mrole,mcontent,user_role)
print(task_response)
```

The output we get is accurate:

```
The sentences "This is a big dog." and "This dog is very big." are
semantically similar. Both sentences convey the idea that the dog
in question is large in size. The difference in wording does not
significantly alter the meaning, as both sentences describe the same
characteristic of the dog.
```

This function will prove to be very useful when we're searching for data that matches the input in a dataset. Let's now run a summarization task.

Summarization

In the following input, GPT-4o can detect the summarization instruction tag and also interpret the maximum length of the response required:

```
input="Summarize this text in 10 words maximum: The group walked in the
forest on a nice sunny day. The birds were singing and everyone was
happy."
mrole = "system"
user_role = "user"
mcontent = "Follow the instructions in the input"
# API function call
task_response = openai_api.make_openai_api_call(
    input,mrole,mcontent,user_role)
print(task_response)
```

The output is once again accurate:

```
Group enjoyed a sunny forest walk with singing birds.
```

The takeaway of this exploration is that no matter which generative AI model we implement, it requires task tags to react as we expect. Next, we'll use this insight to implement semantic textual similarity in our orchestrator for processing task tags.

Implementing the orchestrator for instruction selection

In this section, we will begin building the orchestrator for two instructions based on task tags, as shown in *Figure 2.9*: sentiment analysis to determine the sentiment of a sentence and semantic analysis to analyze the facts in a sentence.

We will make the system more complex by asking the generative AI model to find the best task tag scenario (sentiment or semantic analysis) based on the input. In other words, the task tag will not be part of the input. We will use GPT-4o's semantic textual similarity features to choose the right task tag itself.

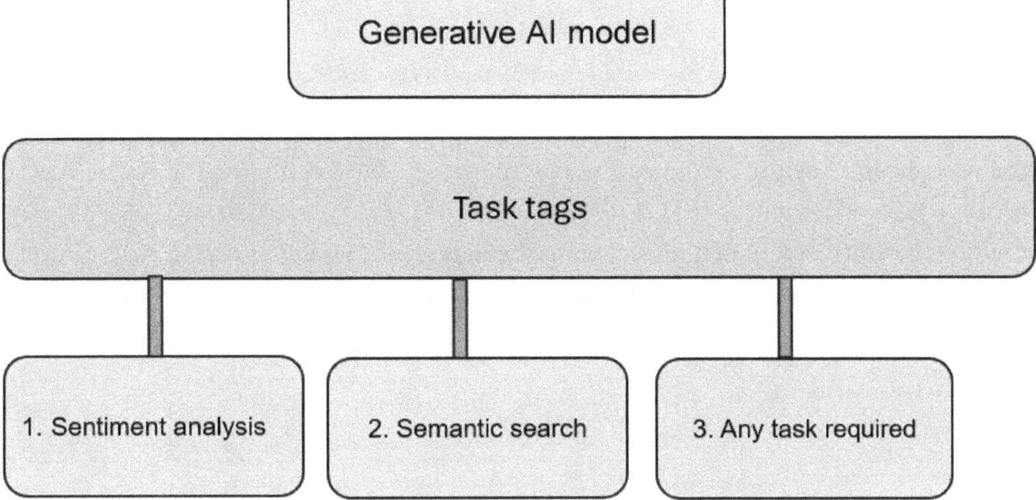

Figure 2.9: Running tasks with implicit task tags

Eventually, our orchestrator will support any task (see **3. Any Task required** in *Figure 2.9*), not just sentiment or semantic analysis.

Setting up the environment is the same as earlier:

```
download("commons","requirements01.py")
download("commons","openai_setup.py")
download("commons","openai_api.py")
```

No additional installations are required for the orchestrator. We will begin by implementing an instruction scenario selection.

Selecting a scenario

The core of an AI controller is to decide what to do when it receives an input (system or human user). The selection of a task opens a world of possible methods that we will explore throughout the book. However, we can classify them into two categories:

- Using an explicit task tag to trigger an instruction. This tag can be a context in a generative AI model and expressed freely in various ways in a prompt.
- The prompt has no task instruction but instead a repository of scenarios from which the AI controller will make decisions based on semantic textual similarity.

Here, we'll explore the second, more proactive approach. We'll test two prompts with no instructions, no task tag, and no clue as to what is expected of the generative AI model. Although we will implement other, more explicit approaches later with task tags, a GenAISys AI controller orchestrator must be able to be proactive in certain situations.

- The first prompt is an opinion on a movie, implying that a sentiment analysis might interest the user:

```
if prompt==1:
    input = "Gladiator II is a great movie although I didn't like
some of the scenes. I liked the actors though. Overall I really
enjoyed the experience."
```

- The second prompt is a fact, implying that a semantic analysis might interest the user:

```
if prompt==2:
    input = "Generative AI models such as GPT-4o can be built into
Generative AI Systems. Provide more information."
```

To provide the AI controller with decision-making capabilities, we will need a repository of instruction scenarios.

Defining task/instruction scenarios

Scenarios are sets of instructions that live in a repository within a GenAISys. While ChatGPT-like models are trained to process many instructions natively, domain-specific use cases need custom scenarios (we'll dive into these starting from *Chapter 5*). For example, a GenAISys could receive a message such as `Customer order #9283444 is late`. The message could be about a production delay or a delivery delay. By examining the sender's username and group (production or delivery department), the AI controller can determine the context and, selecting a scenario, take an appropriate decision.

In this notebook, the scenarios are stored in memory. In *Chapter 3*, we will organize the storage and retrieval of these instruction sets in Pinecone vector stores.

In both cases, we begin by creating a repository of structured scenarios (market, sentiment, and semantic analysis):

```
scenarios = [
    {
        "scenario_number": 1,
        "description": "Market Semantic analysis.You will be provided with
a market survey on a give range of products.The term market must be in the
user or system input. Your task is provide an analysis."
    },

    {
        "scenario_number": 2,
        "description": " Sentiment analysis  Read the content and
classify the content as an opinion  If it is not opinion, stop there  If
it is an opinion then your task is to perform a sentiment analysis on
these statements and provide a score with the label: Analysis score:
followed by a numerical value between 0 and 1  with no + or - sign.Add an
explanation."
    },
    {
        "scenario_number": 3,
        "description": "Semantic analysis.This is not an analysis but a
semantic search. Provide more information on the topic."
    }
 ]
```

We will also add a dictionary of the same scenarios, containing simple definitions of the scenarios:

```
# Original list of dictionaries
scenario_instructions = [
    {
        "Market Semantic analysis.You will be provided with a market
survey on a give range of products.The term market must be in the user or
system input. Your task is provide an analysis."
```

```
    },
    {
        "Sentiment analysis  Read the content return a sentiment analysis
on this text and provide a score with the label named : Sentiment analysis
score followed by a numerical value between 0 and 1  with no + or - sign
and  add an explanation to justify the score."
    },
    {
        "Semantic analysis.This is not an analysis but a semantic search.
Provide more information on the topic."
    }
]
```

We now extract the strings from the dictionary and store them in a list:

```
# Extract the strings from each dictionary
instructions_as_strings = [
    list(entry)[0] for entry in scenario_instructions
]
```

At this point, our AI controller has everything it needs to recognize intent—matching any incoming prompt to the best-fitting scenario.

Performing intent recognition and scenario selection

We first define the parameters of the conversational AI agent just as we did in the *Conversational AI agent* section:

```
# Define the parameters for the function call
mrole = "system"
mcontent = "You are an assistant that matches user inputs to predefined
scenarios. Select the scenario that best matches the input. Respond with
the scenario_number only."
user_role = "user"
```

The orchestrator's job is to find the best task for any given input, making the AI controller flexible and adaptive. In some cases, the orchestrator may decide not to apply a scenario and just follow the user's input. In the following example, however, the orchestrator will select a scenario and apply it.

We now adjust the input to take the orchestrator's request into account:

```
# Adjust `input` to combine user input with scenarios
selection_input = f"User input: {input}\nScenarios: {scenarios}"
print(selection_input)
```

GPT-4o will now perform a text semantic similarity search as we ran in the *Semantic Textual Similarity Benchmark (STSB)* section. In this case, it doesn't just perform a plain text comparison, but matches one text (the user input) against a list of texts (our scenario descriptions):

```
# Call the function using your standard API call
response = openai_api.make_openai_api_call(
    selection_input, mrole, mcontent, user_role
)
```

Our user input is as follows:

```
User input: Gladiator II is a great movie
```

Then, the scenario is chosen:

```
# Print the response
print("Scenario:",response )
```

The scenario number is then chosen, stored with the instructions that go with it, and displayed:

```
scenario_number=int(response)
instructions=scenario_instructions[scenario_number-1]
print(instructions)
```

For our *Gladiator II* example, the orchestrator correctly picks the sentiment analysis scenario:

```
{'Sentiment analysis  Read the content return a sentiment analysis on this
text and provide a score with the label named : Sentiment analysis score
followed by a numerical value between 0 and 1  with no + or - sign and
add an explanation to justify the score.'}
```

This autonomous task-selection capability—letting GenAISys choose the right analysis without explicit tags—will prove invaluable in real-world deployments (see *Chapter 5*). The program now runs the scenarios with the generative AI agent.

Running scenarios with the generative AI agent

Now that the AI controller has identified the correct `scenario_number`, it's time to execute the selected task. In this notebook, we'll walk through that process step by step. We first print the input:

```
print(input)
```

Using the `scenario_number` value, we access the scenario description from our `instructions_as_strings` list:

```
# Accessing by line number (1-based index)
line_number = scenario_number
instruction = instructions_as_strings[line_number - 1]  # Adjusting for
0-based indexing
print(f"Instruction on line {line_number}:\n{instruction}")
mrole = "system"
user_role = "user"
mcontent = instruction
```

The orchestrator is now ready to run a sentiment analysis.

Sentiment analysis

We append the description of the scenario to the original user prompt and send the combined request to GPT-4o:

```
Instruction on line 2:
Sentiment analysis  Read the content return a sentiment analysis nalysis
on this text and provide a score with the label named : Sentiment analysis
score followed by a numerical value between 0 and 1  with no + or - sign
and  add an explanation to justify the score.
```

```
# API function call
sc_input=instruction +" "+ input
print(sc_input)
task_response = openai_api.make_openai_api_call(
    sc_input,mrole,mcontent,user_role
)
print(task_response)
```

For our *Gladiator II* example, the response might look like this:

```
Sentiment analysis score 0.75
The text expresses a generally positive sentiment towards the movie
"Gladiator II." The use of words like "great movie," "liked the actors,"
and "really enjoyed the experience" indicates a favorable opinion.
However, the mention of not liking some of the scenes introduces a slight
negative element. Despite this, the overall enjoyment and positive remarks
about the actors and the movie as a whole outweigh the negative aspect,
resulting in a sentiment score leaning towards the positive side.
```

The response shows that the orchestrator found a scenario that matches the input and produces an acceptable output. Now, let's go back, change the prompt, and see whether the orchestrator finds the right scenario.

Semantic analysis

The goal now is to verify, without changing a single line of code, whether the orchestrator can access another scenario. The orchestrator will rely on GPT-4o's native ability to perform semantic text similarity searches.

We will now activate prompt 2:

```
prompt=2
…
if prompt==2:
    input = "Generative AI models such as GPT-4o can be built into
Generative AI Systems. Provide more information."
```

This input clearly calls for a semantic analysis rather than sentiment analysis. We then reuse the exact same code as our sentiment analysis search:

```
# Accessing by line number (1-based index)
line_number = scenario_number
instruction = instructions_as_strings[line_number - 1]  # Adjusting for
0-based indexing
print(f"Instruction on line {line_number}:\n{instruction}")
mrole = "system"
user_role = "user"
mcontent = instruction
```

The output shows that the right scenario was found:

```
Instruction on line 3:
Semantic analysis.This is not an analysis but a semantic search. Provide
more information on the topic.
```

The task response is displayed:

```
print(task_response)
```

The output shows that the orchestrator produces a coherent semantic analysis:

```
Generative AI models, like GPT-4, are advanced machine learning models
designed to generate human-like text based on the input they receive….
```

This demonstrates that in some cases, the orchestrator will be able to find the right scenarios without task tags. This will prove useful when we tackle more complex workflows, such as advanced production and support.

Summary

The first takeaway from this chapter is the central role of humans in a GenAISys. Human design drove the creation of both our conversational agent and orchestrator. We started developing these two complex components with simply an OpenAI API and Python, yet we *humans* designed the initial levels of the AI controller that powers our custom GenAISys. The basic GenAISys rule will always apply: no human roles, no GenAISys. We design AI systems, implement them, maintain them, and evolve them based on ongoing feedback.

The second takeaway is how our conversational AI agent goes beyond a small-scale ChatGPT-like structure. We not only built short-term context and memory retention for a full-turn dialogue, but we also added long-term memory across multiple users and multiple topics. Our dialogue included three users (John, Myriam, and Bob) and two topics (geology and surfing). As we progress through the book, we will expand the scope of these multi-user, multi-topic sessions to use cases where team cooperation is essential.

The third takeaway concerns our AI controller orchestrator. We gave the orchestrator a small scenario dataset containing custom instructions that we can expand for a domain-specific use case, and then leveraged GPT-4o to both select the appropriate scenario and execute the task itself.

At this point, we have a conversational agent and a nascent AI controller orchestrator. When we assemble our AI controller, they will together form a unique multi-user, multi-domain customized GenAISys. To build our multi-user, multi-domain GenAISys AI controller, we will now build a Pinecone vector store in the next chapter.

Questions

1. A ChatGPT-like GenAISys only needs a generative AI model such as GPT-4o. (True or False)
2. A ChatGPT-like GenAISys doesn't require an AI controller. (True or False)
3. Human roles are critical when building and running GenAISys. (True or False)
4. Generally, not always, a generative AI model such as GPT-4o contains a task tag in one form or the other. (True or False)
5. Sometimes, not always, a generative model can find the most probable task to perform without a task tag. (True or False)
6. Semantic text similarity cannot be natively performed by GPT-4o. (True or False)
7. A full-turn generative AI conversation loop with an OpenAI API AI requires coding. (True or False)
8. Long-term memory AI conversation sessions are never necessary. (True or False)
9. Summarizing a text can only be done in English by GPT-4o. (True or False)
10. An AI controller orchestrator is sentient. (True or False)

References

- Raffel, C., Shazeer, N., Roberts, A., Lee, K., Narang, S., Matena, M., Zhou, Y., Li, W., & Liu, P. J. (2020). *Exploring the Limits of Transfer Learning with a Unified Text-to-Text Transformer.* `https://arxiv.org/abs/1910.10683`
- Ren, J., Sun, Y., Du, H., Yuan, W., Wang, C., Wang, X., Zhou, Y., Zhu, Z., Wang, F., & Cui, S. (2024). *Generative Semantic Communication: Architectures, Technologies, and Applications.* `https://doi.org/10.48550/arXiv.2412.08642`

Further reading

- Koziolek, H., Gruener, S., & Ashiwal, V. (2023). *ChatGPT for PLC/DCS Control Logic Generation.* `https://doi.org/10.48550/arXiv.2305.15809`

Subscribe for a Free eBook

New frameworks, evolving architectures, research drops, production breakdowns—*AI_Distilled* filters the noise into a weekly briefing for engineers and researchers working hands-on with LLMs and GenAI systems. Subscribe now and receive a free eBook, along with weekly insights that help you stay focused and informed.

Subscribe at `https://packt.link/TRO5B` or scan the QR code below.

3

Integrating Dynamic RAG into the GenAISys

A business-ready **generative AI system (GenAISys)** needs to be flexible and ready to face the rapidly evolving landscape of the AI market. The AI controller acts as an adaptive orchestrator for e-marketing, production, storage, distribution, and support, but to satisfy such a range of tasks, we need a **retrieval-augmented generation (RAG)** framework. In the previous chapter, we built a conversational AI agent and a function for similarity search for instruction scenarios (AI orchestrator) for a generative AI model. In this chapter, we will enhance that foundation and build a scalable RAG in a Pinecone index to integrate both instruction scenarios and classical data, which the generative AI model will connect to.

We make a clear distinction in this chapter between **instruction scenarios**—expert-crafted prompt fragments (or *task tags*, as explained in the previous chapter) that tell the model *how* to reason or act—and **classical data**—the reference material the RAG system retrieves to ground its answers.

Why do we need this dynamic and adaptive RAG framework with vectorized scenarios of instructions on top of classical data? Because the global market affects entities internally and externally. This framework is an advanced form of context engineering, where we dynamically retrieve not just data, but also instructions to shape the AI's reasoning. For example, a hurricane can cause electricity shortages, putting the supply chain of businesses in peril. Businesses might have to relocate supply routes, production, or distribution. General-purpose AI cloud platforms might do some of the job. But more often than not, we will need to provide custom, domain-specific functionality. For that reason, we need a dynamic set of instructions in a vector store repository as we do for RAG data.

We will begin by defining the architecture scenario-driven task executions for a generative AI model, in this case, GPT-4o, through a Pinecone index. We will carefully go through the cost-benefits of investing in intelligent scenarios for the generative model through similarity search and retrieval. We will introduce a dynamic framework to produce ChatGPT-like capabilities that we will progressively introduce in the following chapters.

Once the architecture is defined, we will first build a Pinecone index to chunk, embed, and upsert instruction scenarios. We will make sure the GenAISys vector store can embed a query and find a relevant instruction scenario. This capability will be a key component in *Chapter 4, Building the AI Controller Orchestration Interface,* when we design the conversational agent's interface and orchestrator. Finally, we will write a program to upsert classical data in a RAG environment to the same Pinecone index alongside the instruction scenarios. Differentiation between scenarios and classical data will be maintained using distinct namespaces. By the end of this chapter, we will have built the main components to link instructions to a generative AI model. We will be ready to design a user interface and AI controller orchestrator in *Chapter 4*.

This chapter covers the following topics:

- Architecting RAG for the dynamic retrieval of instructions and data
- The law of diminishing returns when developing similarity searches
- Examining the architecture of a hybrid GenAISys CoT
- Creating a Pinecone index by chunking, embedding, and upserting instruction scenarios
- Enhancing a Pinecone index with classical data
- Querying the Pinecone index

Our first task is to architect a RAG framework for dynamic retrieval.

Architecting RAG for dynamic retrieval

In this section, we will define a Pinecone index that stores both instruction scenarios and classical data. This structure gives GenAISys dynamic, cost-effective retrieval: the instruction scenarios steer the generative AI model (GPT-4o in our example), while the classical data supplies the factual context used by the RAG pipeline.

We will go through the following components:

- **Scenario-driven task execution**: Designing optimized instructional prompts ("scenarios") that we will upsert to the Pinecone index.

- **Cost-benefit strategies**: Considering the law of diminishing returns to avoid overinvesting in automation.
- **Partitioning Pinecone with namespaces**: Using Pinecone index namespaces to clearly differentiate instruction scenarios from classical data.
- **Hybrid retrieval framework**: Implementing implicit vector similarity searches but also triggering explicit instructions for the generative AI model (more on this in the *Scenario-driven task execution* section).
- **CoT loops**: Explaining how the flexibility of the scenario selection process will lead to loops of generative AI functions before finally producing an output.
- **GenAISys framework**: Laying the groundwork for the advanced GenAISys framework we are building throughout the book.

Let's first dive deeper into scenario-driven task execution.

Scenario-driven task execution

In the previous chapter, we saw two complementary ways the AI controller can pick what to do next:

- **Implicit selection**: The controller embeds the user's prompt, runs a semantic similarity search across its scenario library, and chooses the closest match without any task tag. This gave us flexible, code-free orchestration (e.g., it automatically chose a sentiment-analysis scenario for the *Gladiator II* review).
- **Explicit selection**: The desired task is spelled out, either as a task tag in the prompt or as a user interface action, such as "Run web search." Here, the controller skips the similarity search and jumps straight to the requested tool or workflow.

That same pattern continues in this chapter, but at a larger scale. Instead of a few hand-picked prompts, we manage hundreds or even thousands of expert-authored instruction scenarios stored in a vector database; instead of single-user experiments, we support many concurrent users and workflows. This scenario-driven (implicit) approach has three advantages:

- Professional experts typically create these advanced prompts/instruction scenarios, often surpassing the expertise level of mainstream users.
- The scenarios can be co-designed by AI specialists and subject-matter experts, covering a wide range of activities in a corporation, from sales to delivery.

- The order of execution of the scenarios is prompt-driven, flexible, and unordered. This dynamic approach avoids hardcoding the order of the tasks, increasing adaptability as much as possible.

However, while implicit task planning maximizes flexibility, as we move toward building business-ready systems, we must balance flexibility with cost-efficiency. In some cases, therefore, *explicit* instructions, such as triggering a web search by selecting the option in the user interface, can significantly reduce the potential costs, as shown in *Figure 3.1*:

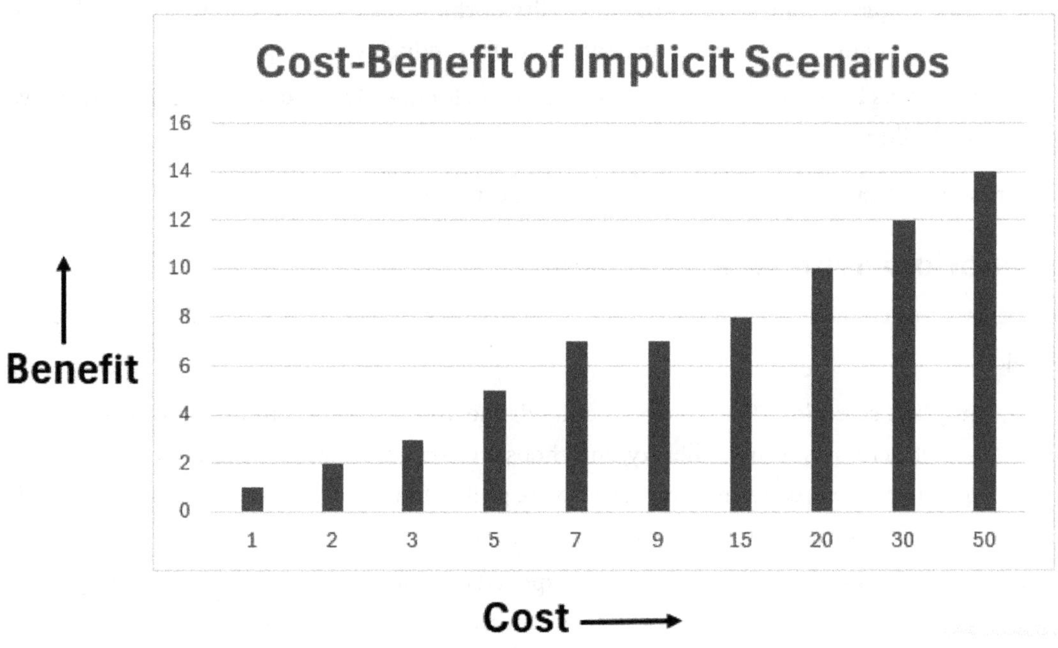

Figure 3.1: Diminishing returns as costs increase

The more we automate implicit scenarios that the generative AI model will select with vector similarity searches in the Pinecone index, the higher the cost. To manage this, we must carefully consider the law of diminishing returns:

$$B = B_0 + (r \times C) - (d \times C^2)$$

In this equation, as illustrated in *Figure 3.1*, in theoretical units, we have the following:

- B represents the overall benefit, which is represented by roughly 15 when the cost reaches 50.

- B_0 represents the initial benefit of storing instruction scenarios in the Pinecone index and asking the generative AI model to select one through vector similarity with the user input. In this case, it is nearly 1 benefit unit for 1 cost unit.
- r is the rate at which the benefit begins to increase as we increase the cost.
- C represents the cost measured in theoretical units (currency, human resources, or computational resources).
- d denotes the rate at which returns diminish as cost increases.

For example, when the cost reaches 7 theoretical units, the benefit reaches 7 theoretical units. This 1 unit of cost generating 1 unit of benefit is reasonable. However, when the benefit reaches 10 units, the cost could double to 14 units, which signals that something is going wrong.

The diminishing d factor has a strong negative impact NI on the benefits through squared costs:

$$NI = -(d\ x\ C^2)$$

We will carefully monitor the factor d as we move through the use cases in this book. We will have to make choices between running implicit automated scenario selections through the Pinecone index and explicitly triggering actions through predefined instructions in the prompt itself.

Let's now explore how we identify instruction scenarios within a Pinecone index.

Hybrid retrieval and CoT

Our first step is teaching the GenAISys framework to distinguish clearly between classical data and instruction scenarios. To achieve this, we will separate the instruction scenarios and the data with two namespaces within the same Pinecone index, named `genai-v1`:

- `genaisys` will contain instruction vectors of information
- `data01` will contain data vectors of information

We will implement `genai-v1` in code with additional explanations in the *Creating the Pinecone index* section of this chapter.

Once the Pinecone index has been partitioned into scenarios and data, we can take the GenAISys to another level with hybrid retrieval, as shown in *Figure 3.2*.

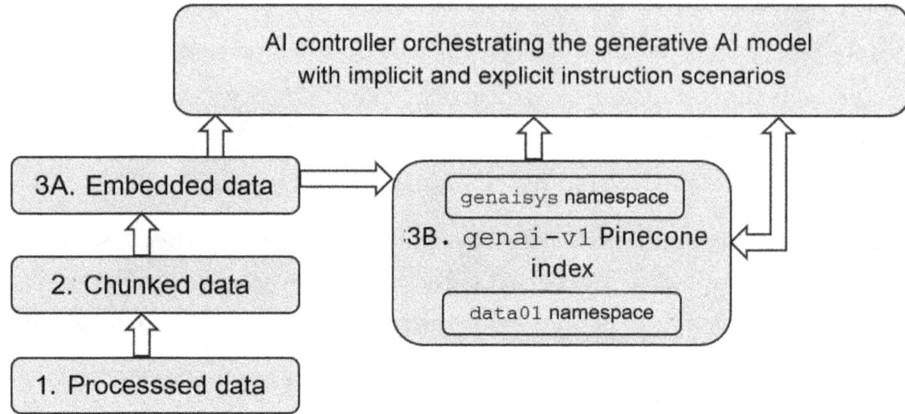

Figure 3.2: AI controller orchestrating GenAISys

The hybrid retrieval framework depicted in the preceding figure will enable GenAISys to do the following:

- Run the generative AI model using processed, chunked, and embedded data memory (see **1–3** in *Figure 3.2*) directly without going through the Pinecone index (see **3B**). This will reduce costs for ephemeral data, for example.
- Run the generative AI model after the chunked and embedded data is upserted to the Pinecone index either as a set of instructions in a scenario or as classical data.
- Create a CoT loop between the Pinecone index (see **3B**) and the generative AI model controller as an orchestrator. For example, the output of the model can serve as input for another CoT cycle that will retrieve scenarios or data from the Pinecone index. ChatGPT-like copilots often present their output and then finish by asking whether you'd like to explore further—sometimes even suggesting ready-made follow-up prompts you can click on.

CoT loops can operate implicitly via vector similarity search or explicitly via direct instruction triggers or task tags (such as "Run a web search"). For example, ChatGPT-like copilots can trigger web searches directly through the user interface or rules in the AI controller.

We'll begin building our GenAISys in this chapter and continue refining it over the next few chapters. Starting from *Chapter 4, Building the AI Controller Orchestration Interface*, we'll use the RAG foundations introduced here to develop the hybrid retrieval framework shown in *Figure 3.2*.

The GenAISys we're building will include dynamic process management—a requisite for keeping pace with the shifting market conditions. Specifically, our GenAISys will do the following:

- Leverage Pinecone's vector database or in-memory chunked and embedded information with similarity searches to optimize retrieval (instructions or data)
- Explicitly trigger direct instructions, such as a web search, and include them in a CoT loop for summarization, sentiment analysis, or semantic analysis
- Break down complex sets of instructions and data retrieval into manageable steps
- Iteratively refine solutions in a human-like thought process before producing an output
- Get the best out of generative AI models, including OpenAI's reasoning models such as o3, by providing them with an optimized instruction scenario

Our initial step in this chapter is building the `genai-v1` Pinecone index, which the AI controller will use to manage instruction scenarios within the `genaisys` namespace. Then, we'll demonstrate how to chunk, embed, and upsert classical data into the `data01` namespace. Let's get moving!

Building a dynamic Pinecone index

We'll focus on creating a Pinecone index designed to manage both instruction scenarios and classical data. In the upcoming sections, we'll begin upserting the instruction scenarios as well as classical data. The workflow breaks down into three straightforward stages:

- Setting up the environment for OpenAI and Pinecone
- Processing the data, chunking it, and then embedding it
- Initializing the Pinecone index

Open the `Pinecone_instruction_scenarios.ipynb` notebook within the Chapter03 directory on GitHub (`https://github.com/Denis2054/Building-Business-Ready-Generative-AI-Systems/tree/main`). Our first task is to set up the environment.

Setting up the environment

As we move through the book, we will continually reuse functions and features implemented in *Chapters 1* and *2*, add new ones for Pinecone, and organize the installations into two parts:

- Installing OpenAI using the same process as in *Chapter 1*. Refer back to that chapter if needed.
- Installing Pinecone for this and following chapters.

To begin, download the files we need by retrieving `grequests.py` from the GitHub repository:

```
!curl -L https://raw.githubusercontent.com/Denis2054/Building-Business-Ready-Generative-AI-Systems/master/commons/grequests.py --output grequests.py
```

To install OpenAI, follow the same steps as in *Chapter 1*. We'll move on to install Pinecone now, which we will refer to in upcoming chapters throughout the book.

Installing Pinecone

Download the Pinecone requirements file that contains the instructions for the Pinecone version we want to use throughout the book. If another version is required, this will be the only file that needs to be updated:

```
download("commons","requirements02.py")
```

Quick tip: Enhance your coding experience with the **AI Code Explainer** and **Quick Copy** features. Open this book in the next-gen Packt Reader. Click the **Copy** button **(1)** to quickly copy code into your coding environment, or click the **Explain** button **(2)** to get the AI assistant to explain a block of code to you.

```
function calculate(a, b) {
  return {sum: a + b};
};
```

Copy (1) Explain (2)

The next-gen Packt Reader is included for free with the purchase of this book. Scan the QR code OR visit `https://packtpub.com/unlock`, then use the search bar to find this book by name. Double-check the edition shown to make sure you get the right one.

The file contains the installation function, which we will call with the following command:

```
# Run the setup script to install and import dependencies
%run requirements02
```

The script is the same as the one for OpenAI described in *Chapter 1*, but adapted to Pinecone. We first uninstall Pinecone and then install the version we need:

```
import subprocess
import sys
def run_command(command):
    try:
        subprocess.check_call(command)
    except subprocess.CalledProcessError as e:
        print(f"Command failed: {' '.join(command)}\nError: {e}")
        sys.exit(1)
# Uninstall the 'pinecone-client' package
print("Uninstalling 'pinecone-client'...")
run_command(
    [sys.executable, "-m", "pip", "uninstall", "-y", "pinecone-client"]
)
# Install the specific version of 'pinecone-client'
print("Installing 'pinecone-client' version 5.0.1...")
run_command(
    [
        sys.executable, "-m", "pip", "install",\
        "--force-reinstall", "pinecone-client==5.0.1"
    ]
)
```

Then, we verify the installation:

```
# Verify the installation
try:
    import pinecone
    print(
        f"'pinecone-client' version {pinecone.__version__} is installed."
)
except ImportError:
    print(
```

```
        "Failed to import the 'pinecone-client' library after
installation."
)
    sys.exit(1)
```

The output shows we have successfully installed the client:

```
Uninstalling 'pinecone-client'...
Installing 'pinecone-client' version 5.0.1...
'pinecone-client' version 5.0.1 is installed.
```

Let's go ahead and initialize the Pinecone API key.

Initializing the Pinecone API key

The program now downloads `pinecone_setup.py`, which we will use to initialize the Pinecone API key:

```
download("commons","pinecone_setup.py")
```

This setup mirrors the Google Colab secrets-based approach we used for OpenAI in *Chapter 1*, but it's adapted here for initializing the Pinecone API.:

```
# Import libraries
import openai
import os
from google.colab import userdata

# Function to initialize the Pinecone API key
def initialize_pinecone_api():
    # Access the secret by its name
    PINECONE_API_KEY = userdata.get('PINECONE_API_KEY')

    if not PINECONE_API_KEY:
        raise ValueError("PINECONE_API_KEY is not set in userdata!")

    # Set the API key in the environment and OpenAI
    os.environ['PINECONE_API_KEY'] = PINECONE_API_KEY
    print("PINECONE_API_KEY initialized successfully.")
```

If Google secrets was set to `True` for OpenAI in the OpenAI section of this notebook, then the Pinecone setup function will be called:

```
if google_secrets==True:
    import pinecone_setup
    pinecone_setup.initialize_pinecone_api()
```

If Google secrets was set to `False`, then you can implement a custom function by uncommenting the code and entering the Pinecone API key with any method you wish:

```
if google_secrets==False: # Uncomment the code and choose any method you
wish to initialize the Pinecone API key
    import os
    #PINECONE_API_KEY=[YOUR PINECONE_API_KEY]
    #os.environ['PINECONE_API_KEY'] = PINECONE_API_KEY
    #openai.api_key = os.getenv("PINECONE_API_KEY")
    #print("Pinecone API key initialized successfully.")
```

The program is now ready to process the data we will upsert to the Pinecone index.

Processing data

Our goal now is to prepare the scenarios for storage and retrieval so that we can then query the Pinecone index. The main steps of the process are represented in *Figure 3.2*, which is only one layer of the roadmap for the following chapters. We will process the data in the following steps:

1. **Data loading and preparation**, in which the data will be broken into smaller parts. In this case, each scenario will be stored in one line of a scenario list, which will prepare the chunking process. We will not always break text into lines, however, as we will see in the *Upserting classical data into the index* section later.
2. **Chunking functionality** to store each line of scenarios into chunks.
3. **Embedding** the chunks of text obtained.
4. **Verification** to ensure that we embedded the corresponding number of chunks.

Let's now cover the first two steps: loading and preparing the data, followed by chunking.

Data loading and chunking

We will use the scenarios implemented in *Chapter 2*. They are stored in a file that we will now download:

```
download("Chapter03","scenario.csv")
```

We will add more scenarios throughout our journey in this book to create a GenAISys. For the moment, our main objective is to get our Pinecone index to work. The program first initializes `start_time` for time measurement. Then we load the lines of scenario instructions directly into `chunks` line by line:

```
import time
start_time = time.time()  # Start timing
# File path
file_path = 'scenario.csv'
# Read the file, skip the header, and clean the lines
chunks = []
with open(file_path, 'r') as file:
    next(file)  # Skip the header line
    chunks = [line.strip() for line in file]  # Read and clean lines as
chunks
```

Then the code displays the number of chunks and the time it took to create the chunks:

```
response_time = time.time() - start_time  # Measure response time
print(f"Response Time: {response_time:.2f} seconds")  # Print response
time
```

```
Total number of chunks: 3
Response Time: 0.00 seconds
```

The program now verifies the first three chunks of scenario instructions:

```
# Optionally, print the first three chunks for verification
for i, chunk in enumerate(chunks[:3], start=1):
    print(chunk)
```

The output shows the three scenarios we will be working on in this chapter:

```
['ID,SCENARIO\n',
 '100,Semantic analysis.This is not an analysis but a semantic search.
Provide more information on the topic.\n',
 '200,Sentiment analysis  Read the content return a sentiment analysis
nalysis on this text and provide a score with the label named : Sentiment
analysis score followed by a numerical value between 0 and 1  with no + or
- sign and  add an explanation to justify the score.\n',
 '300,Semantic analysis.This is not an analysis but a semantic search.
Provide more information on the topic.\n']
```

The chunks of data are now ready for embedding. Let's proceed with embedding.

Embedding the dataset

To embed the dataset, we will first initialize the embedding model and then embed the chunks. The program first initializes the embedding model.

Initializing the embedding model

We will be using an OpenAI embedding model to embed the data. To embed our data with an OpenAI model, we can choose one of three main models:

- `text-embedding-3-small`, which is fast and has a lower resource usage. This is sufficient for real-time usage. It is a smaller model and is thus cost-effective. However, as the vector store will increase in size with complex scenarios, it might be less accurate for nuanced tasks.
- `text-embedding-3-large`, which provides high accuracy and nuanced embeddings and will prove effective for complex semantic similarity searches. It requires more resources and costs more.
- `text-embedding-ada-002`, which is cost-effective for good-quality embeddings. However, it's slightly slower than models such as `text-embedding-3-small` and `text-embedding-3-large`.

You can consult the OpenAI documentation at `https://platform.openai.com/docs/guides/embeddings` for more info.

To import a limited number of scenarios in this chapter, we will use `text-embedding-3-small` to optimize speed and cost. The program initializes the model while the others are commented for further use if needed:

```
import openai
import time
embedding_model="text-embedding-3-small"
#embedding_model="text-embedding-3-large"
#embedding_model="text-embedding-ada-002"
```

We initialize the OpenAI client:

```
# Initialize the OpenAI client
client = openai.OpenAI()
```

An embedding function is then created that will convert the text sent to it into embeddings. The function is designed to produce embeddings for a batch of input texts (`texts`) with the embeddings model of our choice, in this case, `text-embedding-3-small`:

```
def get_embedding(texts, model="text-embedding-3-small")
```

The function first cleans the text by replacing newline characters in each text with spaces:

```
texts = [text.replace("\n", " ") for text in texts]
```

Then, the function makes the API embedding call:

```
response = client.embeddings.create(input=texts, model=model)
```

The embeddings are extracted from the response:

```
embeddings = [res.embedding for res in response.data]  # Extract
embeddings
```

Finally, the embeddings are returned:

```
return embeddings
```

The program is now ready to embed the chunks.

Embedding the chunks

The program first defines a function to embed the chunks:

```
def embed_chunks(
    chunks, embedding_model="text-embedding-3-small",
    batch_size=1000, pause_time=3
):
```

The parameters of the function are:

- `chunks`: The parts of text to embed
- `embedding_model`: Defines the model to use, such as `text-embedding-3-small`

- `batch_size`: The number of chunks the function can process in a single batch, such as `batch_size=1000`
- `pause_time`: A pause time in seconds, which can be useful for rate limits

We then initialize the timing function, `embeddings` variable, and counter:

```
start_time = time.time()  # Start timing the operation
embeddings = []  # Initialize an empty list to store the embeddings
counter = 1  # Batch counter
```

The code is now ready to process the chunks in batches:

```
# Process chunks in batches
    for i in range(0, len(chunks), batch_size):
        chunk_batch = chunks[i:i + batch_size]  # Select a batch of chunks
```

Each batch is then sent to the embedding function:

```
        # Get the embeddings for the current batch
        current_embeddings = get_embedding(
            chunk_batch, model=embedding_model
        )
```

The embedded batch is appended to the embeddings list:

```
# Append the embeddings to the final list
        embeddings.extend(current_embeddings)
```

The number of batches is monitored and displayed and the pause is activated:

```
   # Print batch progress and pause
        print(f"Batch {counter} embedded.")
        counter += 1
        time.sleep(pause_time)  # Optional: adjust or remove this
depending on rate limits
```

Once all the batches are processed, the total time is displayed:

```
   # Print total response time
    response_time = time.time() - start_time
    print(f"Total Response Time: {response_time:.2f} seconds")
```

The embedding function is ready to be called with the chunks list:

```
embeddings = embed_chunks(chunks)
```

The output shows that the scenario data has been embedded:

```
Batch 1 embedded.
Total Response Time: 4.09 seconds
```

The first embedding is displayed for verification:

```
print("First embedding:", embeddings[0])
```

The output confirms that the embeddings have been generated:

```
First embedding: [0.017762450501322746, 0.041617266833782196,
-0.024105189368128777,…
```

The final verification is to check that the number of embeddings matches the number of chunks:

```
# Check the lengths of the chunks and embeddings
num_chunks = len(chunks)
print(f"Number of chunks: {num_chunks}")
print(f"Number of embeddings: {len(embeddings)}")
```

The output confirms that the chunking and embedding process is most probably successful:

```
Number of chunks: 3
Number of embeddings: 3
```

The chunks and embeddings are now ready to be upserted into the Pinecone index.

Creating the Pinecone index

The `genai-v1` Pinecone index we will create will contain two namespaces, as shown in *Figure 3.3*:

- `genaisys`: A repository of instruction scenarios. These prompts drive generative AI behavior and can also trigger traditional functions such as web search.
- `Data01`: The embedded classical data that the RAG pipeline queries.

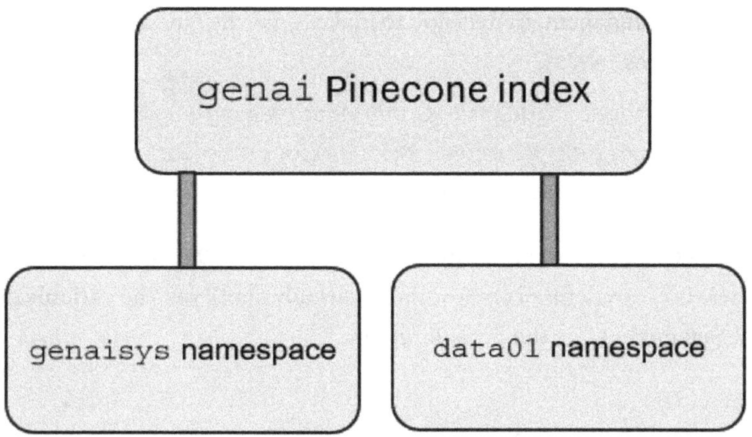

Figure 3.3: Partitioning the Pinecone index into namespaces

We begin by importing two classes:

```
from pinecone import Pinecone, ServerlessSpec
```

The `Pinecone` class is the primary interface to interact with the Pinecone index. We will use this class to configure Pinecone's serverless services.

Before going further, you will need to set up a Pinecone account and obtain an API key. Make sure to verify the cost of these services at `https://www.pinecone.io/`. This chapter is self-contained, so you can begin by reading the content, comments, and code before deciding on creating a Pinecone account.

Once our account is set up, we need to retrieve and initialize our API key:

```
# Retrieve the API key from environment variables
api_key = os.environ.get('PINECONE_API_KEY')
if not api_key:
    raise ValueError("PINECONE_API_KEY is not set in the environment!")

# Initialize the Pinecone client
pc = Pinecone(api_key=api_key)
```

We now import the specification class, define the name of our index (`genai-v1`), and initialize our first namespace (`genaisys`) for our scenarios:

```
from pinecone import ServerlessSpec
index_name = "genai-v1"
namespace="genaisys"
```

We now have a project management decision to make—use the Pinecone cloud to host our index or **Amazon Web Services** (**AWS**)?

```
cloud = os.environ.get('PINECONE_CLOUD') or 'aws'
region = os.environ.get('PINECONE_REGION') or 'us-east-1'
spec = ServerlessSpec(cloud=cloud, region=region)
```

The code first checks whether an environment variable (`PINECONE_CLOUD`) is set to use the Pinecone cloud. If there is no predefined environment variable check set, the variable defaults to AWS with `'aws'` and `'us-east-1'` as the default region.

For more information, refer to the Pinecone Python SDK documentation at `https://docs.pinecone.io/reference/python-sdk`.

In this case, AWS was chosen for the following reasons:

- **Market leadership and reliability**: AWS has a market share of over 30% of the global infrastructure market. As such, it is deemed reliable by a large number of organizations.
- **Compliance and security standards**: AWS has over 140 security standards for data security and privacy, including PCI-DSS and HIPAA/HITECH, FedRAMP, GDPR, FIPS 140-2, and NIST 800-171.
- **Scalability**: AWS has a global network of data centers, making scalability seamless.

Alternatively, you can create an index manually in your Pinecone console to select the embedding model and the host, such as AWS or **Google Cloud Platform** (**GCP**). You can also select your pod size from x1 to more, which will determine the maximum size of your index. Each choice depends on your project and resource optimization strategy.

In any case, we need metrics to monitor usage and cost. Pinecone provides detailed usage metrics accessible via your account, allowing you to manage indexes efficiently. For example, we might want to delete information you don't need anymore, add targeted data, or optimize the usage per user.

Pinecone provides three key metrics:

- **Serverless storage usage**: Measured in **gigabyte-hours** (**GB-hours**). The cost is calculated at 1 GB of storage per hour. Carefully monitoring the amount of data we store is an important factor in any AI project.

- **Serverless write operations units:** Measures the resources consumed by write operations to the Pinecone database that contains our index.
- **Serverless read operations units:** Measures the resources consumed by read operations.

You can download detailed information on your consumption by going to your Pinecone account, selecting **Usage**, and then clicking on the **Download** button, as shown here:

Figure 3.4: Downloading Pinecone usage data

The download file is in CSV format and contains a detailed account of our Pinecone usage, such as `BillingAccountId` (account identifier), `BillingAccountName` (account name), `OrganizationName` (organization name), `OrganizationId` (organization ID), `ProjectId` (project identifier), `ProjectName` (project name), `ResourceId` (resource identifier), `ResourceName` (resource name), `ChargePeriodStart` (charge start date), `ChargePeriodEnd` (charge end date), `BillingPeriodStart` (billing start date), `BillingPeriodEnd` (billing end date), `SkuId` (SKU identifier), `SkuPriceId` (SKU price ID), `ServiceName` (service name), `ChargeDescription` (charge details), `CloudId` (cloud provider), `RegionId` (region), `Currency` (currency type), `PricingQuantity` (usage quantity), `PricingUnit` (usage unit), `ListCost` (listed cost), `EffectiveCost` (calculated cost), `BilledCost` (final cost), and `Metadata` (additional data).

As AI slowly enters its industrial age, straying away from the initial excitement of the early 2020s, continuous monitoring of these metrics becomes increasingly critical.

We will now check whether the index we selected exists or not. The program imports the `pinecone` and `time` classes to insert a sleep time before checking whether the index exists:

```
import time
import pinecone
# check if index already exists (it shouldn't if this is first time)
if index_name not in pc.list_indexes().names():
```

If the index exists, the following code will be skipped to avoid creating duplicate indexes. If not, an index is created:

```
    # if does not exist, create index
    pc.create_index(
```

```
        index_name,
        dimension=1536,  # dimension of the embedding model
        metric='cosine',
        spec=spec
    )
    # wait for index to be initialized
    time.sleep(1)
```

The parameters are as follows:

- `index_name`, which is the name of our Pinecone index, `genai-v1`
- `dimension=1536`, the dimensionality of the embedding vectors
- `metric='cosine'`, which sets the distance metric for similarity searches to cosine similarity
- `spec=spec`, which defines the region and the serverless specification we defined previously for the cloud services
- `time.sleep(1)`, which makes the program wait to make sure the index is fully created before continuing

If the index has just been created, the output shows its details with `total_vector_count` set to `0` (if you see a number other than `0`, the notebook has likely already been run):

```
{'dimension': 1536,
 'index_fullness': 0.0,
 'namespaces': {},
 'total_vector_count': 0}
```

If the index already exists, the statistics will be displayed, including `index_fullness` to monitor the space used in your index pod from 0 to 1:

```
Index stats
{'dimension': 1536,
 'index_fullness': 0.0,
 'namespaces': {'genaisys': {'vector_count': 3}},
 'total_vector_count': 3}
```

In this case, we haven't populated the index yet. We can connect to the index we just created and display its statistics before populating it:

```
# connect to index
index = pc.Index(index_name)
# view index stats
index.describe_index_stats()
```

The output displays the information, confirming that we are connected:

```
{'dimension': 1536,
 'index_fullness': 0.0,
 'namespaces': {'genaisys': {'vector_count': 0}},
 'total_vector_count': 0}
```

The selected embedding model must match Pinecone's index dimension (`1536`). We will create the parameters of a Pinecone index interactively when we begin working on use cases in *Chapter 5*. Here, we are using `embedding_model="text-embedding-3-small` with its 1,536 dimensions, which matches the dimension of the Pinecone index.

Note also that the `'genaisys'` namespace we initialized is taken into account. This ensures that when we upsert the scenarios we designed, they will not be confused with the classical data that is in another namespace of the same index. We are now ready to upsert the data to our Pinecone index.

Upserting instruction scenarios into the index

Upserting embedded chunks into a Pinecone index comes with a cost, as explained at the beginning of this section. We must carefully decide which data to upsert. If we upsert all the data, we might do the following:

- Overload the index and make retrieval challenging, be it instruction scenarios or classical data
- Drive up the cost of write and read operations
- Add more noise than is manageable and confuse the retrieval functions

If we choose not to upsert the data, we have two options:

- **Querying in real time in memory**: Loading chunked, embedded data into memory and querying the information in real time could alleviate the data store and be a pragmatic way to deal with ephemeral information we don't need to store, such as the daily weather forecast. However, we must also weigh the cost/benefit of this approach versus upserting at each step for the use cases we'll be working on starting from *Chapter 5*.
- **Fine-tuning data**: This comes with the cost of building training datasets, which requires human and computing resources. In the case of fast-moving markets, we might have to fine-tune regularly, which entails high investments. The cost/benefit will be up to the project management team to consider. A cost-benefit analysis of fine-tuning versus RAG will be explored in *Chapter 5*.

We first initialize the libraries and start a timer to measure how long it takes to run the script:

```
import pinecone
import time
import sys
start_time = time.time()  # Start timing before the request
```

The program must then calculate the maximum size of the batch we send to Pinecone. It is set to 400,000 bytes, or 4 MB, to play it safe. If the limit is reached, the batch size is returned:

```
# Function to calculate the size of a batch
def get_batch_size(data, limit=4000000):  # Limit set to 4MB to be safe
    total_size = 0
    batch_size = 0
    for item in data:
        item_size = sum([sys.getsizeof(v) for v in item.values()])
        if total_size + item_size > limit:
            break
        total_size += item_size
        batch_size += 1
    return batch_size
```

We now need an `upsert` function that takes the batch size into account when called:

```
# Upsert function with namespace
def upsert_to_pinecone(batch, batch_size, namespace="genaisys"):
    """
    Upserts a batch of data to Pinecone under a specified namespace.
```

```
    """
    try:
        index.upsert(vectors=batch, namespace=namespace)
        print(
            f"Upserted {batch_size} vectors to namespace '{namespace}'."
        )
    except Exception as e:
        print(f"Error during upsert: {e}")
```

In production, we would typically exit on error, but for this educational notebook, printing helps us observe without stopping execution.

Note that we will upsert the instruction scenarios into the namespace, `genaisys`, within the Pinecone index. We can now define the main batch upsert function:

```
def batch_upsert(data):
```

The function begins by determining the total length of the data and then prepares batches that match the batch size that it will calculate with the `get_batch_size` function. Then, it creates a batch and sends it to the `upsert_to_pinecone` function we defined:

```
# Function to upsert data in batches
def batch_upsert(data):
    total = len(data)
    i = 0
    while i < total:
        batch_size = get_batch_size(data[i:])
        batch = data[i:i + batch_size]
        if batch:
            upsert_to_pinecone(batch, batch_size, namespace="genaisys")
            i += batch_size
            print(f"Upserted {i}/{total} items...")  # Display current
progress
        else:
            break
    print("Upsert complete.")
```

When the upsert is completed, the output will display a success message, signaling that we are ready to prepare the upsert process. A Pinecone index requires an ID that we will now create:

```
# Generate IDs for each data item
ids = [str(i) for i in range(1, len(chunks) + 1)]
```

Once each embedded chunk has an ID, we need to format the data to fit Pinecone's index structure:

```
# Prepare data for upsert
data_for_upsert = [
    {"id": str(id), "values": emb, "metadata": {"text": chunk}}
    for id, (chunk, emb) in zip(ids, zip(chunks, embeddings))
]
```

The data is now formatted with an ID, values (embeddings), and metadata (the chunks). Let's call the `batch_upsert` function that will call the related functions we created:

```
# Upsert data in batches
batch_upsert(data_for_upsert)
```

When the upserting process is finished, the number of vectors upserted to the namespace and the time it took are displayed:

```
Upserted 3 vectors to namespace 'genaisys'.
Upserted 3/3 items...
Upsert complete.
Upsertion response time: 0.45 seconds
```

We can also display the statistics of the Pinecone index:

```
#You might have to run this cell after a few seconds to give Pinecone
#the time to update the index information
print("Index stats")
print(index.describe_index_stats(include_metadata=True))
```

Note that you might have to wait a few seconds to give Pinecone time to update the index information.

The output displays the information:

```
Index stats
{'dimension': 1536,
 'index_fullness': 0.0,
 'namespaces': {'genaisys': {'vector_count': 3}},
 'total_vector_count': 3}
```

The information displayed is as follows:

- `'dimension': 1536`: Dimension of the embeddings.
- `'index_fullness': 0.0`: A value between 0 and 1 that shows how full the Pinecone index is. We must monitor this value to optimize the data we are upserting to avoid having to increase the size of the storage capacity we are using. For more information, consult the Pinecone documentation at `https://docs.pinecone.io/guides/get-started/overview`.
- `'namespaces': {'genaisys': {'vector_count': 3}}`: Displays the namespace and vector count.
- `'total_vector_count': 3}`: Displays the total vector count in the Pinecone index.

We are now ready to upload the classical data into its namespace.

Upserting classical data into the index

Building a GenAISys involves teams. So that each team can work in parallel to optimize production times, we will upsert the classical data in a separate program/notebook. One team can work on instruction scenarios while another team works on gathering and processing data.

Open `Pinecone_RAG.ipynb`. We will be reusing several components of the `Pinecone_instruction_scenarios.ipynb` notebook built in the *Building a dynamic Pinecone index* section of this chapter. Setting up the environment is identical to the previous notebook. The Pinecone index is the same, `genai-v1`. The namespace for source-data upserting is `data01`, as we've already established in earlier sections, to make sure the data is separated from the instruction scenarios. So, the only real difference is the data we load and the chunking method. Let's get into it!

Data loading and chunking

This section embeds chunks using the same process as for instruction scenarios in `Pinecone_instruction_scenarios.ipynb`. However, this time, GPT-4o does the chunking. When importing lines of instruction scenarios, we wanted to keep the integrity of the scenario in one chunk to be able to provide a complete set of instructions to the generative AI model. In this case, we will leverage the power of generative AI and chunk raw text with GPT-4o.

We begin by downloading data, not scenarios, and setting the path of the file:

```
download("Chapter03","data01.txt")
# Load the CSV file
file_path = '/content/data01.txt'
```

Now, the text file is loaded as one big chunk in a variable and displayed:

```
try:
    with open(file_path, 'r') as file:
        text = file.read()
    text
except FileNotFoundError:
    text = "Error: File not found. Please check the file path."
print(text)
```

While a production application would typically exit on a critical `FileNotFoundError`, for this educational notebook, printing the error allows us to observe the outcome without interrupting the learning flow.

You can comment `print(text)` or only print a few lines. In this case, let's verify that we have correctly imported the file. The output shows that we did:

```
The CTO was explaing that a business-ready generative AI system (GenAISys)
offers functionality similar to ChatGPT-like platforms…
```

The text contains a message from the CTO of the company whose data we are uploading to our custom RAG database. A company might have thousands of such internal messages—far too many (and too volatile) to justify model fine-tuning. Storing only the key chunks in Pinecone gives us searchable context without flooding the index with noise.

The `text` variable is not ready yet to be chunked by GPT-4o. The first step is to create an OpenAI instance and give the GPT-4o model instructions:

```
# Import libraries
from openai import OpenAI
# Initialize OpenAI Client
client = OpenAI()
# Function to chunk text using GPT-4o
def chunk_text_with_gpt4o(text):
    # Prepare the messages for GPT-4o
```

```
    messages = [
        {"role": "system", "content": "You are an assistant skilled at 
splitting long texts into meaningful, semantically coherent chunks of 50-
100 words each."},
        {"role": "user", "content": f"Split the following text into 
meaningful chunks:\n\n{text}"}
    ]
Now we send the request to the API:
    # Make the GPT-4o API call
    response = client.chat.completions.create(
        model="gpt-4o",  # GPT-4o model
        messages=messages,
        temperature=0.2,  # Low randomness for consistent chunks
        max_tokens=1024  # Sufficient tokens for the chunked response
    )
```

We need to keep an eye on the `max_tokens=1024` setting: GPT-4o will stop generating once it hits that limit. For very large documents, you can stream the text in smaller slices—then let GPT-4o refine each slice. We can also use ready-made chunking functions that will break the text down into optimized *chunks* to obtain more nuanced and precise results when retrieving the data. However, in this case, let's maximize the usage of GPT-4o; we send the entire file in one call with a low temperature so we can watch the model partition a real-world document from end to end.

Now we can retrieve the chunks from the response, clean them, store them in a list of chunks, and return the chunks variable:

```
    # Extract and clean the response
    chunked_text = response.choices[0].message.content
    chunks = chunked_text.split("\n\n")  # Assume GPT-4o separates chunks 
with double newlines
    return chunks
```

Now, we can call the chunking function. We don't have to display the chunks and can comment the code in production. However, in this case, let's verify that everything is working:

```
# Chunk the text
chunks = chunk_text_with_gpt4o(text)
# Display the chunks
print("Chunks:")
for i, chunk in enumerate(chunks):
```

```
    print(f"\nChunk {i+1}:")
    print(chunk)
```

The output shows that the chunks were successfully created:

```
Chunks:
Chunk 1:
The CTO was explaining that …
Chunk 2:
GenAISys relies on a generative AI model…
Chunk 3:
We defined memoryless, short-term, long-term…
```

The remaining embedding and upsert steps are identical to those in `Pinecone_instruction_scenarios.ipynb`—just remember to use `namespace="data01"` when writing the vectors. After that, we're ready to query the index and verify retrieval.

Querying the Pinecone index

As you know, our vector store now has two logical areas—`genaisys` for instruction scenarios and `data01` for classical data. In this section, we'll query each area interactively to prove the retrieval code works before we wire it into the multi-user interface in *Chapter 4*. We will query these two namespaces in the Pinecone index, as shown in *Figure 3.5*:

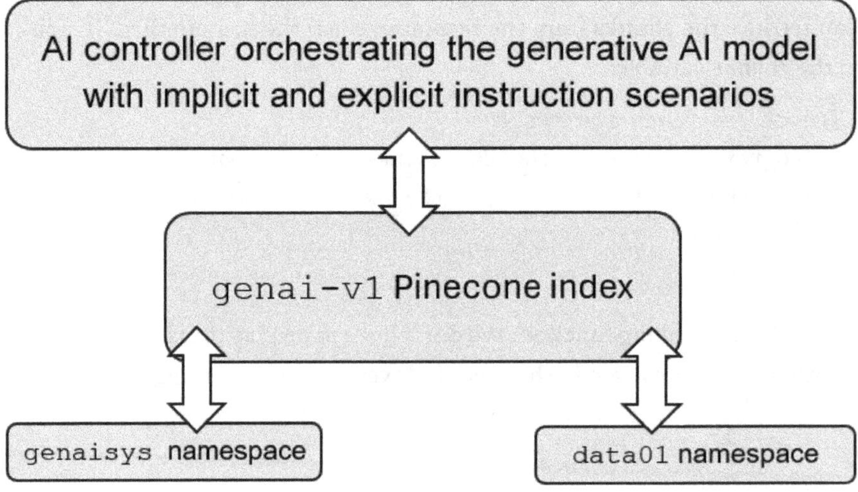

Figure 3.5: Generative AI model querying either the instruction scenarios or the data

Open `Query_Pinecone.ipynb` to run the verification queries. The next steps are the same as those in the *Setting up the environment* and *Creating the Pinecone index* sections, except for two minor differences:

- The namespace is not provided when we connect to the Pinecone index, only its name: `index_name = 'genai-v1'`. This is because the querying function will manage the choice of a namespace.
- The `Upserting` section of the notebook has been removed because we are not upserting but querying the Pinecone index.

The `Query` section of the notebook is divided into two subsections. The first subsection contains the querying functions and the second one the querying requests. Let's begin with the querying functions.

Querying functions

There are four querying functions, as follows:

- QF1: `query_vector_store(query_text, namespace)`, which receives the query, sends the request to QF2, and returns the response. It will use QF4 to display the results.
- QF2: `get_query_results(query_text, namespace)`, which receives the query from QF1, sends it to QF3 to be embedded, makes the actual query, and returns a response to QF1.
- QF3: `get_embedding(text, model=embedding_model)`, which receives text to embed from QF2 and sends the embedded text back to QF2.
- QF4: `display_results(query_results)`, which receives the results from QF1, processes them, and returns them to QF1.

We can simplify the representation as shown in *Figure 3.6* by creating two groups of functions:

- A group with QF1, `query_vector_store`, and QF4, `display_results`, in which QF1 queries the vector store through QF2 and returns the results to display.

- A group with QF2, `get_query_results`, queries the vector store after embedding the query with QF3, `get_embedding`, and returns the results to QF1.

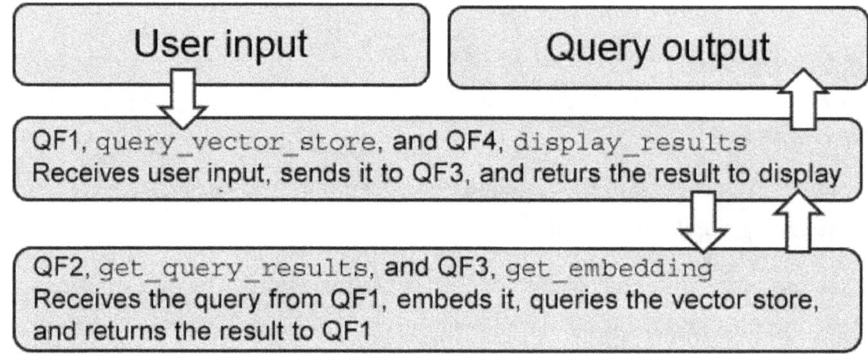

Figure 3.6: Querying the vector store with two groups of functions

Let's begin with the first group of functions.

Querying the vector store and returning results

The first function, QF1, receives the user input:

```
def query_vector_store(query_text, namespace):
    print("Querying vector store...")
```

Then, the function calls QF2, `query_results`:

```
    # Retrieve query results
    query_results = get_query_results(query_text, namespace)
```

QF2 then returns the results in `query_results`, which, in turn, is sent to `display_results` to obtain the text and target ID:

```
    # Process and display the results
    print("Processed query results:")
    text, target_id = display_results(query_results)
    return text, target_id
```

`display_results` processes the query results it receives and returns the result along with metadata to find the text obtained in the metadata of the Pinecone index. When it is found, the function retrieves the ID:

```
def display_results(query_results):
    for match in query_results['matches']:
```

```
        print(f"ID: {match['id']}, Score: {match['score']}")
        if 'metadata' in match and 'text' in match['metadata']:
            text=match['metadata']['text']
            #print(f"Text: {match['metadata']['text']}")
            target_id = query_results['matches'][0]['id']  # Get the ID 
from the first match
            #print(f"Target ID: {target_id}")
        else:
            print("No metadata available.")
    return text, target_id
```

The text and ID are returned to QF1, `query_vector_store`, which, in turn, returns the results when the function is called. Note that for educational purposes, this function assumes `query_results` will always contain at least one match with `'metadata'` and `'text'` fields. Let's now see how the query is processed.

Processing the queries

The program queries the Pinecone index with `get_query_results` with the input text and namespace provided. But first, the input text must be embedded to enable a vector similarity search in the vector store:

```
def get_query_results(query_text, namespace):
    # Generate the query vector from the query text
    query_vector = get_embedding(query_text)  # Replace with your method 
to generate embeddings
```

Once the input is embedded, a vector search is requested with the vectorized input within the namespace specified:

```
    # Perform the query
    query_results = index.query(
        vector=query_vector,
        namespace=namespace,
        top_k=1,  # Adjust as needed
        include_metadata=True
    )
```

Note that k is set to 1 in this example to retrieve a single top result for precision, and also, the metadata is set to True to include the corresponding text. The results are returned to QF2,query_results:

```
    # Return the results
    return query_results
```

The embedding function is the same as what we used to upsert the data in the Pinecone index:

```
import openai
client = openai.OpenAI()
embedding_model = "text-embedding-3-small"
def get_embedding(text, model=embedding_model):
    text = text.replace("\n", " ")
    response = client.embeddings.create(input=[text], model=model)
    embedding = response.data[0].embedding
    return embedding
```

Make sure to use the same model to embed queries as you did to embed the data you upserted so that the embedded input is in the same vector format as the embedded data stored. This is critical for similarity search to make accurate similarity calculations.

We're now ready to run two tests: an instruction scenario query (namespace genaisys) and a source data query (namespace data01).

Retrieval queries

To retrieve an instruction scenario, we will enter a user input and the namespace to let the system find the closest instruction to perform:

```
# Define your namespace and query text
namespace = "genaisys"  # Example namespace
query_text = "The customers like the idea of travelling and learning.
Provide your sentiment."
```

The system should detect the task briefly asked for and return a comprehensive instruction scenario. For that, we'll call the entry point of the functions, query_vector_store, and display the output returned:

```
# Call the query function
text, target_id = query_vector_store(query_text, namespace)
# Display the final output
print("Final output:")
print(f"Text: {text}")
print(f"Target ID: {target_id}")
```

The output is satisfactory and is ready to be used in *Chapter 4* in a conversational loop:

```
Querying vector store...
Processed query results:
ID: 2, Score: 0.221010014
Querying response time: 0.54 seconds
Final output:
Text: 200,Sentiment analysis  Read the content return a sentiment analysis
nalysis on this text and provide a score with the label named : Sentiment
analysis score followed by a numerical value between 0 and 1  with no + or
- sign and  add an explanation to justify the score.
Target ID: 2
```

The program now retrieves data from the Pinecone index. The query functions are identical since the namespace is a variable. Let's just look at the query and output. The query is directed to the data namespace:

```
# Define your namespace and query text
namespace = "data01"  # Example namespace
query_text = "What did the CTO say about the different types of memory?"

The result is printed:
# Display the final output
print("Final output:")
print(f"Text: {text}")
print(f"Target ID: {target_id}")
```

The output is satisfactory:

```
Querying vector store...
Processed query results:
ID: 3, Score: 0.571151137
Querying response time: 0.45 seconds
Final output:
Text: We defined memoryless, short-term, long-term memory, and cross-topic
memory. For the hybrid travel marketing campaign, we will distinguish
semantic memory (facts) from episodic memory (personal events in time, for
example). The CTO said that we will need to use episodic memories of past
customer trips to make the semantic aspects of our trips more engaging.
Target ID: 3
```

We have thus populated a Pinecone vector store and queried it. Let's summarize the implementation of the Pinecone index before we move on to adding more layers to our GenAISys.

Summary

In this chapter, we pushed our GenAISys project another step forward by moving beyond ordinary RAG. First, we layered expert-written instruction scenarios on top of the source data corpus, turning a static RAG pipeline into a dynamic framework that can fetch not only facts but also the exact reasoning pattern the model should follow. The global market is accelerating so quickly that users now expect ChatGPT-level assistance the moment a need arises; if we hope to keep pace, our architecture must be flexible, cost-aware, and capable of near-real-time delivery.

We began by laying out that architecture, then introduced the law of diminishing returns to determine when an implicit similarity search is worth its compute bill and when a direct, explicit call—such as a simple web search—will do the job more cheaply. With the theory in place, we wrote a program to download, chunk, embed, and upsert the instruction scenarios into a dedicated namespace inside a Pinecone index. Next, we enlisted GPT-4o to perform the same chunk-and-embed routine on the source documents, storing those vectors in a second namespace. Once both partitions were in place, we verified the retrieval layer: a single query function now routes any prompt to the correct namespace and returns the best match along with its metadata.

With scenarios and data cleanly separated yet instantly searchable, the GenAISys has the retrieval backbone it needs. In the next chapter, we will plug these components into the conversational loop and let the system demonstrate its full, business-ready agility.

Questions

1. There is no limit to automating all tasks in a generative AI system. (True or False)
2. The law of diminishing returns is of no use in AI. (True or False)
3. Chunking is the process of breaking data into smaller parts to retrieve more nuanced information. (True or False)
4. There is only one embedding model you should use. (True or False)
5. Upserting data to a Pinecone index means uploading data to a database. (True or False)
6. A namespace is the name of a database. (True or False).
7. A namespace can be used to access different types of data. (True or False)
8. Querying the Pinecone index requires the user input to be embedded. (True or False)
9. Querying the Pinecone index is based on a metric such as cosine similarity. (True or False)
10. The Pinecone index and the query functions are the only components of a GenAISys. (True or False)

References

- OpenAI embeddings documentation: `https://platform.openai.com/docs/guides/embeddings`
- Pinecone Python SDK documentation: `https://docs.pinecone.io/reference/python-sdk`
- Pinecone documentation: `https://docs.pinecone.io/guides/get-started/overview`

Further reading

- *AI Development Cost: Learn What Makes Developing an AI Solution*: `https://www.spaceo.ai/blog/ai-development-cost/`

4

Building the AI Controller Orchestration Interface

Businesses today need to design, produce, and deliver goods and services at a speed never attained before. Responsiveness has become key in nearly every field, from online cloud services to delivering food, medication, clothing, and so on. Such an event-driven economy produces an endless stream of tasks, and only an equally event-driven, human-centered **generative AI system (GenAISys)** can keep pace.

Human judgment still anchors even the most automated workflows: when fires break out, storms destroy infrastructure, or supply chains falter, teams—not algorithms alone—must act. An advanced GenAISys that leaves people out of the loop is a myth. This chapter, therefore, begins by outlining an architecture that tears down the walls between users and AI to create a collaborative, multi-user chatbot.

First, we sketch the event-driven GenAISys interface at a high level, showing how the building blocks from earlier chapters—short-term, episodic, and long-term memory, the multi-turn conversational agent, and twin RAG pipelines for instruction scenarios and data—fit together. To then implement the responsive system, we must code the processes of the GenAISys and then the conversational agent that will manage the generative AI agent. Once our GenAISys interface is built, we will run a multi-user, multi-turn conversation with three users working in an online travel agency. Their online meeting will include a conversational AI agent as a participant.

These users will be able to have an online meeting with or without the AI agent. They will be able to utilize RAG to find instruction scenarios or simply ask the generative AI agent to answer a question. By the end of the chapter, we will have a fully working GenAISys interface ready for the multimodal chain-of-thought extensions in *Chapter 5*.

In a nutshell, this chapter covers the following topics:

- A high-level view of the architecture of an event-driven GenAISys interface
- A low-level, hands-on flowchart of the GenAISys interface
- Implementing response widgets for inputs, the AI agent, and active users
- The chatbot's event-driven flow in a multi-turn conversation
- Multi-user GenAISys conversation with an AI agent as a participant
- The response RAG features of the conversational agent
- The orchestration capabilities of the GenAISys interface and AI agent

Our first task is to define an event-driven GenAISys interface.

Architecture of an event-driven GenAISys interface

Our event-driven GenAISys interface integrates the functionality we built in the previous chapters. The interface will leverage the flexibility of IPython widgets to create a reactive event-driven environment in which the following apply:

- The high-level tasks will be event-driven, triggered by user inputs
- Generative AI tasks will trigger generative AI agent functions

We will first examine the program we are building at a high level, as represented in *Figure 4.1*:

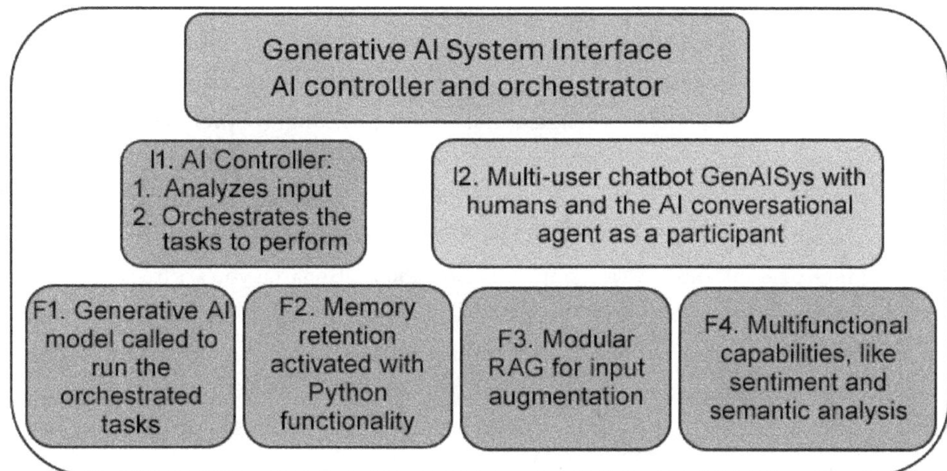

Figure 4.1: High-level architecture of the GenAISys interface

Let's go through the functions we have already built in the previous chapters and also list the key ones we are adding in this chapter:

- **I1 – AI controller**: This chapter's main new component is the generative AI Python interface with responsive widgets, which will be run as an AI controller and orchestrator
- **I2 – Multi-user chatbot**: The chat surface through which several users interact concurrently
- **F1 – Generative AI model**: Inherited from all the previous chapters, especially *Chapter 3*, in which we ran generative AI calls with GPT-4o
- **F2 – Memory retention**: Inherited from *Chapter 1*, which introduced different types of memory
- **F3 – Modular RAG**: The instruction-and-data pipelines inherited from *Chapter 3*
- **F4 – Multifunctional capabilities:** Semantic and sentiment analysis from *Chapters 2* and *3*, to be expanded in *Chapter 5* with image, audio, web search, and ML features

To build this architecture, we will do the following:

- Build the processes of an event-driven GenAISys interface
- Implement the conversational agent with GPT-4o and an OpenAI embedding model
- Run a multi-user, multi-turn session exploring the main features of the GenAISys AI controller and orchestrator

The decision to present the main components of the GenAISys architecture (in this chapter and the next) without arrows is a deliberate choice designed to convey a core concept: modularity and architectural flexibility. The figure is not a rigid blueprint but rather a conceptual toolkit. It shows you the powerful components at your disposal—**I1. AI controller, I2. Multi-user chatbot, F1. Generative AI model, F2. Memory retention, F3. Modular RAG**, and **F4. Multifunctional capabilities**—as independent, interoperable blocks. This empowers you, illustrating that you are free to design your own system architecture. For instance, a user could choose to run some functional components, such as **F4. Multifunctional capabilities**, as independent, distributed agents that are called upon by the controller. Alternatively, they could implement a completely different interface or even run the system headlessly without one.

However, the focus of this architecture is on demonstrating a human-centered GenAISys. In this configuration, **I1. AI controller** (the generative AI IPython interface) serves as the central hub and orchestrator. This human-centered architecture guarantees full control and transparency. This is essential to build trust in risk-averse corporate environments. The control flow, while not drawn with arrows, is implicit: user interactions from **I2. Multi-user chatbot** are managed by the AI controller, which then strategically delegates tasks to the various functional components (**F1** to **F4**) to generate responses, access memory, perform RAG, or execute specific functions. This approach provides a clear, stable, and explainable pathway to building a business-ready generative AI system.

Let's first explore scenario-driven task execution.

Building the processes of an event-driven GenAISys interface

Let's begin by building the GenAISys interface shown in *Figure 4.2*, using IPython widgets to create a responsive, event-driven environment. The result will be a dynamic multi-user chat surface with drop-down menus, text-input fields, and a checkbox—everything needed for real-time collaboration between people/users and the generative AI agent.

Open `Event-driven_GenAISys_framework.ipynb` notebook within the Chapter04 directory on GitHub (`https://github.com/Denis2054/Building-Business-Ready-Generative-AI-Systems/tree/main`). Setting up the environment is the same as described in the previous chapters:

- To set up OpenAI, refer to *Chapter 1*, including the custom OpenAI API call used here: `openai_api.make_openai_api_call`

- Refer to *Chapter 3* for setting up Pinecone, connecting to the index, and querying it

An additional package (`ipython`) is required for the notebook environment. IPython is pre-installed in Google Colab; if needed, install it using the following:

```
!pip install ipython
```

The code we'll build demonstrates core concepts such as event-driven interactions, dynamic content updating, and modular function organization. By the end of this section, you will have learned how to bridge the gap between AI functionality and end user engagement.

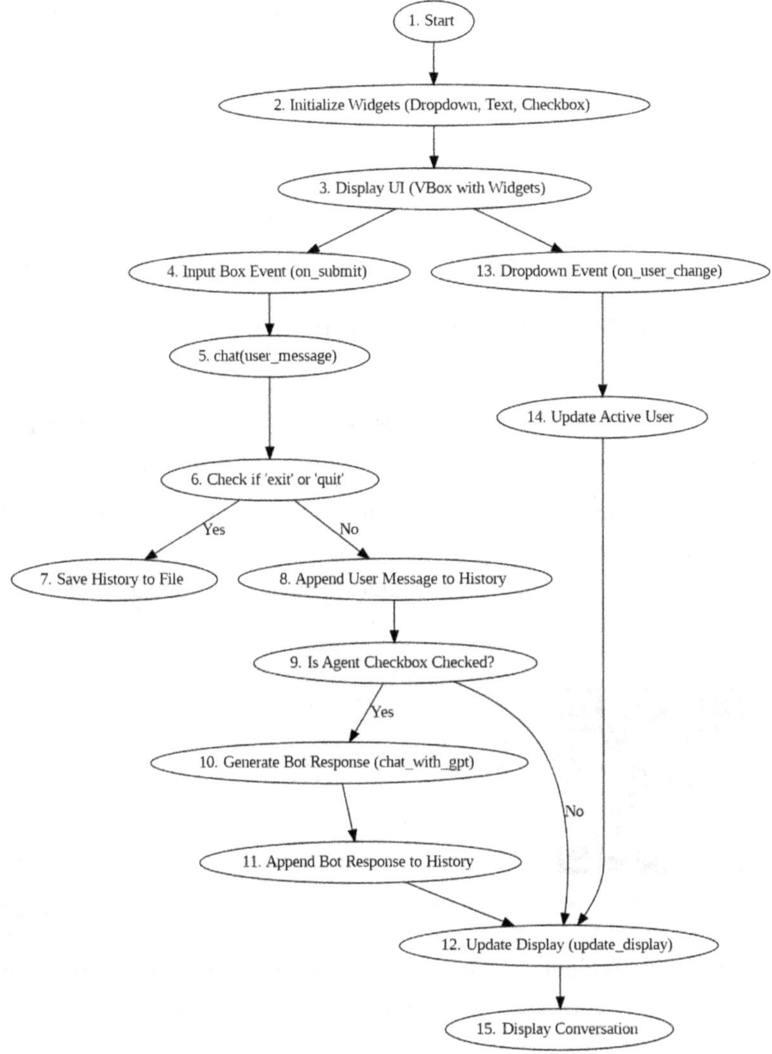

Figure 4.2: The flowchart of an event-driven GenAISys interface

The main groups of functions required to build this interface are the following:

- Initializing widgets
- Handling user input and selection changes
- Processing chat messages, including triggering functions and exit commands
- Generating and processing AI responses
- Updating the UI dynamically
- Saving the conversation history

Before diving into the code from a developer's perspective, let's keep the user's point of view in mind. We must build an intuitive interface that can seamlessly execute the flow outlined in *Figure 4.2*.

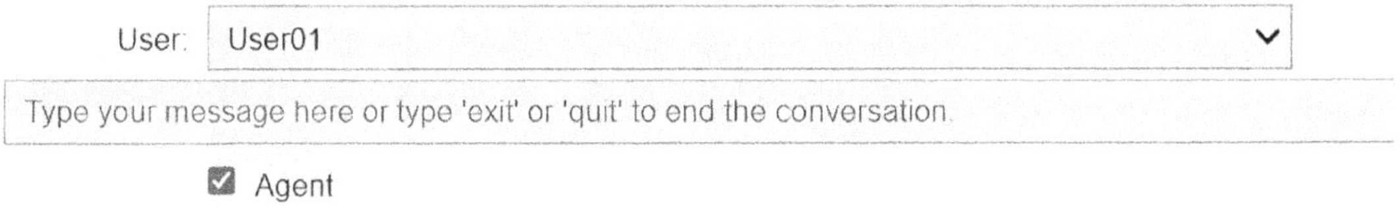
User: User01
Type your message here or type 'exit' or 'quit' to end the conversation.
Agent

Figure 4.3: GenAISys from a user's perspective

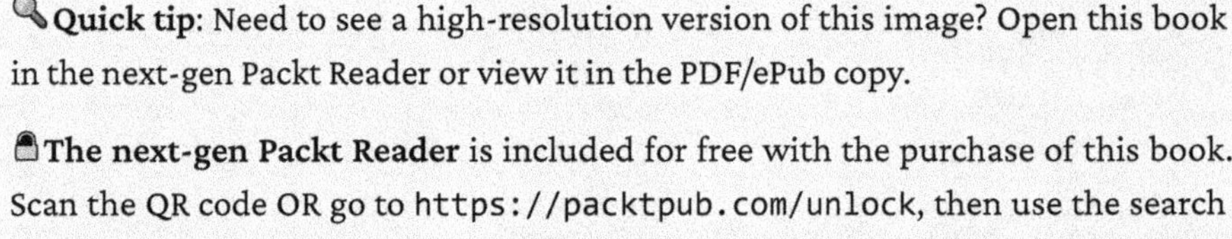

Quick tip: Need to see a high-resolution version of this image? Open this book in the next-gen Packt Reader or view it in the PDF/ePub copy.

The next-gen Packt Reader is included for free with the purchase of this book. Scan the QR code OR go to `https://packtpub.com/unlock`, then use the search bar to find this book by name. Double-check the edition shown to make sure you get the right one.

The UI contains only three widgets: an input box for entering prompts, a drop-down list for selecting active users, and a checkbox for activating and deactivating the conversational AI agent.

Let's walk through the process of setting up and running this interactive GenAISys environment.

1. Start

The program starts from the *Multi-user conversation with the agent as a participant* cell. We first import the modules and libraries we need, starting with `IPython`:

```
from IPython.display import display, HTML, clear_output
```

Let's go through each functionality we will be implementing in Google Colab:

- `display` and `HTML` to display objects such as widgets, images, and rich HTML outputs
- `clear_output` to clear the output of a cell

Then, we import `ipywidgets` managed by the Jupyter project:

```
from ipywidgets import Dropdown, Text, Checkbox, VBox, Layout
```

`ipywidgets` is the core component of the interactive interface in this notebook, in which we will use the following widgets:

- `Dropdown`: A drop-down widget to select a value from a list of options
- `Text`: A widget for text input from a user
- `Checkbox`: A widget for Boolean checked/unchecked input
- `Vbox`: A container widget to arrange child widgets in a vertical box layout
- `Layout`: To customize the style of the widgets with layout properties such as width, height, and margin

Finally, we import JSON, used to store multi-user conversation histories:

```
import json
```

We then initialize the conversation histories for all users, define the first active user, and set the active conversation to `True`:

```
# Initialize conversation histories for all users and active user
user_histories = {"User01": [], "User02": [], "User03": []}
active_user = "User01"  # Default user
conversation_active = True
```

We are thus, from the start, initializing a multi-user collaborative GenAISys in which the users can be human prompts and system prompts. For example, a "user" could be a message from another system and triggered in this interface by an event that reads pending messages. The user list can be expanded, stored in variables, or utilized in any user management system that suits a project's needs, including access rights, passwords, and roles for various applications. Next, we initialize the widgets themselves.

2. Initialize widgets

The code now sets up the `Dropdown`, `Text`, and `Checkbox` widgets we need. The widgets are also linked to event handlers. The `Dropdown` widget for the users defines the three users initialized at the start of the conversation:

```
# Create a dropdown to select the user
user_selector = Dropdown(
    options=["User01", "User02", "User03"],
    value=active_user,
    description='User:',
    layout=Layout(width='50%')
)
```

The selector has four parameters:

- `options` lists the available users that can be expanded and can access any user management repository as needed for your project.
- `value` determines the active user. The program started with `User01` as the initial user. This can be automated when an authorized user first connects to the GenAISys.
- `description` provides a label for the drop-down list that will be displayed.
- `layout` sets the width of the widget that will be displayed.

Note that we are creating a core GenAISys, not a platform. The goal is to grasp the inner workings of a GenAISys. Once it works as expected, we can then add the classical layers of user management (names, roles, and rights). In this case, we are remaining focused on the flexible core concepts of GenAISys, not how they will be encapsulated in a specific platform and framework. We are learning how to be generative AI agentic architects, not operators of a specific framework.

The next step is to insert an event handler. In this case, it is an event listener that will detect when the value of `user_selector` changes. When another user is selected, the `on_user_change` function is automatically called, and `value` becomes the new user:

```
user_selector.observe(on_user_change, names='value')
```

This dynamic change in users within a GenAISys conversation represents a major evolution from the one-on-one chatbots. It introduces a whole new dimension to collaborative teamwork with AI as a co-participant.

The second widget to activate is the input widget:

```
# Create the input box widget
input_box = Text(placeholder="Type your message here or type 'exit' to end
the conversation.", layout=Layout(width='100%'))
```

The input can be any text and will occupy 100% of the UI layout. The conversation ends when a user enters `exit` or `quit`. When the text is typed and the *Enter* button is pressed, the event handler takes over:

```
input_box.on_submit(handle_submit)  # Attach the on_submit event handler
```

`on_submit` is a method of the `input_box` widget. `handle_submit` is a callback function that we can write as we wish and will be described later in this section.

The third widget is the checkbox for the AI conversational agent:

```
# Create a checkbox to toggle agent response
agent_checkbox = Checkbox(
    value=True,
    description='Agent',
    layout=Layout(width='20%')
)
```

The checkbox displays the description label, which is an agent in this case. The layout will occupy 20% of the UI. If `value` is set to `True`, then the conversational AI agent will be activated. We will build the AI agent in the *Conversational agent* section of this chapter. The AI agent will also be event-driven.

The UI box is now ready to be displayed.

3. Display the UI

The UI container widget now combines the three event-driven widgets we defined in VBox (V stands for vertical; i.e., in a vertical box). The three widgets are in brackets:

```
# Display the initial interface
display(
    VBox(
        [user_selector, input_box, agent_checkbox],
        layout=Layout(
            display='flex', flex_flow='column',
            align_items='flex-start', width='100%'
        )
    ))
```

The layout is then defined:

```
layout=Layout(
    display='flex', flex_flow='column',
    align_items='flex-start', width='100%'
)))
```

The parameters of this responsive UI are the following:

- `display='flex'` activates the CSS flexbox model for layouts dynamically without specifying the sizes of the items
- `flex_flow='column'` arranges the child widgets vertically
- `align_items='flex-start'` aligns the widgets to the start (left side) of the UI (left side) container
- `width='100%'` makes the container take up the full width of the available space

With that, the UI is ready. We can choose to begin with any of the three widgets. The user selector can be run before the input, as well as the AI agent checkbox. In this case, the user selector was set to a default value, `User01`, and the AI agent checkbox was set to the default value, `True`.

The three widgets and their processes can be built into any classical web or software interface, depending on your project's needs. Since there is no default value for the input, let's continue with the input widget.

4. Input box event

The input text is managed by the UI described in the previous section, which triggers input_box.on_submit(handle_submit) when a user enters text. In turn, the submit method calls the handle_submit function:

```
# Function to handle the submission of the input
def handle_submit(sender):
    user_message = sender.value
    if user_message.strip():
        sender.value = ""  # Clear the input box
        chat(user_message)
```

Now, the function does three things:

- user_message = sender.value processes the text received from the input widget
- if user_message.strip() checks whether there is a message and clears the input box for the next input with sender.value = "" # Clear the input box
- chat(user_message) is called if there is a message

chat(user_message) is the next process and a key event processing hub for the GenAISys. Let's go through it.

5. chat(user_message) function

The chat(user_message) function is an *orchestrator* component of our event-driven GenAISys. It should remain human-centered for critical human control. Once the system has gained the trust of the users and after careful consideration, some of the actions it manages can be triggered by system messages. The orchestrator contains important decisions when it processes the user message it receives from the handle_submit(sender) function. It encapsulates several choices and functions, as represented in *Figure 4.2*: deciding whether to continue the conversation, appending or saving the conversation history to a file, determining whether to call the AI conversational agent, and updating the UI display.

It first inherits the global status of the conversation variable (conversation_active = True) we initialized at the start of the conversation (in node **1** of *Figure 4.2*):

```
# Function to handle user input and optional bot response
def chat(user_message):
    global conversation_active
```

It continues to determine whether the multiple-turn conversation is over or not by checking whether the user has exited or quit the conversation (see **6** in *Figure 4.2*):

```
        if user_message.lower() in ['exit', 'quit']:
```

Let's see what happens if the user chooses to exit the conversation.

6. If 'exit' is chosen

Suppose the user enters `exit` or `quit`; then the `conversation_active` variable we set to `True` at the start of the conversation (in node **1** of *Figure 4.2*) will now be set to `False`. The system now knows that there is no need to update the display anymore. It then tells the `clear_output` function to wait until the next conversation turn to clear the output to avoid flickering effects:

```
clear_output(wait=True)
```

The exit process continues by displaying a message signaling the end of the conversation and indicating that the conversation history is being saved:

```
display(HTML("<div style='color: red;'><strong>Conversation ended. Saving
history...</strong></div>"))
```

The exit process ends by calling the *save* function of the conversation, which will save all history to a file (see node **7** in *Figure 4.2*):

```
save_conversation_history()
```

The conversation is thus saved at the end of the session for further use (for a new session or a meeting summary), as shown in node **7** of *Figure 4.2*:

```
# Function to save conversation history to a file
def save_conversation_history():
    filename = "conversation_history.json"  # Define the filename
    with open(filename, 'w') as file:
        json.dump(user_histories, file, indent=4)  # Write the user
histories dictionary to the file in JSON format
    display(HTML(f"<div style='color: green;'><strong>Conversation history
saved to {filename}.</strong></div>"))
```

Now, let's go through the process when the user(s) chooses to continue the conversation.

7. If user(s) continue the conversation

If the user input does not contain `exit` or `quit`, then the multi-turn, multi-user conversation will continue. We have some big decisions to make with this function, however. Do we append it to each user request or not? If we append it to each user request, at some point, the context window will be complete, but the number of tokens we send through the API will increase processing time and costs.

The first step is to append the history of the conversation we initialized at the start (in node **1** of *Figure 4.2*):

```
# Append user message to the active user's history
user_histories[active_user].append(
    {"role": "user", "content": user_message}
)
```

This step is a direct example of context engineering, as we are actively managing the information in the model's context window to maintain conversational continuity.

So, in the hybrid scenario of this notebook, at this point, we will save the user history in memory until the end of the session, and we will thus augment each user's input with their input history, as seen in node **11** of *Figure 4.2*. If the user input does not contain `exit` or `quit`, then the multi-turn, multi-user conversation will continue. It will append the user message to the history (in node **8** of *Figure 4.2*) of the user.

However, if we don't want to append a user request to it but still want to keep a record of the entire conversation for context, we can also summarize the conversation at the midpoint or the end. If we summarize it during the conversation, we can add a function to append it to the user input each time. If we summarize after the end of a session, we can continue with a new, fresh session with a summary of the previous session's history.

In this notebook, we will implement a hybrid short- and long-term memory process. We can continue the conversation by not entering `'quit'` or `'exit'`. Now, the `chat(user_message)` function will check the conversational agent's checkbox value:

```
if agent_checkbox.value:
```

This verification is shown in node **9** in *Figure 4.2*. If the checkbox is checked, then the functions we created in the previous chapters are activated by calling `chat_with_gpt`:

```
        response = chat_with_gpt(user_histories[active_user],
            user_message)
```

Once the response is returned, it is appended to the history of the response described previously:

```
        user_histories[active_user].append(
            {"role": "assistant", "content": response}
        )
```

We now have an entry-point memory framework. The program then calls `update_display()`, another key function that is shown in node **14** of *Figure 4.2*. If the agent checkbox is checked, `chat_with_gpt` will be called.

8. Generate bot response

The `chat_with_gpt` function assembles the work we did in the previous chapters to create a conversational AI agent with the Pinecone-based RAG functionality. We will fully implement this integration in the *Conversational agent* section of this chapter.

`chat_with_gpt` orchestrates the AI conversational agent by providing information, enabling it to be dynamic and responsive. The user history of this conversation and the user message are sent to the `chat_with_gpt` conversational agent function:

```
response = chat_with_gpt(user_histories[active_user], user_message)
```

Once the response is returned, the `update_display` function is called from `chat(user_message)`.

9. Update display

The `update_display` function refreshes the UI with the updated conversation history and also displays the status of the widgets. It first tells the UI to wait until a new output arrives by setting `wait` to `True`:

```
def update_display():
    clear_output(wait=True)
```

The function then filters and displays the active user's history (see node **15** of *Figure 4.2*):

```
    for entry in user_histories[active_user]:  # Show only the active
user's history
        if entry['role'] == 'user':
            display(HTML(f"<div style='text-align: left; margin-left:
20px; color: blue;'><strong>{active_user}:</strong> {entry['content']}</
div>"))
```

```
        elif entry['role'] == 'assistant':
            display(HTML(f"<div style='text-align: left; margin-left:
20px; color: green;'><strong>Agent:</strong> {entry['content']}</div>"))
```

If the conversation is active, the UI VBox is displayed along with the status of the widgets:

```
    if conversation_active:
        display(VBox([user_selector, input_box, agent_checkbox]))  # Keep
input box, selector, and checkbox visible if active
```

The input box is cleared, the agent checkbox has been checked independently by the user, and the system has verified its status. The active user will be displayed based on the independent decision of the user. In this case, the active user, `active_user`, who was initialized at the start **(1)** of the conversation, remains the same. If the user changed, the `on_user_change` drop-down event **(13)** would have been triggered by the `observe` method of the `user_selector` widget:

```
user_selector.observe(on_user_change, names='value')
```

In that case, `user_selector.observe` will independently call the `update active_user` function **(14)** and first make sure that the active user is a global variable:

```
def on_user_change(change):
    global active_user
```

Then, it will make the new user the active user:

```
    active_user = change['new']
```

Finally, it will call the `update_display` function we built in this subsection:

```
    update_display()
```

Now that we have our dynamic UI and event-driven functions in place, let's implement the conversational agent logic called by `chat_with_gpt`.

Conversational agent

We implemented an AI conversational agent in *Chapters 1* and *2* and built the query Pinecone functionality in *Chapter 3*. Go to the *Conversational agent* section of the notebook. If needed, take the time to revisit those chapters before proceeding. In this section, it's time we integrate those components, preparing our GenAISys conversational agent for multi-user sessions.

We begin by importing OpenAI and initializing the client:

```
from openai import OpenAI
# Initialize the OpenAI client
client = OpenAI()
```

Next, we make a decision to store or not to store all of the user's conversation history for each call to optimize context window size for cost and clarity:

```
user_memory = True # True=User messages are memorized False=User messages
are not memorized
```

The memory setting should be strategically monitored in production environments. For example, here we set `user_memory` to `True`, but we avoid applying it during RAG queries, as historical context could confuse the Pinecone similarity searches. We then define the `chat_with_gpt` function, which is called in node **10** of *Figure 4.2*:

```
def chat_with_gpt(messages, user_message):
```

The function first searches the input text for a keyword to trigger a RAG retrieval from the Pinecone index as implemented in `Query_Pinecone.ipynb` and described in *Chapter 3*. The code first determines the namespace:

```
    try:
      namespace=""
      if "Pinecone" in user_message or "RAG" in user_message:
        # Determine the keyword
        if "Pinecone" in user_message:
            namespace="genaisys"
        elif "RAG" in user_message:
            namespace="data01"
        print(namespace)
…
```

If the user message contains "Pinecone," the query will target the `genaisys` namespace, which contains the instruction scenarios. The `genaisys` namespace implementation departs from static data retrieval and takes us into agentic, dynamic decision-making to trigger an instruction or a task. If the user message contains "RAG," the query will target the `data01` namespace, which contains static data. The queries and content of the Pinecone index are those implemented in *Chapter 3*:

```
        #define query text
        query_text=user_message
```

```
        # Retrieve query results
        query_results = get_query_results(query_text, namespace)
        # Process and display the results
        print("Processed query results:")
        qtext, target_id = display_results(query_results)
        print(qtext)
```

Once the query result is returned, we append the user message to it to augment the input:

```
        #run task
        sc_input=qtext + " " + user_message
        mrole = "system"
        mcontent = "You are an assistant who executes the tasks you are
asked to do."
        user_role = "user"
```

The message parameters and the OpenAI API call are described in the *Setting up the environment* section of *Chapter 1*. The OpenAI response is stored in `task response`:

```
        task_response = openai_api.make_openai_api_call(
            sc_input,mrole,mcontent,user_role
        )
        print(task_response)
```

The response returned by the OpenAI API call, augmented with the result of the Pinecone query, is stored in `aug_output`:

```
        aug_output=namespace + ":" +task_response
```

If the user message does not contain a keyword to trigger the RAG function, the user request will be sent directly to the OpenAI API call, and the response will be stored in `aug_output`. However, the system must first check whether `user_memory` is `True` or not. The system must also extract the text content of `user_message`:

```
    else:
        if user_memory:
                # Extract ALL user messages from the conversation history
                user_messages_content = [
                    msg["content"] for msg in messages
                    if msg["role"] == "user" and "content" in msg
                ]
```

```
            # Combine all extracted user messages into a single string
            combined_user_messages = " ".join(user_messages_content)

            # Add the current user_message to the combined text
            umessage = f"{combined_user_messages} {user_message}"
```

In this case, `umessage` now contains a concatenation of the conversation history of the active user extracted and stored in `combined_user_messages` and the user message itself in `user_message`. The generative AI model now has complete context about the dialogue with this user.

The strategy for managing conversation history will depend heavily on each real-world use case. For example, we might choose to extract the history of all users involved in a session or only specific users. Alternatively, a team could decide to use a single shared username throughout an entire conversation. Generally, the best practice is to organize workshops with end users to define and configure the conversation-memory strategies that best fit their workflow.

In some cases, we might decide to ignore the conversation history altogether. In that scenario, we set the `user_memory` parameter to `False`, and the system disregards prior exchanges:

```
        else:
            umessage = user_message
```

The `umessage` variable is now ready to be sent directly to the generative AI model:

```
        mrole = "system"
        mcontent = "You are an assistant who executes the tasks you are
asked to do."
        user_role = "user"
        task_response =openai_api.make_openai_api_call(
            umessage,mrole,mcontent,user_role
        )
        aug_output=task_response
```

The response from the OpenAI API call is then returned to the `chat_with_gpt` function (in node **10** of *Figure 4.2*):

```
        # Return the augmented output
        return aug_output
```

If the OpenAI API call fails, an exception is raised and returned:

```
    except Exception as e:
        # Return the error message in case of an exception
        return f"An error occurred: {str(e)}"
```

And with that, we have assembled the generative AI functionalities developed across the previous three chapters. At this stage, we've built a responsive GenAISys interface and integrated a generative agent, together forming a cohesive AI controller and orchestrator. Let's now put our GenAISys into motion.

Multi-user, multi-turn GenAISys session

We now have a responsive, event-driven GenAISys capable of executing multiple tasks in diverse ways, as illustrated in *Figure 4.4*. We will explore the flexibility of this GenAISys interface we built using IPython and assemble the OpenAI and Pinecone components from previous chapters.

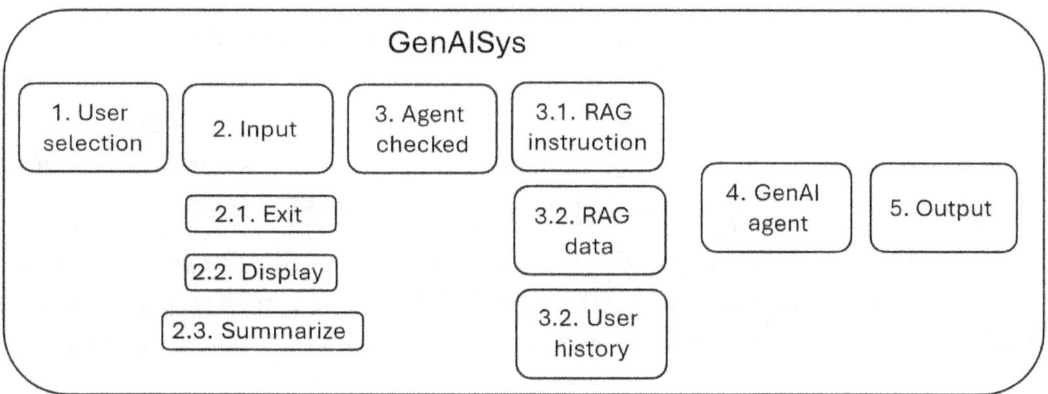

Figure 4.4: Summing up the components we have built and assembled in this chapter

Since the functions within GenAISys are event-driven, a user (human or system) or a group of users can leverage this framework to address multiple cross-domain tasks. The system is human-centric, creating a collaborative, frictionless environment between humans and a generative AI agent. Importantly, there is no competition between humans and AI in this framework. Teams can maintain human social relationships with co-workers while using the GenAISys to boost their performance and productivity exponentially. This human-centric approach is one I have always advocated throughout my decades of experience providing AI-driven automation solutions for global corporations, mid-sized businesses, and smaller organizations. When teams adopt AI as a collaborative tool rather than a competitor, it fosters a positive atmosphere that leads to quick wins—demonstrating the combined effectiveness of teamwork and technology.

If we look deeper into how the GenAISys framework can be leveraged in teamwork scenarios, we can establish several fundamental sequences of events typically needed in real-world projects:

a. User selection => Input => Agent checked => RAG instruction => GenAI agent => Output
b. User selection => Input => Agent checked => RAG data => GenAI agent => Output
c. User selection => Input => Agent checked => User history => GenAI agent => Output
d. User selection => Input => Agent checked => No user history => GenAI agent => Output
e. User selection => Input => Agent unchecked => Output

These basic sequences constitute a set of sequences, ***S***:

$$\boldsymbol{S} = \{\boldsymbol{a}, \boldsymbol{b}, \boldsymbol{c}, \boldsymbol{d}, \boldsymbol{e}\}$$

To achieve a goal for a single user or a group of users, the sequences can be assembled as follows:

- {a, b}: Running a sentiment analysis with RAG, followed by the retrieval of an episodic memory of a past meeting.
- {d, e}: Running an OpenAI API request and then making a comment for other users. The novelty in this case is that the AI agent remains a co-worker in a team and sometimes doesn't express itself, allowing the team to ponder the ideas it suggested.

These sequences can be arranged into longer session flows as required by the specific tasks and scenarios. Because sequences can repeat themselves, we have an indefinite number of possible dynamic combinations. For instance, here's a glimpse into the flexibility that this provides:

- Set of three members, such as {a, c, e}, {b, d, e}, {a, b, c}
- Set of four members, such as {a, b, c, d}, {b, c, d, e}, {a, c, d, e}
- Set of five members, such as {a, b, c, d, e}

We could add exiting the session and summarizing to these sequences, as well as reloading a saved file and continuing the session. There can also be a repetition of sets, sets with different users, and sets with more functions. In the following chapters, we will add new features, including image generation, audio, web search, and ML, that will expand the scope of the GenAISys framework we have built.

In this section, however, we will run a session with two users in a simple sequence of events. Then, we will run a scenario with multiple users and some basic sequences. Let's begin with a straightforward sequence of events.

A session with two users

In this example session, two users collaborate to brainstorm ideas for attractive travel destinations they could recommend to customers on their online travel website. We start by running an interface session, then display the conversation history, and finally summarize the discussion. To begin the session, open `Event-driven_GenAISys_framework.ipynb` and run these sections of cells:

- **Setting up the environment**: Run all cells
- **Conversational agent**: Run the single cell
- **Running the interface in the GenAISys IPython interface**: This will initialize the conversation

Due to the stochastic nature of generative AI models, the outputs might vary slightly with each execution. Likewise, minor differences may appear between this notebook and the printed chapter, as multiple runs are performed during quality control.

With the conversation initialized, let's now run the interactive session.

The interactive conversation

The conversation starts with `User01` by default, displaying the input box and the activated agent checkbox. The sequence of events and functions triggered in this scenario is illustrated in *Figure 4.5*.

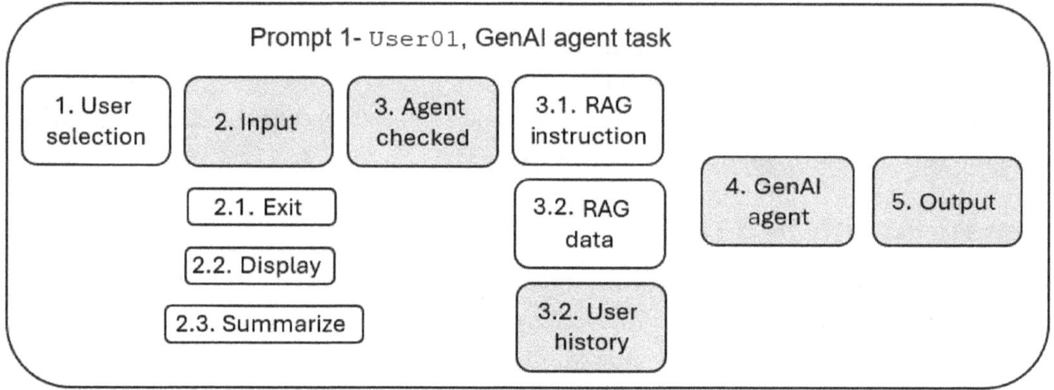

Figure 4.5: The GenAI agent performs a task with the user's history

The flow follows this sequence:

User selection => Input => Agent checked => User history => GenAI agent => Output

To the user, this process is seamless, as illustrated in *Figure 4.6*. However, the underlying functions required careful design and development to produce this smooth effect.

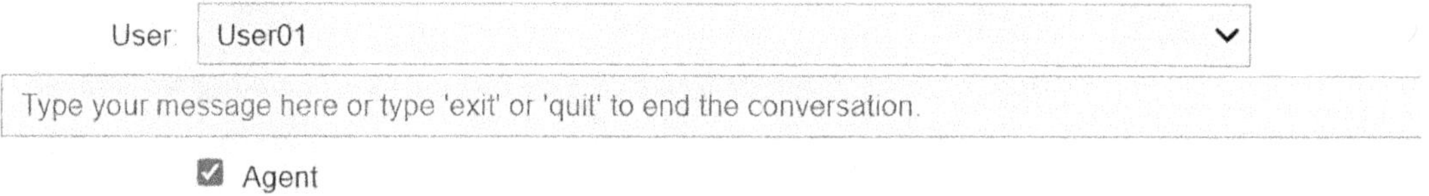

Figure 4.6: UI with the GenAI agent checked

From the user's perspective, the process is straightforward. `User01` types a prompt into the input box: `What is the capital of France?`.

Figure 4.7: User entering a simple prompt

The output will be displayed above the input widget, as shown here:

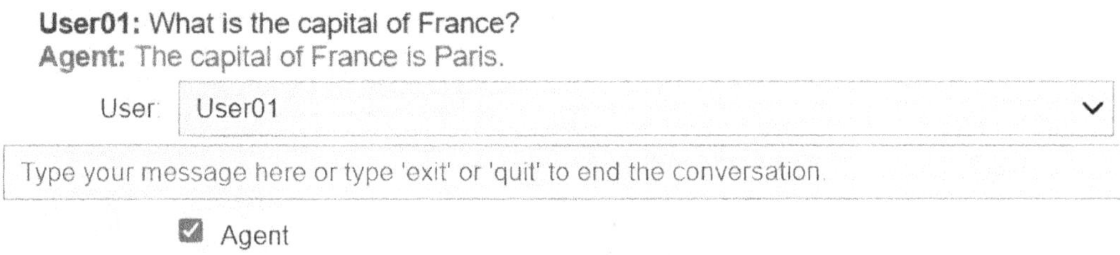
User01: What is the capital of France?
Agent: The capital of France is Paris.
User: User01
Type your message here or type 'exit' or 'quit' to end the conversation.
Agent

Figure 4.8: Output is displayed above the input box

At this point, even an untrained user can intuitively run the GenAISys with the basic information provided in a one-page document. `User01` continues with a follow-up question:

```
User01: What is there to visit?
```

Since the user history option is set to `True` in the conversational agent, the agent knows that the question is about Paris and provides information on some iconic locations to visit:

```
Agent: The capital of France is Paris. There are numerous attractions
to visit in Paris, including: 1. **Eiffel Tower** - An iconic symbol of
Paris, offering stunning views of the city. 2. **Louvre Museum** - The
world's largest art museum, home to the Mona Lisa…
```

`User02` now enters the dialogue by selecting **User02** in the drop-down menu, as shown in *Figure 4.9*. **User02** then asks what the capital of Spain is, the agent responds, and the output is displayed above the input box.

User02: What is the capital of Spain?
Agent: The capital of Spain is Madrid.

User: User02

Type your message here or type 'exit' or 'quit' to end the conversation.

Agent

Figure 4.9: Input: Another user can enter the conversation

At this point, the GenAISys core framework is set, and we will progressively enhance it in subsequent chapters by addressing features such as security, user access controls, and data privacy. `User02` goes further and asks about the places worth visiting. The agent responds correctly because the user history option is activated:

```
User02: What is there to visit?
```

```
Agent: The capital of Spain is Madrid. There are many attractions to visit
in Madrid, including: 1. **The Royal Palace of Madrid** - The official
residence of the Spanish royal family, although it is now used only for
state ceremonies. 2. **Prado Museum** …
```

Now, a key feature comes into play. `User02` deactivates the AI agent, as shown in *Figure 4.10*, and directly addresses `User01`, similar to an interaction on social media or remote collaboration platforms.

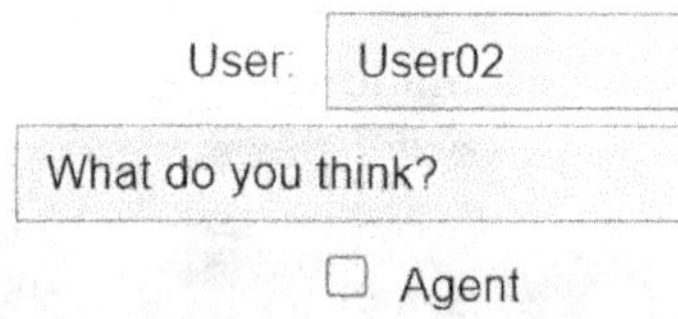

Figure 4.10: Direct communication in a team without an AI agent

`User01` responds informally: "Yes, that's a good idea, but let's think it over." Subsequently, `User01` ends the session by typing `exit`. This capability, as illustrated in *Figure 4.11*, takes our GenAISys to a new level for the use cases we will explore in this book, such as the following configurations:

- A user can communicate with GenAISys alone in a one-to-one conversation
- A team can work together, enhancing their performance with the AI agent as a collaborator
- The users can be AI agents playing the role of managers from different locations when the human managers are not available

- The users can be systems providing information in real-time to human users

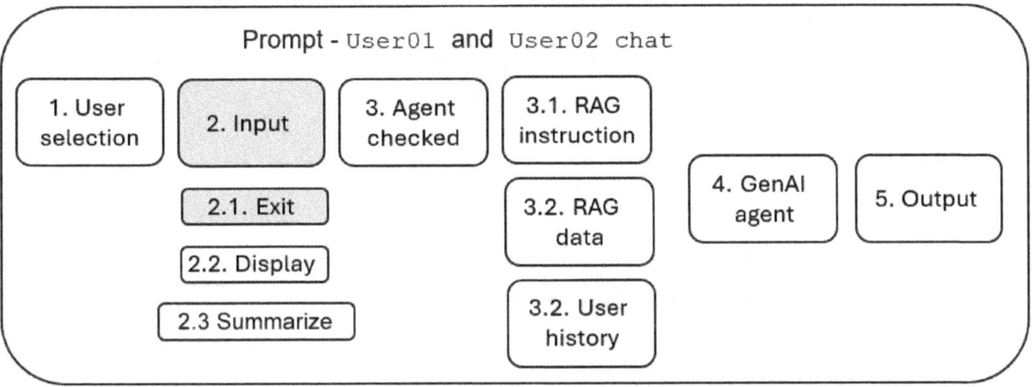

Figure 4.11: A team communicates directly and then ends the session

Upon exiting, the session ends, and the conversation history is saved to the `conversation_history.json` file:

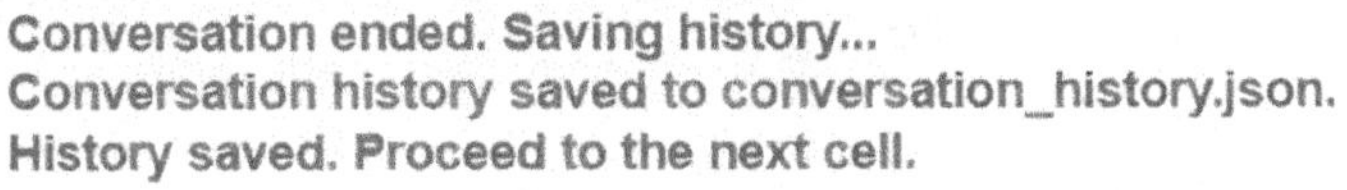

```
Conversation ended. Saving history...
Conversation history saved to conversation_history.json.
History saved. Proceed to the next cell.
```

Figure 4.12: Saving and concluding the conversation

Like all other features in this framework, the exit behavior can be customized for individual projects. Take the following examples:

- The conversation history can be saved or not
- Only parts of the conversation history can be saved
- The name of the saved conversation history file can contain a timestamp
- Going `"…to the next cell"` is optional

These are decisions to make for each use case. They will not modify the overall framework of the GenAISys but allow for a high level of customization.

In this case, the team wants to display the conversation they just had.

Loading and displaying the conversation

The code for this function is a standard IPython display function to convert the JSON file, `conversation_history.json`, into Markdown format. Let's first check whether the conversation history parameter and/or the summary parameter is activated:

```
display_conversation_history=True
summary=True
```

In this case, the conversation history and the summary function are both activated. Now, we will check whether a conversation history file is present or not:

```
import json
from IPython.display import display, Markdown
import os

if display_conversation_history == True or summary==True:
    # File path
    file_path = 'conversation_history.json'

    # Check if the file exists
    if os.path.exists(file_path):
        display_conversation_history=True
        summary=True
        print(f"The file '{file_path}' exists.")
    else:
        display_conversation_history=False
        summary=False
        print(f"The file '{file_path}' does not exist.")
        print("The conversation history will not be processed.")
```

If a file exists, `display_conversation_history` will be set to `True` and `summary=True` (even if it was set to `False` previously). A message will signal that the file exists:

```
The file 'conversation_history.json' exists.
```

If `display_conversation_history==True`, then the conversation will be displayed:

```
# Display option
if display_conversation_history==True:
  # File path
  file_path = 'conversation_history.json'

  # Open the file and read its content into the 'dialog' variable
  with open(file_path, 'r', encoding='utf-8') as file:
      dialog = json.load(file)  # Parse JSON content
…
# Function to format JSON content as markdown
def format_json_as_markdown(data, level=0):
    html_output = ""
    indent = "  " * level
…
return html_output
# Format the JSON into markdown
formatted_markdown = format_json_as_markdown(dialog)
# Display formatted JSON as Markdown
display(Markdown(formatted_markdown))
```

The output is nicely formatted:

```
…
User01:
role:
user
content:
What is the capital of France?
role:
assistant
content:
The capital of France is Paris.
…
Content:
The capital of Spain is Madrid.
role:
user
```

```
content:
What is there to visit?
role:
assistant
content:
The capital of Spain is Madrid. There are many attractions to visit in
Madrid, including:
The Royal Palace of Madrid - …
```

The team has displayed the conversation but wants to take the process further and summarize this online meeting that included an AI agent as a participant.

Loading and summarizing the conversation

The conversation we are summarizing shows how to merge an AI agent into an existing human team to boost productivity. In some cases, the GenAISys will have worked on automated tasks alone. In other cases, the GenAISys will be the copilot of one or several users. In others, in the many critical moments of the life of an organization, teams of humans and AI agents will be able to work together to make decisions.

In this section, we will ask the AI agent to summarize the conversation. We will integrate this feature as a function in the GenAISys in the following chapters. For the moment, we will run it separately after displaying the conversation, as shown in *Figure 4.13*.

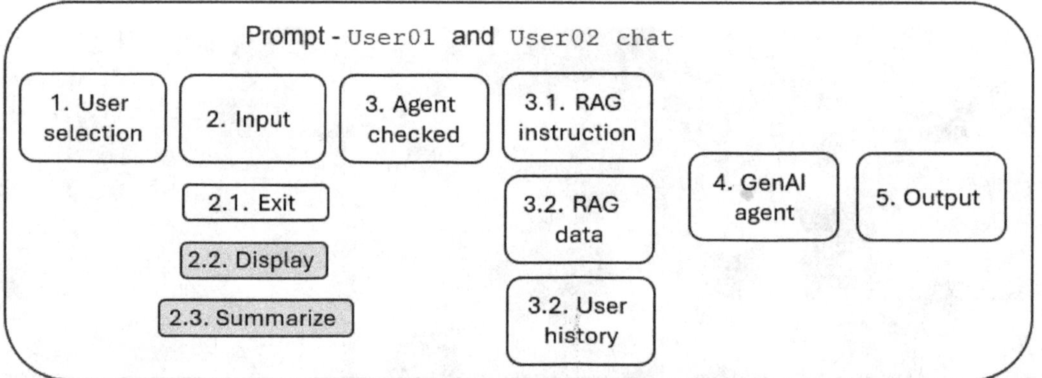

Figure 4.13: Displaying and summarizing a conversation

The code first loads the `conversation_history.json` file as in the display function. Then, we define a function that converts the conversation history content into an optimal format for the OpenAI API:

```
# Function to construct dialog string from the JSON conversation history
def construct_dialog_for_summary(conversation_history_json):
    dialog = ""
    for user, messages in conversation_history_json.items():
        dialog += f"\n{user}:\n"
        for message in messages:
            role = message["role"]
            content = message["content"]
            dialog += f"- {role}: {content}\n"
    return dialog
```

The function to construct the full conversation history is called:

```
# Construct the full dialog from the JSON history
formatted_dialog = construct_dialog_for_summary(conversation_history_json)
```

Now, we prepare the complete message for the custom GenAISys API call built for the system and imported in the *OpenAI* subsection of the *Setting the environment* section in our notebook:

```
# Task to summarize the conversation
mrole = "system"
mcontent = "Your task is to read this JSON formatted text and summarize
it."
user_role = "user"
task = f"Read this JSON formatted text and make a very detailed summary of
it with a list of actions:\n{formatted_dialog}"
```

Finally, we call the GenAISys OpenAI function:

```
# The make_openai_api_call function is called
task_response = openai_api.make_openai_api_call(
    task, mrole, mcontent, user_role
)
```

The API response code will be displayed in Markdown format:

```
from IPython.display import Markdown, display
# Display the task response as Markdown
display(Markdown(task_response))
```

Now, everything is ready. We can call the summarizing function if `summary==True`:

```
if summary==True:
    # File path to the JSON file
    file_path = '/content/conversation_history.json'
    # Check if the file exists before calling the function
    if os.path.exists(file_path):
        summarize_conversation(file_path)
    else:
        print(f"File '{file_path}' does not exist. Please provide a valid
file path.")
```

Note that in Google Colab, `/content/` is the default directory. So, the following file paths point to the same directory:

```
  file_path = '/content/conversation_history.json' or
file_path = 'conversation_history.json'
```

In another environment, you may need absolute paths.

The output is a summary of the conversation history that contains an introduction and then a detailed summary. The prompt for this summary can be modified to request shorter or longer lengths. We can also design a prompt asking the generative AI model to target part of the conversation or design any other specific prompt for a given project. In this case, the output is satisfactory:

```
The JSON formatted text contains interactions between users and an
assistant, where users inquire about the capitals of France and Spain and
seek recommendations for attractions to visit in these cities. Below is a
detailed summary with a list of actions:
User01 Interaction:
1. Question about the Capital of France:
    User01 asks for the capital of France.
    The assistant responds that the capital of France is Paris.
2. Inquiry about Attractions in Paris:
```

```
User01 asks what there is to visit in Paris.
The assistant provides a list of notable attractions in Paris:
1. Eiffel Tower - Iconic symbol and must-visit landmark.
2. Louvre Museum - Largest art museum, home to the Mona Lisa….
```

By running through the many possible sequences of tasks and events, we have seen the flexibility that the GenAISys offers us. Let's run a more complex multi-user session.

Multi-user session

In this section, we will run a technical session that activates the main functions we have built in the previous chapters and this chapter:

- Semantic and sentiment analysis
- RAG for episodic memory retrieval
- A dialogue without an AI conversational agent
- Loading, displaying, and summarizing the conversation history

If you haven't interrupted the previous session, then simply run the *Running the interface section in the GenAISys IPython interface* cell again in our notebook, which will start a new conversation.

If you are starting from scratch, then to start the session, open `Event-driven_GenAISys_framework.ipynb` and run the following sections of cells:

- Setting up the environment: All the cells
- Conversational agent: Contains one cell
- Running the interface in the GenAISys IPython interface: Will start the conversation

We are ready to explore some advanced features of the GenAISys. We will highlight the events and functions that are activated by each prompt. The first sequence in the session is semantic and sentiment analysis.

Semantic and sentiment analysis

To perform semantic and sentiment analysis, we will need to run the following sequence orchestrated by the GenAISys as shown in *Figure 4.14*:

- **1. User selection** is not activated because `User01` is the default user at the beginning of a session. We could call this user the "host" if we wish, depending on the use case.
- `User01` enters an input at **2. Input** triggering **3. Agent checked**, which is checked as the default value when the session starts.

- The AI conversational AI controller takes over, parses the prompt, finds the `Pinecone` keyword in the prompt, triggers a Pinecone query in the instruction scenario namespace, augments the prompt, and triggers **4. GenAI agent**.
- **4. GenAI agent** triggers an API call to GPT-4o and returns the response.
- **5. Output** triggers the updating of the display. The system is ready for a new input.

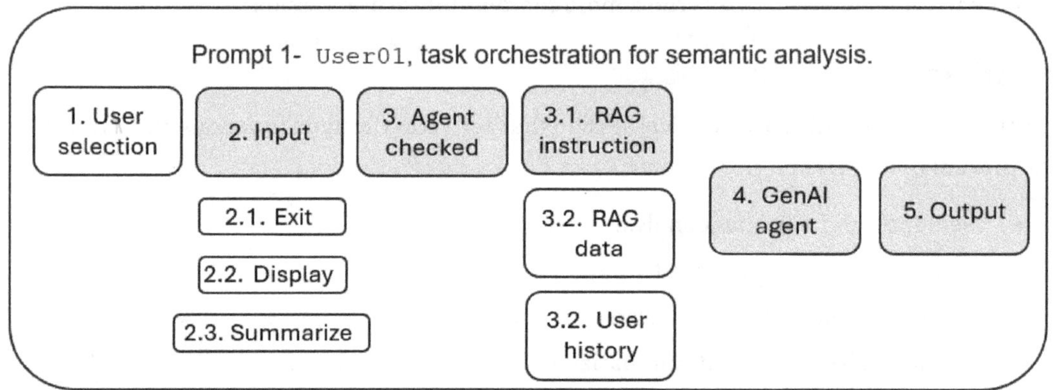

Figure 4.14: The sequence of events and functions to perform semantic and semantic analysis

The prompt that triggers this sequence of functions and events is as follows:

```
A customer said that our travel agency was pretty good but should have
more activities. Let's ask Pinecone for ideas.
```

The generative AI controller correctly identified `Pinecone` as a trigger to query the instruction scenario namespace, which GPT-4o used to produce a satisfactory response:

```
Agent: genaisys:To enhance your travel agency's offerings based on the
customer's feedback, … 3. **Gather Data**: Collect data from various
sources such as travel blogs, customer reviews, social media, and travel
forums. This data can be used to train or query your semantic search
model. 4. **Generate Ideas**: Use the results from the semantic search to
generate a list of potential activities. For example, if you're looking
to expand adventure activities, the search might suggest zip-lining, rock
climbing, or guided hiking tours….
```

Note that the AI agent begins the response with `genaisys`, signaling that the proper namespace was queried. This is an optional quality control feature that is good practice to implement when developing the GenAISys.

The prompt 2 sentiment analysis sequence represented in *Figure 4.15* is identical to the semantic analysis sequence with two differences:

- The **1. User selection** event is activated because `User02` is selected
- The prompt will contain an allusion to sentiment analysis

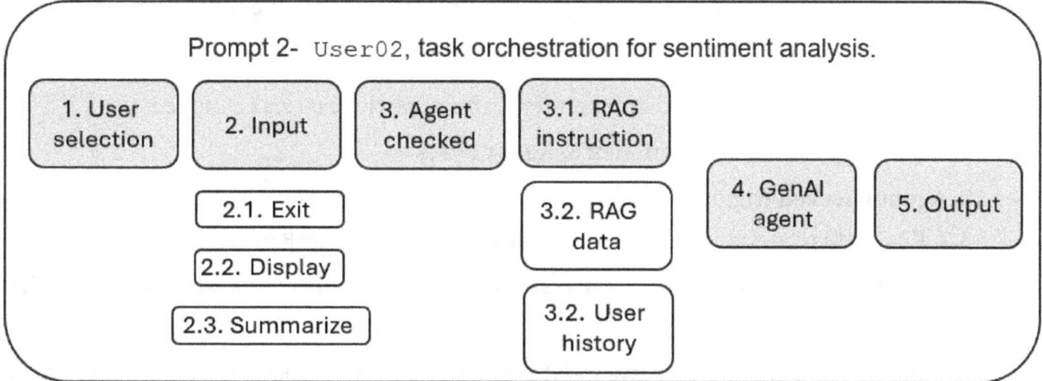

Figure 4.15: Task orchestration for sentiment analysis

The prompt contains the `Pinecone` keyword to activate a Pinecone query and the word `sentiment`, which is an indicator for similarity search:

```
`A customer said that our travel agency was worse than our competition and
should have better service. Let's ask Pinecone what its sentiment is.`
```

We could add an explicit drop-down list for all the tasks requested, and we will in some cases. But in this case, we are relying on implicit keyword searches. The balance between explicit (choosing the task in a drop-down list, for example) and implicit (using a keyword or relying on the content of the prompt) should be decided in workshops with the users.

In this case, the `Pinecone` keyword `genaisys` namespace is selected, and the output is satisfactory:

```
Agent: genaisys:Sentiment analysis score: 0.2 Explanation: The text
expresses a negative sentiment towards the travel agency. The customer
explicitly states that the agency is "worse than our competition" and
suggests that it "should have better service." These statements indicate
dissatisfaction and a negative comparison to competitors, which justifies
a low sentiment score.
```

The session continues to trigger episodic memory retrieval.

RAG for episodic memory retrieval

The sequence for episodic memory retrieval will search the Pinecone index for a personal memory of the team members. The memory is related to a past meeting that was saved and upserted to the Pinecone index in the `data01` namespace. `User03` will trigger the following sequence of functions and events, as represented in *Figure 4.16*:

- **1. User selection** is triggered independently because the user has changed.
- **2. Input** triggers the verification of **3. Agent checked**.
- **3. Agent checked** triggers the conversational AI agent controller, which in turn triggers **3.2. RAG data** retrieval in the `data01` namespace based on the `RAG` keyword in the prompt.
- Finally, the generative AI agent, GPT-4o, runs a request with the augmented input and returns an output.
- **4. Output** will trigger a display update, and the system is ready for a new turn in the conversation.

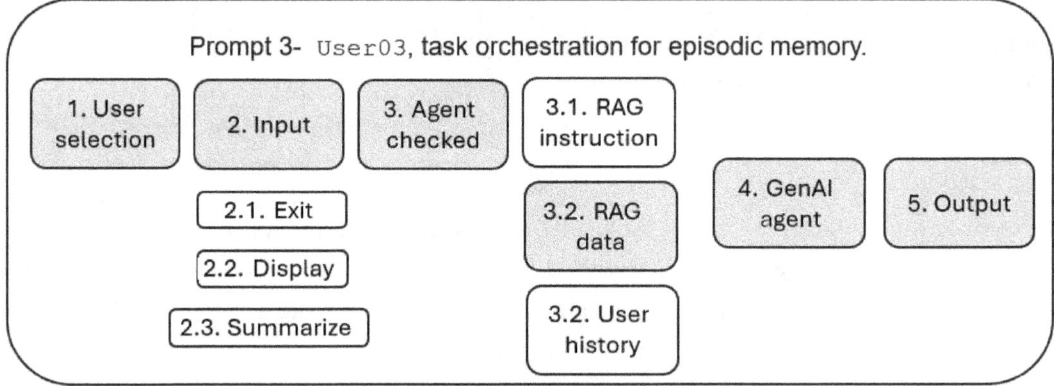

Figure 4.16: Episodic memory retrieval

The prompt clearly refers to a past event that is related to the personal experience of the team. The trace of this event is an episodic memory stored in the Pinecone index:

```
The CTO was talking about leveraging different kind of memories the other
day. What did he mean by that? Let's search RAG.
```

The agent's response is satisfactory:

```
Agent: data01:RAG, or Retrieval-Augmented Generation, is a framework that
combines retrieval-based methods with generative models to enhance the
capabilities of AI systems. When your CTO mentioned leveraging different
kinds of memories, they were likely referring to the various ways in which
information can be stored, accessed, and utilized within such a system.
Here are some types of "memories" that might be relevant in this context:
1. **Short-term Memory**:
```

We can see that the agent correctly found the namespace and also uses the phrase `your CTO mentioned`, recognizing that this is an event linked to the personal experience of the team, not impersonal semantic data.

The session continues with ideation.

Generative AI agent for ideation

The session continues with ideation. We want to leverage the power of the generative AI model, GPT-4o, in this case, to give us ideas. As such, the sequence will not use the Pinecone index to retrieve instructions or data, as shown in *Figure 4.17*:

- The user remains unchanged, **2. Input** goes directly to **3. Agent checked**.
- The system then ignores the Pinecone index but takes **3.2. User history** into account.
- Finally, **4. GenAI agent** triggers the generative AI call and returns the output.
- **5. Output** triggers the display update and the system is ready for another conversation turn.

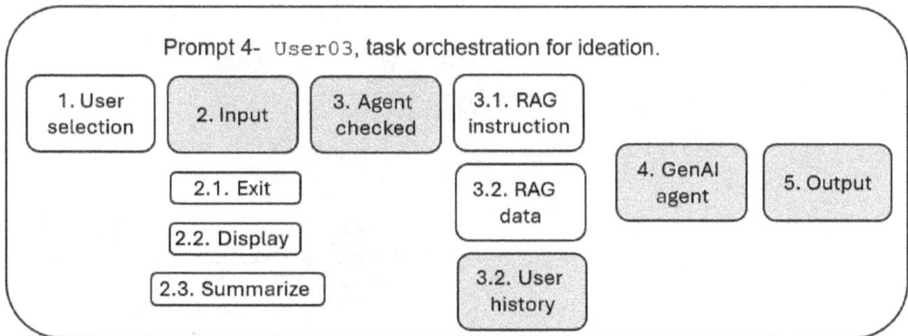

Figure 4.17: GenAISys as an ideation generator

The prompt asks the AI agent for help to get some ideas:

```
But what do you, the AI Agent, suggest we do to leverage these types of
memories in our traveling promotion campaigns?
```

The AI agent inherits episodic memory since it refers to the CTO's reflections in the conversation history and now gives its suggestions based on the history of the conversation:

```
When the CTO mentioned leveraging different kinds of memories, they were
likely referring to the concept of using various types of memory systems
in artificial intelligence and computing to enhance performance and
capabilities.…
…consider the following strategies: 1. **Personalized Recommendations**:
Use short-term and long-term memory to analyze customer preferences
and past interactions to provide personalized travel recommendations
and offers. 2. **Dynamic Content**: Utilize episodic memory to tailor
marketing content based on past customer interactions and experiences,
making the promotions more relevant and engaging. …
```

The GenAISys has provided ideas for the team. Now, the team wants to think these ideas over.

Dialogue without an AI conversational agent

The team has now had a short conversation that could have continued as long as they needed to. `User01` takes over and communicates directly with the team. The GenAISys is now used as a collaborative remote meeting tool, as shown in *Figure 4.18*:

1. User selection is triggered because `User01` is stepping in.
2. `User01` unchecks the **Agent** widget.
3. A message is entered, but the prompt is for other users, not the AI agent
4. Then, `User01` ends the conversation, which is saved.

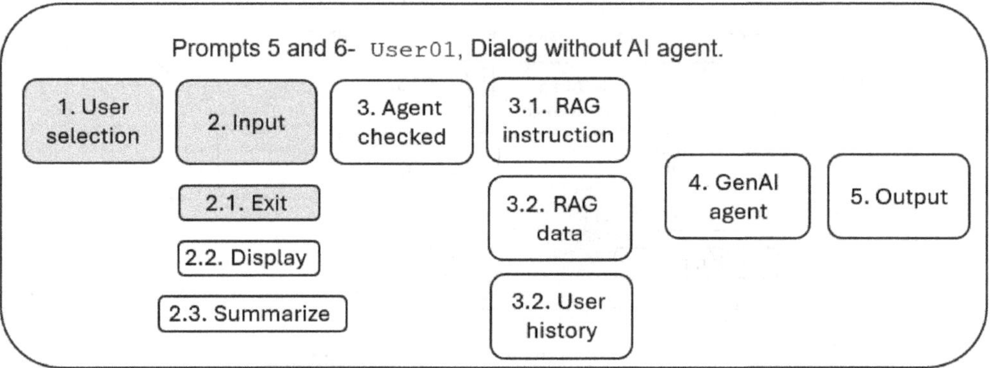

Figure 4.18: A dialogue without an AI agent

`User01` enters a message for the others:

```
OK. Let's stop here, get a summary, and go see the manager to get some
green lights to move ahead.
```

Figure 4.19 shows that `User01` has unchecked the AI agent to send the message and is now ready to end the session by entering `exit`.

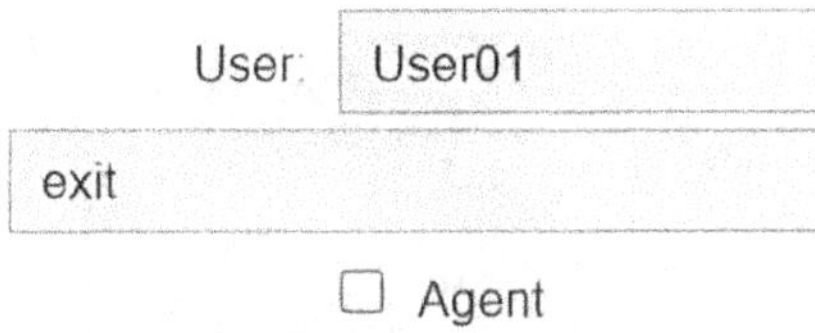

Figure 4.19: The user ends the conversation

The GenAISys displays the *conversation ended* message, as shown in *Figure 4.20*.

Conversation ended. Saving history...
Conversation history saved to conversation_history.json.
History saved. Proceed to the next cell.

Figure 4.20: Conversation ends

The message instructs the users to proceed to the next cell to display and summarize the conversation.

Loading, displaying, and summarizing the conversation

The display and summarization of a conversation will be integrated into the functions of the GenAISys framework in *Chapter 5, Adding Multimodal, Multifunctional Reasoning with Chain of Thought*.

In this notebook, we will proceed to the next cells as described in the *A session with two users* section.

The output of the display function provides Markdown text of the conversation:

```
…assistant
content:
When the CTO mentioned leveraging different kinds of memories, they were likely referring to the concept of…
Episodic Memory: This involves storing information about specific events or experiences. In AI, episodic memory can be used to recall past interactions or events to inform future decisions…
…To leverage these types of memories in your travel promotion campaigns, consider the following strategies:
Personalized Recommendations: Use short-term and long-term memory to analyze customer preferences and past interactions to provide personalized travel recommendations and offers.
Dynamic Content: Utilize episodic memory to tailor marketing content based on past customer interactions and experiences, making the promotions more relevant and engaging….
The summary is interesting because it provides useful suggestions for this online travel agency:

AI Suggestion for Travel Promotion:
1.Personalized Recommendations: Use short-term and long-term memory for personalized travel offers.
2.Dynamic Content: Utilize episodic memory for tailored marketing content.
3.Knowledge-Based Insights: Leverage semantic memory for travel tips and destination information.
4.Real-Time Engagement: Use working memory for real-time customer interactions.
5.Feedback and Improvement: Implement long-term memory systems to analyze feedback and improve campaigns.
```

We built the fundamental structure of the GenAISys framework we will be enhancing throughout the next chapters. We also ran some basic conversations. Let's summarize this chapter and move up to the next level.

Summary

A complex, event-driven, fast-moving economy requires powerful automation for the hundreds of tasks generated by just-in-time consumer needs. A GenAISys can satisfy those requirements with a responsive interface and generative AI capabilities. The challenge is providing a dynamic, intuitive system. No matter how generative AI automates tasks—and they can be tremendously automated—the final decisions will be made by humans. Humans need to communicate in meetings, whether they are organized physically or online. The challenge then evolves to provide an organization with multi-user GenAISys.

In this chapter, we first explored a high-level framework to build multi-user, multi-turn, multi-functional, and RAG features. The framework includes real-time memory features and long-term knowledge stored in a vector store. The overall ChatGPT-like system requires a response interface and conversational agent that we will enhance in the following chapters.

We then build an event-driven GenAISys response interface with IPython. The interface was seamless for an end user who can use the system with three widgets. The first widget managed the users' input, the second one the active user, and the third an agent checkbox to activate or deactivate the AI conversational agent built with GPT-4o.

Finally, we ran a multi-user, multi-turn GenAISys session centered on traveling for an online travel agency team. The first goal was to run a seamless GenAISys for the users with three widgets. The second goal was to explore the scope of short-term, long-term, semantic, and episodic memory. The third goal was to run RAG to retrieve instructions and data. Finally, the goal was to let the users communicate with or without the AI agent. We concluded the session by saving and summarizing it.

We now have a framework that we can configure and enhance in the following chapters, starting by adding multimodal functions and external extensions to the GenAISys in *Chapter 5, Adding Multimodal, Multifunctional Reasoning with Chain of Thought.*

Questions

1. The interface of a GenAISys must be seamless for the users. (True or False)
2. IPython is the only tool available to build a GenAISys interface. (True or False)
3. The AI conversational AI agent built with GPT-4o must be enhanced with RAG. (True or False).

4. GPT-4o can provide sufficient information and perform tasks quite well. (True or False)
5. Pinecone can be used to retrieve instruction scenarios. (True or False)
6. A namespace is only for data in Pinecone. (True or False)
7. A vector store such as Pinecone is a good way to store episodic memory. (True or False)
8. We don't need an agent checkbox option. (True or False)
9. Querying Pinecone is done by the user in a GenAISys. (True or False)
10. GenAISys is a complex system that should be seamless for the user. (True or False)

References

- IPython documentation: `https://ipython.org/`
- OpenAI multi-turn conversations: `https://platform.openai.com/docs/guides/audio/multi-turn-conversations/`
- Google Colab functionality: `https://colab.research.google.com/`

Further reading

- Liu, J., Tan, Y. K., Fu, B., & Lim, K. H. (n.d.). *Balancing accuracy and efficiency in multi-turn intent classification for LLM-powered dialog systems in production*: `https://arxiv.org/abs/2411.12307`

5

Adding Multimodal, Multifunctional Reasoning with Chain of Thought

At this point in our journey, we've built the core framework of our GenAISys. We have a responsive, small-scale, ChatGPT-like interactive interface. We expanded beyond typical one-to-one copilot interactions, creating a collaborative multi-user environment where an AI agent actively participates in discussions. We further extended this human-centric design by integrating RAG, giving our AI agent access to a Pinecone index capable of managing both instruction scenarios and data. Finally, we built a flexible GenAISys that allows users to activate or deactivate the AI agent during collaborative meetings. In short, we have created a human-centric AI system that augments human teams rather than attempting to replace people with machine intelligence.

However, despite its human-centric nature, the exponential growth of global transcontinental supply chains and the vast daily flow of goods, services, and digital content require significant levels of automation. For example, we cannot realistically expect social media platforms such as Meta, X, or LinkedIn to employ millions of people to moderate billions of messages—including images, audio, and video files—every day. Similarly, companies such as Amazon cannot manage millions of online transactions and physical deliveries exclusively through human efforts. Automation is essential to augment human decision-making and reasoning, particularly for critical tasks at scale. Therefore, in this chapter, we will extend the GenAISys framework by adding multimodal capabilities and reasoning functionalities. To address the challenges of cross-domain automation, we will implement image generation and analysis and begin integrating machine learning. Our objective is to build a new agentic AI layer into our GenAISys.

We will begin by outlining features that we are integrating into our existing GenAISys framework. Given the broadening scope of our GenAISys, we will introduce **chain-of-thought (CoT)** reasoning processes to orchestrate and manage complex tasks effectively. We will then incorporate computer vision capabilities. This includes building an image generation function with DALL-E and an image analysis function using GPT-4o. Next, we will add audio functionality for those who prefer voice interactions—using **speech to text** (**STT**) for input prompts and **text to speech** (**TTS**) for responses. Lastly, we'll introduce a decision tree classifier as a machine learning endpoint within the GenAISys, capable of predicting activities. By the end of this chapter, we will have successfully extended the GenAISys into a fully interactive, multimodal reasoning platform ready to tackle complex cross-domain use cases.

In all, this chapter covers the following topics:

- The architecture of the additional functions for our GenAISys
- Implementing a widget image file processing
- Implementing a widget to enable voice dialogues
- Image generation with DALL-E
- Image analysis with GPT-4o
- Building an endpoint for machine learning with a decision tree classifier
- Implementing CoT reasoning

Let's begin by designing an enhanced interface for our GenAISys with additional AI capabilities.

Enhancing the event-driven GenAISys interface

So far, the GenAISys framework we've developed is event-driven, activated by user inputs (human- or system-generated) that trigger specific AI agent functions. In this chapter, we'll expand the GenAISys by adding several new capabilities:

- **Voice interaction**, allowing users to manage the GenAISys through speech
- A new **machine learning endpoint** using a decision tree classifier for predictive tasks
- **Multimodal functionality**, including image generation with DALL-E and image analysis using GPT-4o
- A **CoT** reasoning orchestrator to coordinate sophisticated, self-reflective instruction scenarios

Let's start by examining the expanded GenAISys architecture shown in *Figure 5.1*:

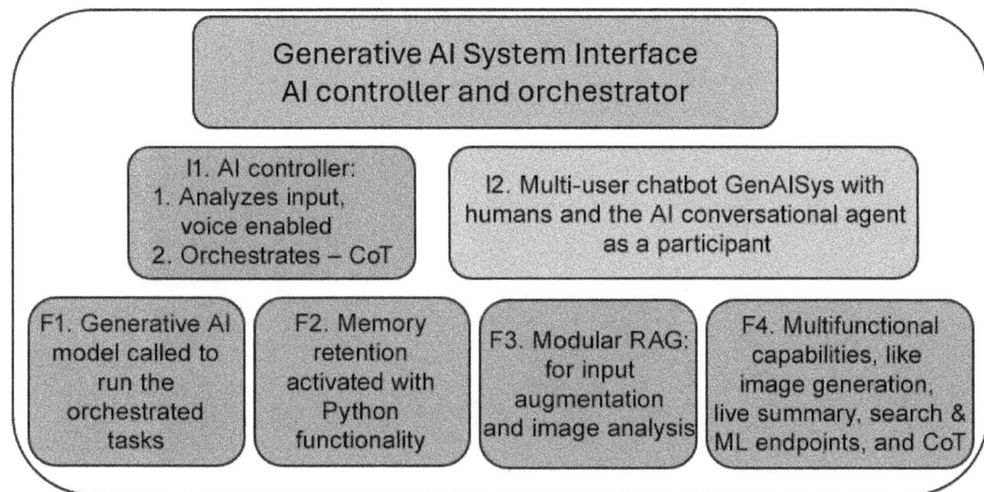

Figure 5.1: Architecture of the enhanced GenAISys interface

This figure (which is an extended version of *Figure 4.1* from the previous chapter) highlights the new capabilities we'll integrate into our GenAISys:

- **I1 – AI controller**: Enhanced with CoT reasoning, enabling automated sequences of tasks as needed and incorporating a widget to manage voice-based user interactions
- **I2 – Multi-user chatbot**: Maintained exactly as designed in previous chapters
- **F1 – Generative AI model**: Extended to handle multimodal tasks
- **F2 – Memory retention**: Continues unchanged from earlier chapters
- **F3 – Modular RAG**: Continues unchanged from earlier chapters
- **F4 – Multifunctional capabilities:** New additions covering audio and image processing, including a decision tree classifier for making predictions

Reminder

The decision to present the main components of the GenAISys architecture without arrows is a deliberate choice designed to convey a core concept: modularity and architectural flexibility. The figure is not a rigid blueprint but rather a conceptual toolkit. It shows you the powerful components at your disposal—**I1. AI controller, I2. Multi-user chatbot, F1. Generative AI model, F2. Memory retention, F3. Modular RAG**, and **F4. Multifunctional capabilities**—as independent, interoperable blocks.

We are expanding the functionality of GenAISys as built in *Chapter 4* by adding new layers rather than replacing existing components. Our emphasis here is on enhancement and seamless integration. The following figure provides a high-level flowchart demonstrating how the additional capabilities will integrate into our existing GenAISys architecture:

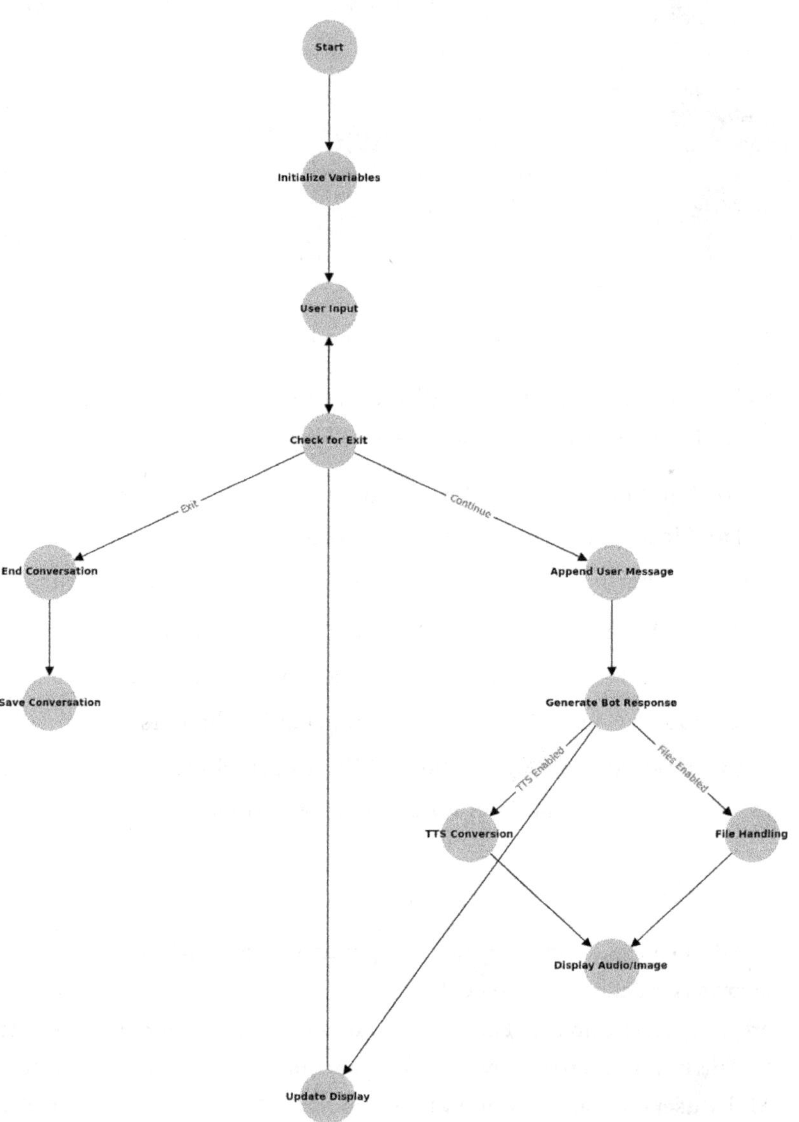

Figure 5.2: Flowchart of additional functions to the GenAISys

The following additional functions will be integrated into our existing GenAISys interface:

- **Start**: Initializes two new widgets—one for TTS functionality and another to handle image files
- **User Input**: Now includes optional voice input, enabled if the user chooses
- **Generate Bot** and **Generate Bot Response**: These processes connect directly to the existing `VBox` interface, displaying reasoning steps clearly whenever the AI agent utilizes CoT logic

To achieve this expanded functionality, we will develop the following key features:

- **STT and TTS**: Integrated using **Google Text-to-Speech (gTTS)**
- **Machine learning endpoint**: Implementing a decision tree classifier for predictive capabilities
- **Image generation and analysis**: Powered by OpenAI's DALL-E and GPT-4o models
- **CoT reasoning**: Orchestrating tasks, functions, and extensions, thus providing GenAISys with explicit machine (not human) reasoning abilities

Although we are adding several new functions, including reasoning functionality (CoT), we will introduce only a single new package installation, gTTS, to minimize complexity in this chapter. Our primary focus remains on building a reliable architecture with optimal dependency management. To begin, let's explore the updated elements of the IPython interface and the enhancements to the AI agent.

IPython interface and AI agent enhancements

The GenAISys architecture we've developed can now be viewed as comprising three interconnected layers, as shown in *Figure 5.3*. These enhancements blur the lines between orchestration, control, and agent functionality, as these roles are now distributed across multiple layers:

- **Layer 1 (IPython interface)** manages user and system inputs through event-driven widgets, orchestrating tasks based on user interactions (inputs and checkboxes).
- **Layer 2 (AI agent)** controls the generative AI models (in our case, OpenAI models) and can trigger a CoT reasoning sequence.

- **Layer 3 (functions and agents)** contains functions triggered by the AI agent. Notably, the CoT function itself acts as an agent, capable of orchestrating generative AI tasks, machine learning, and additional functions as needed.

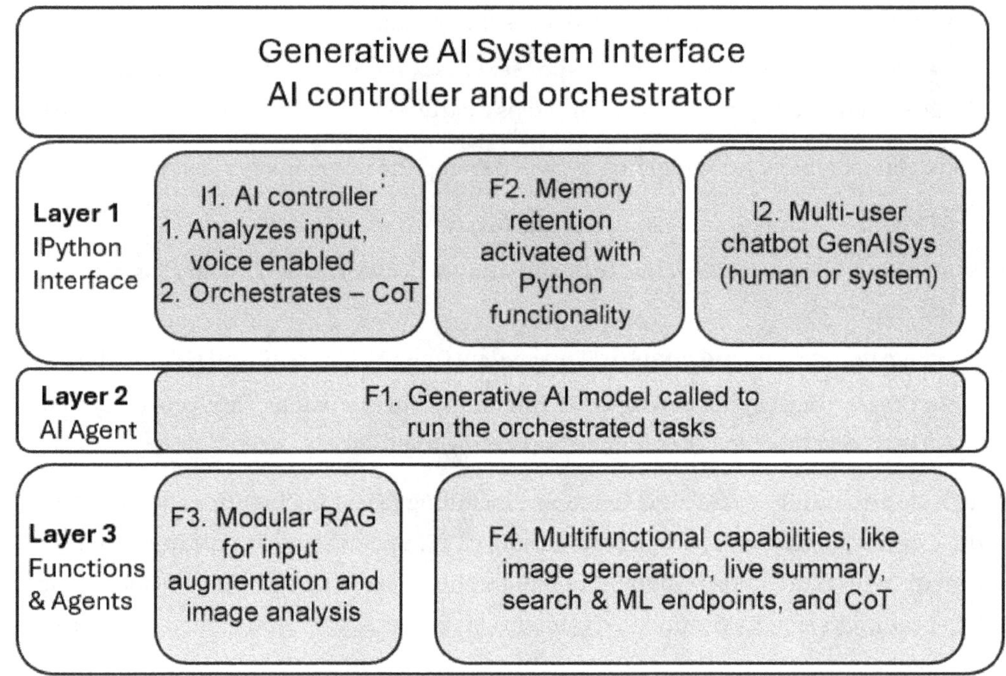

Figure 5.3: The three layers of the event-driven GenAISys

This high-level architecture integrates orchestrators, controllers, and agents, each broken down into specific Python functionalities. Let's start by exploring **Layer 1**, the IPython interface, from a functional standpoint.

Layer 1: IPython interface

The IPython interface now incorporates three new features (highlighted in yellow in *Figure 5.4*): a voice widget, a file-handling widget, and a dedicated reasoning interface triggered by user inputs and AI agent activities. These enhancements bring the interface total to six interactive widgets and functions.

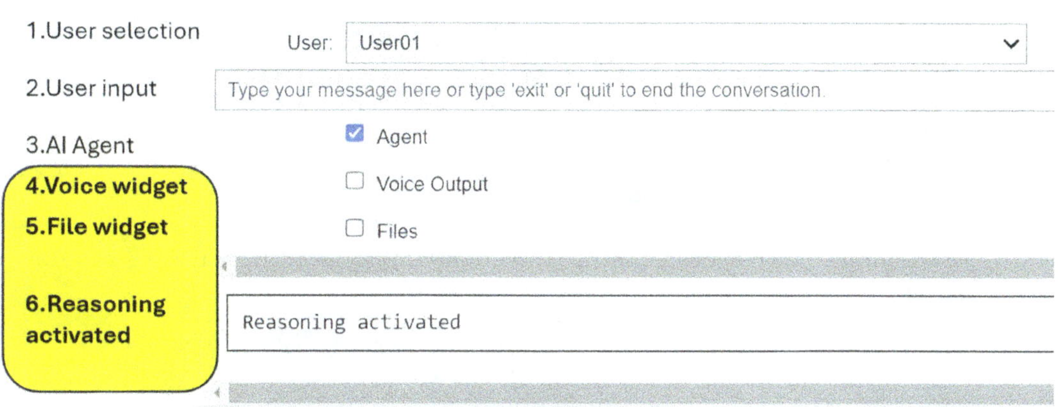

Figure 5.4: Voice, file, and reasoning features are added to the IPython interface

Let's go through each widget and function:

1. **User selection** remains as designed in *Chapter 4*. It is central to the collaborative design of the GenAISys and remains unchanged.
2. **User input** is also retained from *Chapter 4* without modification; this widget remains central for capturing user prompts.
3. The **AI agent**, as described in *Chapter 4*, activates or deactivates the generative AI agent (`chat_with_gpt`).
4. The **voice widget** enables voice-based interactions through STT and TTS. We're using cost-free, built-in functionality for STT:
    - **Windows**: Press the Windows key + *H*
    - **macOS**: Enable **Dictation** under **Keyboard settings** and choose a custom shortcut

    For TTS, the gTTS service is utilized and controlled via a checkbox set to `False` by default:

    ```
    # Create a checkbox to toggle text-to-speech
    tts_checkbox = Checkbox(
        value=False,
        description='Voice Output',
        layout=Layout(width='20%')
    )
    ```

If the AI agent's checkbox is checked, then the TTS function is called:

```
if agent_checkbox.value:
...
if tts_checkbox.value:
            text_to_speech(response)
```

The resulting MP3 file (`response.mp3`) is automatically played in the `update_display()` function:

```
def update_display():
...
#Audio display
    if os.path.exists("/content/response.mp3"):
      display(Audio("/content/response.mp3", autoplay=True))
      !rm /content/response.mp3
```

5. **The files widget** is a new widget that activates file management. It will display images generated and saved by the generative AI model (DALL-E) triggered in the AI agent function, `chat_with_gpt`. It is controlled via another checkbox, initially set to `False`:

```
# Create a checkbox to toggle agent response
files_checkbox = Checkbox(
    value=False,
    description='Files',
    layout=Layout(width='20%')
)
```

If an image exists, it is displayed with the **Python Image Library** (**PIL**) in the `update_display()` function:

```
    if os.path.exists("/content/c_image.png") and files_checkbox.
value==True:
    # Open the image using PIL
    original_image = PILImage.open("/content/c_image.png")
    # Resize the image to 50% of its original size
    new_size = (original_image.width //2, original_image.height//2)
    resized_image = original_image.resize(new_size)
    # Display the resized image
    display(resized_image)
```

6. **Reasoning activated** is another new widget of the GenAISys. The user input will trigger an event in the AI agent, and that, in turn, will trigger a CoT reasoning process. The reasoning interface will display the thought process of the CoT in real time. The reasoning output widget is created at the start of a session:

   ```
   # Create an output widget for reasoning steps
   reasoning_output = Output(
       layout=Layout(border="1px solid black", padding="10px",
           margin="10px", width="100%")
   )
   ```

   The widget will receive outputs from the CoT process and display them independently from `VBox` and persistently in the `update_display()` function:

   ```
   def update_display():
   …
   # Display reasoning_output persistently
       display(reasoning_output)…
   ```

The `VBox` interface now contains all interactive widgets, including the newly added TTS and files widgets:

```
if conversation_active:
        display(
            VBox(
                [user_selector, input_box, agent_checkbox,
                tts_checkbox, files_checkbox],
                layout=Layout(display='flex', flex_flow='column',
                    align_items='flex-start', width='100%')
            )
        )
```

Given the length and complexity of responses from the AI agent (especially during CoT processes), we introduced an enhanced formatting feature using Markdown. The `update_display()` function now formats entries clearly, calling a dedicated formatting function:

```
def update_display():
    clear_output(wait=True)
    for entry in user_histories[active_user]:
        formatted_entry = format_entry(entry)
        display(Markdown(formatted_entry))
```

The `format_entry(entry)` function formats the user's (blue) and assistant's (green) responses, ensuring readability:

```
def format_entry(entry):
    """Format the content of an entry for Markdown display."""
    if entry['role'] == 'user':
        formatted_content = format_json_as_markdown(entry['content'])
            if isinstance(entry['content'], (dict, list))
            else entry['content']
        formatted_content = formatted_content.replace("\n", "<br>")  #
Process newlines outside the f-string
        return f"**<span style='color: blue;'>{active_user}:</span>**
{formatted_content}"
…
    elif entry['role'] == 'assistant':
        formatted_content = format_json_as_markdown(entry['content'])
        …
        return f"**<span style='color: green;'>Agent:</span>** {formatted_
content}"
```

This design emphasizes that the IPython interface (**Layer 1**) is purely to orchestrate user interactions and trigger underlying layers of functions and agents. This architecture ensures that you have the flexibility you need if you want to call the functions and agents directly without a user interface.

With the IPython interface described, let's explore the enhanced capabilities in **Layer 2**, the AI agent.

Layer 2: AI agent

The AI agent invoked by the IPython interface in **Layer 1** remains the `chat_with_gpt` function, reinforcing the conversational nature of GenAISys. With the introduction of reasoning capabilities, the conversation can now occur directly between AI agents as well.

The `chat_with_gpt` function has been expanded with several new features. If necessary, review the core functionalities described in *Chapter 4*.

Let's explore the new enhancements added to the AI agent:

- `continue_functions=True` has been introduced at the beginning of the function to ensure that only one requested task is executed at a time.
- `continue_functions` is set to `False` at the end of the Pinecone query process, triggered by the presence of the `Pinecone` keyword in the user message. This stops any additional unintended task executions.
- The new function, `reason.chain_of_thought_reasoning`, described later, in the *Reasoning with CoT* section, is called under specific conditions:

```
if "Use reasoning" in user_message and "customer" in user_message
and "activities" in user_message and continue_functions==True:
```

The `continue_functions==True` condition ensures the reasoning function is called with the initial user query. A sample customer activities file is also downloaded as part of this process:

```
initial_query = user_message
download("Chapter05","customer_activities.csv")
reasoning_steps = reason.chain_of_thought_reasoning(initial_query)
```

In the example use case for this chapter, a team can automatically access and query a regularly updated customer activity data source. The sample file provided contains 10,000 records of historical customer activities, including customer IDs, locations, activity types, and activity ratings:

CUSTOMER_ID	LOCATION	ACTIVITY	RATING
729	Paris	Montmartre	7
970	Athens	Acropole	5
356	Athens	Acropole	7
340	Paris	Eiffel Tower	3
394	Paris	Parliament	5
980	Washington	Senate	3
633	Athens	restaurants	4
101	Rome	walking	6
954	Washington	statues	3
350	Athens	restaurants	2

Figure 5.5: The customer ratings of historical sites

A decision tree classifier later utilizes this dataset within the CoT reasoning function to predict the most popular customer activity. Once the response is generated, it is added to the output, and `continue` is set to `False`:

```
aug_output=reasoning_steps
continue_functions=False
```

- The new function, `reason.generate_image`, that we will implement in the *Image generation and analysis* section has also been integrated. It is called as follows:

```
prompt = user_message
image_url = reason.generate_image(prompt, model="dall-e-3",
    size="1024x1024", quality="standard", n=1)
```

The generated image URL is returned, and the image itself is downloaded and saved locally for display or further processing:

```
# Save the image locally
save_path = "c_image.png"
image_data = requests.get(image_url).content
with open(save_path, "wb") as file:
    file.write(image_data)
```

A corresponding message is added to the output, and the `continue` flag is set to `False`:

```
aug_output="Image created"
continue_functions=False
```

- The function previously known as `openai_api.make_openai_api_call` is now renamed `reason.make_openai_api_call`. It maintains the same functionality as in *Chapter 4* but is now part of the GenAISys reasoning library. The memory management `if user_memory`... `else` condition, which takes the complete user history or just the present user message into account, has been updated with explicit conditions that check both the state of `user_memory` and the `continue_functions` flag:

```
if user_memory==False and continue_functions==True:
…
if user_memory==True and continue_functions==True: …
```

The AI agent thus acts as an intermediate orchestrator, calling and managing the execution of lower-layer functions rather than executing them directly. The Pinecone interface remains the top layer that invokes the AI agent, which in turn interacts with the specific functions within **Layer 3**.

Layer 3: Functions

In this layer, our focus is on the new functionalities introduced to enable advanced reasoning through the CoT cognitive agent. Pinecone indexing and standard OpenAI calls remain as implemented in *Chapter 4*. The primary additions in this chapter are as follows:

- **Image generation and analysis** using DALL-E and GPT-4o, respectively
- **CoT reasoning**, which introduces a cognitive agent capable of orchestrating tasks
- **Voice interaction capabilities** enabled through gTTS
- **A machine learning endpoint** leveraging a decision tree classifier

We will explore these functionalities in the upcoming sections of this chapter, as follows:

- The environment setup and initialization for gTTS and machine learning are detailed in the *Setting up the environment* section
- Image functionalities are covered in the *Image generation and analysis* section
- The reasoning orchestration is built in the *Reasoning with CoT* section

By the end of this chapter, our enhanced three-layer GenAISys will have new, robust capabilities designed to expand even further in subsequent chapters. Let's now dive deeper into these enhancements, beginning with the environment setup.

Setting up the environment

In this section, we will enhance, expand, and rearrange the environment previously built to finalize the GenAISys framework. These changes are essential for the upcoming use cases in subsequent chapters. Open the `Multimodal_reasoning_with_Chain_of_Thought.ipynb` notebook within the Chapter05 directory on GitHub (`https://github.com/Denis2054/Building-Business-Ready-Generative-AI-Systems/tree/main`).

Regarding package installations, the *Setting up the environment* section in the notebook remains largely unchanged from the previous chapter (`Event-driven_GenAISys_framework.ipynb`), with just one new addition: *Google Text-to-Speech (gTTS)*.

However, several significant updates have been made to support the CoT generative AI reasoning features. Let's examine each of these updates, starting with the *OpenAI* section.

OpenAI

The first two files we download remain the same as in previous chapters. The third and fourth files, however, are new and have been added to support advanced functionality:

```
from grequests import download
download("commons","requirements01.py")
download("commons","openai_setup.py")
download("commons","reason.py")
download("commons","machine_learning.py")
```

`reason.py` now contains the generative AI library with the functions built in the previous chapters and the ones we are adding in this chapter. These functions in the generative AI library and their status are as follows:

- `make_openai_api_call(input, mrole,mcontent,user_role)` is a general-purpose OpenAI API call described in the *Setting up the environment* section of *Chapter 1*. It is now imported as follows:

  ```
  from reason import make_openai_api_call
  ```

- `image_analysis` is the image analysis function that can describe an image or use the image as a starting point to generate content such as a story. This function is described in the *Image generation and analysis* section of this chapter.
- `generate_image` is a new function that generates images with DALL-E, detailed in the *Image generation and analysis* section of this chapter.
- `chain_of_thought_reasoning` is a new CoT logic function of the GenAISys we are building. We will implement it in the *Reasoning with CoT* section of this chapter. It can call functions from other libraries, such as `machine_learning`.

`machine_learning.py` will now contain a decision tree classifier in a function named `ml_agent`. The function takes two arguments:

```
ml_agent(ml_agent(feature1_value, feature2_column)
```

In our example use case, `feature1_value` will represent a customer location, and `feature2_column` will represent customer activities. The `ml_agent` classifier will predict the most popular customer activity for a specific location based on historical data.

We import `ml_agent` from `machine_learning.py` as follows:

```
# Import the function from a custom machine learning file
import os
import machine_learning
from machine_learning import ml_agent
```

The remaining OpenAI setup subsections, including package installation and API key initialization, remain identical to previous chapters. Let's now initialize our new functionalities.

Initializing gTTS, machine learning, and CoT

We will initialize the following new functions:

- **gTTS** is installed with `!pip install gTTS==2.5.4`, which is an open source, free TTS library that fits prototyping purposes: `https://pypi.org/project/gTTS/`. `` `click` ``, a command-line library, is required for gTTS. The first cell checks if we wish to use gTTS by setting `use_gtts` to `True`:

  ```
  use_gtts = True #activates Google TTS in Google Colab if True and
  deactivates if False
  ```

  The second cell of the notebook will check for and set up the correct `` `click` `` version if `use_gtts` is set to `True`. If an update is needed, it will then display a clear message in the notebook output prompting you to manually restart the runtime. After restarting, simply click `` `Run All` `` to continue. The code will display an HTML message to restart if the version is updated:

  ```
  import importlib.metadata
  from IPython.display import display, HTML # Required for the message
  # ... (define required_click_version, current_click_version, and
  html_message as in your code) ...
  if current_click_version != required_click_version:
      # --- Commands to uninstall and install 'click' would go here
  ---
      # Example: !pip uninstall -y click
      # Example: !pip install click==8.1.8
      # Display the styled message prompting for manual restart
      display(HTML(html_message))
  ```

```
    # Stop the Python cell execution gracefully, prompting restart
          SystemExit("Please restart the Colab runtime to apply
changes.")

    print(f"--- 'click' is already at the correct version
({required_click_version}). No action needed. ---")
```

If use_gtts is set to True, we install gTTS and define a TTS conversion function:

```
# use_gtts activates Google TTS in Google Colab if True and
deactivates if False
   use_gtts:
  !pip install gTTS==
      gtts        gTTS
      IPython.display        Audio
    text_to_speech(text):
    # Convert text to speech and save as an MP3 file
      use_gtts:
            isinstance(text, str):
          text = str(text) # Making sure the text is a string not a
list
      tts = gTTS(text)
      tts.save("response.mp3")
```

This function will be activated in the IPython interface when the AI agent returns a response, as explained earlier in the *Layer 1: IPython interface* section.

- **The ml_agent algorithm endpoint** is imported from `machine_learning.py`:

```
# Import the function from the custom OpenAI API file
import os
import machine_learning
from machine_learning import ml_agent
```

This decision tree classifier function will predict popular customer activities based on historical data, enhancing our GenAISys's predictive capabilities.

- **The CoT reasoning** framework is imported from `reason.py`:

```
# Import the function from the custom OpenAI API file
import os
import reason
from reason import chain_of_thought_reasoning
```

The Pinecone installation, initialization, and queries are then defined as explained in *Chapters 3* and *4*. Take some time to revisit those chapters if needed, as we will reuse the functions previously developed. We're now prepared to build the image generation and analysis functions.

Image generation and analysis

In this section, we will begin by creating a flexible image generation function using OpenAI's DALL-E model. Following that, we'll build a function for image analysis. The objective is to enhance GenAISys with computer vision capabilities while preserving its responsive, event-driven functionality, as illustrated in *Figure 5.6*:

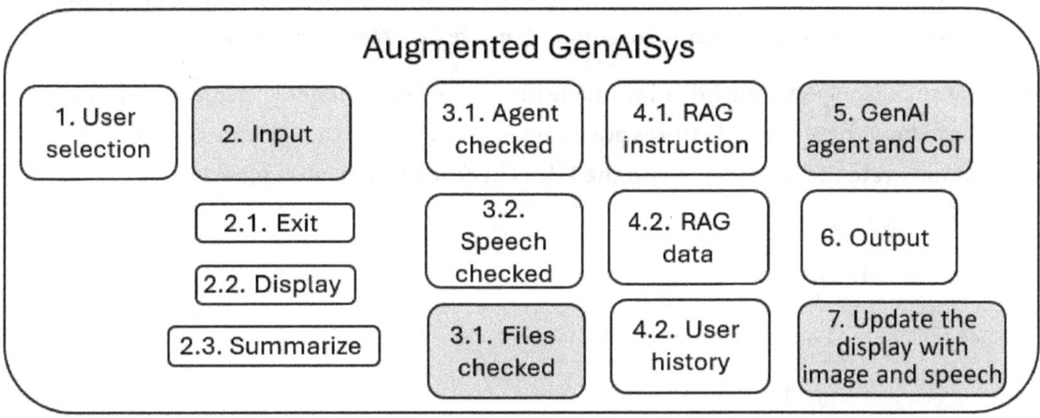

Figure 5.6: Generating images with flexible event-driven triggers

The preceding figure is an evolution of the architecture we first developed in *Chapter 4*. It has been augmented to include new capabilities: activation of speech (voice) features, management of image files, enhanced display functionality, and reasoning through CoT. In this section, our focus will specifically be on integrating and demonstrating computer vision capabilities alongside the enhanced display functionality.

The image generation and analysis processes are designed to be flexible:

- No mandatory selection or explicit widget activation is required for image generation or analysis. We could easily add explicit widgets labeled **Image Generation** or **Image Analysis** if a use case demands it. However, the approach we're adopting here is intentionally flexible, paving the way for integration within more complex, automated reasoning workflows such as CoT.
- The **Files** checkbox widget serves two distinct purposes:
    - If *unchecked*, an image is generated by DALL-E, saved to a file, but not displayed. This allows images to be generated quietly in the background for later use or storage.
    - If *checked*, the generated or analyzed image will be displayed in the user interface, as illustrated in *Figure 5.7*.
- The AI conversational agent automatically activates image generation or analysis based on user prompts. These vision capabilities can also trigger automated reasoning processes, enabling the system to execute comprehensive CoT tasks seamlessly.

Note that the display will only display image files if the **Files** widget is checked. Let's now dive deeper into how these vision features are integrated within the GenAISys interface. Specifically, we'll demonstrate the scenario where the **Files** checkbox is activated (checked), as depicted in *Figure 5.7*:

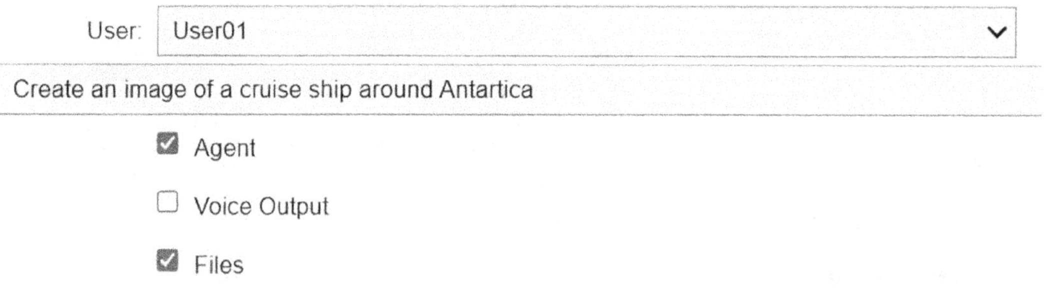

Figure 5.7: The files checkbox is checked so that the image will be displayed

With the **Files** checkbox selected, the image generated by DALL-E in response to the user's prompt will be immediately displayed, as shown in *Figure 5.8*:

User01: Create an image of a cruise ship near Antartica

Agent: Image created

Figure 5.8: Entering a prompt and displaying the image generated

If the **Files** option is not checked, the image will be generated and saved but not displayed. Similarly, image display functionality also applies to analyzing images downloaded from external sources. When the **Files** checkbox is unchecked, the analysis runs without visually displaying the image. We are now ready to examine the implementation details of the image generation function.

Image generation

The function to generate an image is located in the custom generative AI library, `reason.py`, in the commons directory. A user prompt or a CoT framework can trigger this function. The name of the function is `generate_image`, and it takes five arguments:

```
def generate_image(
    prompt, model="dall-e-3", size="1024x1024", quality="standard", n=1
):
```

The five arguments are as follows:

- `prompt`: The query related to the image that is provided by the user or the system.
- `model`: The OpenAI model to use. In this case, the default value is `gpt-4o`.
- `size`: The size of the image. The default size of the image is `1024x1024`.
- `quality`: Defines the quality of the image. The default value is `standard`, which costs less than the higher-quality `hd` option.
- `n`: Defines the number of images to generate. The default value is `1`.

The function returns the URL of the generated image. The code first initializes the OpenAI client:

```
def generate_image(
    prompt, model="dall-e-3", size="1024x1024", quality="standard", n=1
):
    # Initialize the OpenAI client
    client = OpenAI()
```

The DALL-E model is then called via the OpenAI API with the specified parameters:

```
    # Generate the image using the OpenAI API
    response = client.images.generate(
        model=model,
        prompt=prompt,
        size=size,
        quality=quality,
        n=n,
    )
```

The parameters are described in detail in *Chapter 1* in the *Setting up the environment* section.

Once the content, messages, and parameters are defined, the OpenAI API is called:

```
    # Make the API call
    response = client.chat.completions.create(
        model=model,
        messages=messages,
        **params  # Unpack the parameters dictionary
    )
```

The URL of the image is extracted from `response` and returned:

```
# Extract and return the image URL from the response
    return response. data[0].url
```

Once an image has been generated or retrieved, we can choose to display or analyze it, depending on our needs.

Image analysis

The function to analyze an image is also located in the custom generative AI library, `reason.py`, in the commons directory. This function, named `image_analysis`, is defined as follows, and takes three arguments:

```
def image_analysis(image_path_or_url, query_text, model="gpt-4o"):
```

The three arguments are as follows:

- `image_path_or_url (str)`: The path to access a local image file or the URL of the image.
- `query_text (str)`: The query related to the image that is provided by the user or the system
- `model (str)`: The OpenAI model to use. In this case, the default value is `gpt-4o`, which possesses vision capabilities(generation and analysis).

The function initializes the content structure for the API call with the provided query text:

```
# Initialize the content list with the query text
    content = [{"type": "text", "text": query_text}]
```

The function then searches for the image in a URL or a local file:

```
    if image_path_or_url.startswith(("http://", "https://")):
        # It's a URL; add it to the content
        content.append({"type": "image_url",
            "image_url": {"url": image_path_or_url}})
    else:
        # It's a local file; read and encode the image data
        with open(image_path_or_url, "rb") as image_file:
            image_data = base64.b64encode(
                image_file.read()).decode('utf-8')
```

If the image is in a URL, it is appended to the content. If the image is a local file, it is encoded in Base64 and formatted as a UTF-8 string. This format enables embedding the image data within text-based systems (such as JSON or HTML). A data URL is then created and appended to the content:

```
# Create a data URL for the image
    data_url = f"data:image/png;base64,{image_data}"
    content.append({"type": "image_url", "image_url": {"url": data_url}})
```

The OpenAI message is created with the context that contains the query information and the image:

```
# Create the message object
    messages = [{"role": "user", "content": content}]
```

The API call includes a set of standard parameters, detailed in *Chapter 1* (in the *Setting up the environment* section):

```
# Define the parameters
    params = {
        "max_tokens": 300,
        "temperature": 0,
        "top_p": 1,
        "frequency_penalty": 0,
        "presence_penalty": 0,
```

Once the content, messages, and parameters are defined, the OpenAI API is called:

```
    # Make the API call
    response = client.chat.completions.create(
        model=model,
        messages=messages,
        **params  # Unpack the parameters dictionary
    )
```

For further integration, particularly with RAG using Pinecone in *Chapter 6*, the response is saved as text in a file. This enables subsequent use and retrieval:

```
# Save the result to a file
    with open("image_text.txt", "w") as file:
        file.write(response.choices[0].message.content)
return response.choices[0].message.content
```

This `image_analysis` function will also be called by the CoT reasoning process built later in this chapter, where `query_text` will be dynamically created and passed into the function:

```
    response = image_analysis(image_url, query_text)
```

We now have fully functional computer vision components integrated into our GenAISys. With these capabilities, we are ready to build the CoT reasoning process.

Reasoning with CoT

The exponential acceleration of global markets has led to billions of micro-tasks being generated daily across platforms such as social media, e-marketing sites, production lines, and SaaS platforms. Without robust automation, keeping pace with these real-time demands is impossible. Speed and efficiency have become paramount, requiring tasks to be executed in real time or near-real time. Recent advances in AI have significantly helped us adapt to these market paradigms, where we must handle an increasing volume of tasks in increasingly shorter timeframes. However, as we increase the number and scope of AI functions to solve problems, it is becoming confusing for users to run complex scenarios with copilots. It is also quite challenging for a team of developers to create a GenAISys that contains the functions they need and includes a clear and intuitive sequence of operations for problem-solving.

In this section, we address these challenges by implementing CoT reasoning. CoT reasoning breaks complex tasks into smaller, more manageable steps where the output of one step becomes the input for the next. This technique, often also known as *context chaining*, mimics (without replacing) human-like reasoning. It reduces cognitive overload for users, allowing them to focus primarily on decision-making. Additionally, CoT reasoning makes the AI agent's internal thought process transparent, providing real-time explainability of each reasoning step.

The goal of this section is to build a CoT reasoning process using Python, leveraging the flexible and interactive GenAISys framework we've developed. Specifically, we will apply CoT to simulate customer-preference analysis for an online travel platform, generate creative suggestions for activities, produce images using DALL-E, and create storytelling narratives based on these images with GPT-4o.

At first glance, a CoT cognitive agent might seem similar to traditional sequences of functions found in classical software development. Hence, let's first clarify the important distinctions between them before we dive into the code.

CoT in GenAISys versus traditional software sequences

Seasoned software developers are used to implementing complex sequences of functions. To bridge the conceptual gap between traditional software sequences and cognitive CoT reasoning (which mimics rather than replaces human cognition), let's first distinguish their purposes clearly:

- A **traditional sequence** of non-AI or AI functions consists of a series of steps executed independently, following a black-box model in which the output of one function serves as the static input of the next.
- In a **CoT reasoning process**, the steps mimic human-like reasoning. Each step goes beyond a simple function and follows a logical progression. Each new process builds on the output of the previous step, as we will see when we implement CoT. This method is a sophisticated form of context engineering, where the output of one step dynamically creates the context for the next. We will observe the GenAISys's "thinking process" displayed in real time through our interactive interface. The process is transparent and explainable, as it is visualized in real time within the IPython interface. We can see what the system is doing and isolate any function to investigate the process if necessary.

Another critical aspect of CoT is its *intermediate reasoning*:

- Each step in a CoT process builds on the previous one, but not all steps are static. For instance, when DALL·E generates an image, it creates something entirely new—not retrieved from a database. This relies on a generative AI model, not pre-programmed content.

- The next step in the process isn't pre-generated, like a fixed list of messages. For example, when DALL-E generates an image, we will ask GPT-4o to perform a storytelling task that it will invent *ex nihilo* based on the input it received. Alternatively, we could ask GPT-4o to simply describe the image—without needing to change or fine-tune the model.

CoT reasoning offers *cognitive alignment* closer to human thinking patterns. We humans break monolithic problems into smaller parts, process each part, and then assemble the intermediate conclusions to reach a global solution. The human-like framework of the CoT process we are building in this chapter makes the GenAISys more intuitive and creative, mimicking (not replacing) human problem-solving methods. In the following chapters, notably in *Chapter 6*, we'll further expand and enhance the CoT reasoning capabilities. The takeaway here is that CoT involves sequences of tasks, but in a more flexible and creative way than classical non-AI or AI sequences. Let's move on and define the cognitive flow of CoT reasoning.

Cognitive flow of CoT reasoning

Instead of the traditional term flowchart, we'll use the term *cognitive flow* to describe the CoT process we are implementing. This term emphasizes the human-like reasoning and dynamic problem-solving capabilities of our AI agent, differentiating clearly from classical software flowcharts. A classic flowchart provides a visual representation of a sequence of functions. A reasoning CoT cognitive flow or cognitive workflow maps the logical progression of the AI agent's thought process from one step to another. The cognitive flow shows how the AI agent mimics human reasoning. This progression is powered by context chaining, where each completed step enriches the context for the subsequent one.

Let's first walk through the cognitive flow we will implement in Python, visualized in *Figure 5.9*. The Python functions we'll use reside in `reason.py`, located in the `commons` directory, and are described in detail in the *OpenAI* subsection of this chapter's *Setting up the environment* section.

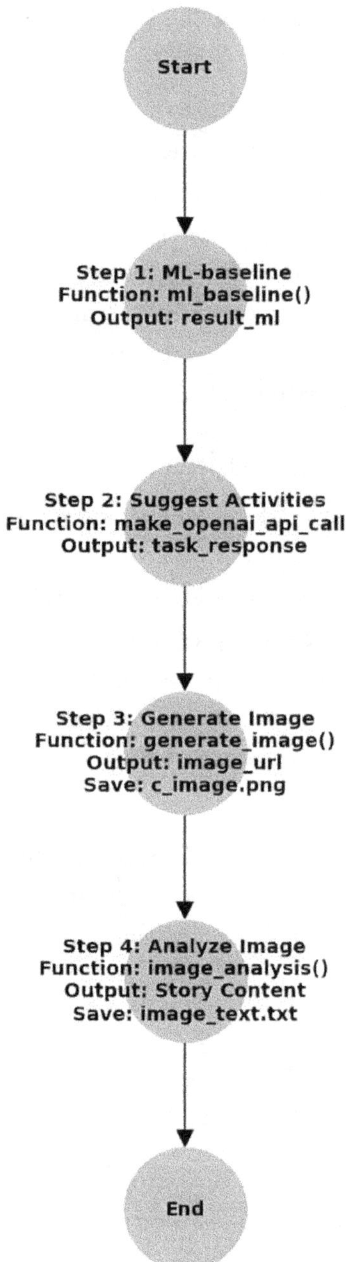

Figure 5.9: Cognitive flow of the CoT process

The cognitive flow for our CoT reasoning process consists of five main phases, orchestrated by the `chain_of_thought_reasoning()` function. This sequence begins with **Start**.

Start

The CoT reasoning process begins when it receives input text provided by the AI agent. The AI agent analyzes the user input and then triggers the CoT function, as described earlier in the *Layer 2: AI agent* section. At the start of the CoT function, two key initializations occur: the reasoning memory (`steps = []`) is initialized, and the reasoning display widget is activated within the IPython interactive interface:

```
steps = []
    # Display the reasoning_output widget in the interface
    display(reasoning_output)
```

`display(reasoning_output)` triggers the `display` widget, which enables real-time updates in the interactive IPython interface, ensuring the CoT process remains transparent and easily interpretable by users.

Step 1: ML-baseline

The first step, **ML-baseline**, activates the machine learning endpoint (`machine_learning.ml_agent()`). It utilizes a decision tree classifier to analyze customer data dynamically and predict activities of interest. The function takes a location (for example, `"Rome"`) and `"ACTIVITY"` as the target column for the prediction:

```
# Step 1: Analysis of the customer database and prediction
    steps.append("Process: Performing machine learning analysis of the
customer database. \n")
    with reasoning_output:
        reasoning_output.clear_output(wait=True)
        print(steps[-1])  # Print the current step
    time.sleep(2)  # Simulate processing time
    result_ml = machine_learning.ml_agent("Rome", "ACTIVITY")
    steps.append(f"Machine learning analysis result: {result_ml}")
```

This block of code is repeated for each reasoning step:

- Each part of the thought process begins with a comment like so: `# Step 1: Analysis of the customer database and prediction`

- `steps.append("Process: Performing machine learning analysis of the customer database. \n")` appends a description of the step to the reasoning memory step list
- `with reasoning_output` initiates a code block for the display widget
- `reasoning_output.clear_output(wait=True)` clears `reasoning_output t`
- `print(steps[-1]) # Print the current step` prints the most recent step added
- `time.sleep(2) # processing time` introduces a two-second delay
- `result_ml =machine_learning.ml_agent("Rome", "ACTIVITY")` calls `ml_agent`
- `steps.append(f"Machine learning analysis result: {result_ml}")` appends the result returned by the machine learning function to the list of steps

Through context chaining, the output from `machine_learning.ml_agent`, which predicts the top customer-preferred activity for the location `"Rome"`, becomes the input for the subsequent step, suggesting creative activities.

Before moving on to the next step, let's briefly explore the underlying decision tree classifier inside `machine_learning.py`.

Decision tree classifier

A decision tree classifier is well suited for our task because it is a machine learning model that makes predictions by splitting data into a tree-like structure based on feature values. It works by recursively choosing the optimal feature at each split until it reaches a defined stopping condition, such as a maximum depth or a minimum sample size per leaf. At each step, the possibilities narrow down until a single prediction emerges.

To run it, we first import the required libraries for handling data and building the decision tree. We also disable warnings to avoid cluttering the IPython output:

```
import pandas as pd
import random
from sklearn.preprocessing import LabelEncoder  # For encoding categorical
variables
from sklearn.tree import DecisionTreeClassifier  # For training the
Decision Tree model
import warnings
warnings.simplefilter(action='ignore', category=UserWarning)
```

Next, we define our classifier function, `ml_agent()`, with two parameters:

```
def ml_agent(feature1_value, feature2_column):
```

The two parameters are the following:

- `feature1_value`: The value of the location we want to predict activities for.
- `feature2_column`: The target column ("ACTIVITY") we want to predict.

The function starts by loading the customer activities dataset into a pandas DataFrame:

```
    # Load the dataset from a CSV file into a DataFrame
    df = pd.read_csv("customer_activities.csv")
```

Then, we encode the categorical variables (`LOCATION` and `ACTIVITY`) for processing:

```
    # Create LabelEncoder objects for encoding categorical variables
    le_location = LabelEncoder()
    le_activity = LabelEncoder()
# Encode categorical values
    df["LOCATION_ENCODED"] = le_location.fit_transform(df["LOCATION"])
    df["ACTIVITY_ENCODED"] = le_activity.fit_transform(df["ACTIVITY"])
```

If no specific location (`feature1_value`) is provided, the function selects the most frequent location by default:

```
# Select default location if feature1_value is empty
    if not feature1_value.strip():  # If empty string or only spaces
        feature1_value = df["LOCATION"].mode()[0]  # Most common location
```

We then prepare the features (`X`) and the target variable (`y`) from our encoded data:

```
    # Select the encoded 'LOCATION' column as the feature (X)
    X = df[["LOCATION_ENCODED"]]
    # Select the encoded 'ACTIVITY' column as the target variable (y)
    y = df["ACTIVITY_ENCODED"]
```

With our data prepared, we train the decision tree model:

```
    # Train a Decision Tree Classifier on the dataset
    model = DecisionTreeClassifier(random_state=42)
    model.fit(X, y)
```

Setting `random_state=42` ensures consistent results each time we run the code. Now, we encode the provided (or default) location input to prepare it for prediction:

```
    # Encode the input location using the same LabelEncoder
    feature1_encoded = le_location.transform([feature1_value])[0]
```

The Python `.transform` method on the `le_location` object converts the categorical string into its unique integer code.

The function is now ready to predict the most probable activity and convert it back to its original label. We will use the Python `.predict` method of our trained model to see what it predicts for this new data point:

```
    # Predict the encoded activity for the given location
    predicted_activity_encoded = model.predict([[feature1_encoded]])[0]
    # Convert the predicted numerical activity back to its original label
    predicted_activity = le_activity.inverse_transform(
        [predicted_activity_encoded]
    )[0]
```

Finally, the function constructs a customer's descriptive output message tailored to the predicted activity:

```
   # Generate output text
    text = (f"The customers liked the {predicted_activity} because it
reminded them of how "
            f"our democracies were born and how it works today. "
            f"They would like more activities during their trips that
provide insights into "
            f"the past to understand our lives.")
```

This descriptive output is returned to the CoT function:

```
    return text
```

To invoke the classifier from the CoT function, we use the following:

```
result_ml = ml_agent("", "ACTIVITY")
print(result_ml)
```

We're letting the classifier find the location and activity. The expected output, in this case, will be the following:

```
Machine learning analysis result: The customers liked the Forum of Rome
because it reminded them of how our democracies were born and how it works
today. They would like more activities during their trips that provide
insights into the past to understand our lives.
```

Let's now use the output of this step to suggest activities.

Step 2: Suggest activities

This step follows the same logic and structure as *Step 1*. The name of the process is as follows:

```
steps.append("Process: Searching for activities that fit the customer
needs. \n")
```

The output from *Step 1* (`result_ml`) becomes part of the instruction sent to GPT-4o to augment the input context. The combined query (`umessage`) for GPT-4o becomes as follows:

```
umessage = (
        "What activities could you suggest to provide more activities and
excitement in holiday trips."
        + result_ml
    )
```

At this stage, the instructions are tailored specifically for our travel-focused domain. In *Chapter 6*, we'll evolve these instructions to become dynamic event-based variables. Here, we continue using the established GenAISys OpenAI API call we built in earlier chapters:

```
mrole = "system"
    mcontent = (
        "You are an assistant that explains your reasoning step by step
before providing the answer. "
        "Use structured steps to break down the query."
    )
    user_role = "user"
    task_response = make_openai_api_call(umessage, mrole, mcontent, user_
role)
```

The output received from GPT-4o (`task_response`) will serve as the input for the next step (*Step 3*). The method of appending and displaying the reasoning steps remains consistent with *Step 1*.

Step 3: Generate image

This step begins by taking the detailed suggestion received from the previous step (`task_response`) and passing it directly as the prompt to DALL-E's image generation function. The structure and logic here are consistent with the previous steps, now focused on generating images:

```
prompt = task_response
image_url = generate_image(prompt)
```

Once generated, the image is downloaded and saved locally as `c_image.png`. This image file will then be displayed through the IPython interface if the **Files** widget is checked, as explained in the *Layer 1: IPython interface* section:

```
…
save_path = "c_image.png"
image_data = requests.get(image_url).content
with open(save_path, "wb") as file:
    file.write(image_data)
steps.append(f"Image saved as {save_path}")
…
```

With the image now generated and saved, the CoT process advances to analyzing this newly created image.

Step 4: Analyze image

The input for this analysis step is the URL of the image generated in *Step 3*, stored as `image_url`. As mentioned earlier, in this notebook, the query text is currently set as a generic, yet travel-specific, request to GPT-4o. In subsequent chapters, this query text will become event-driven and more dynamic.

For our image analysis, we instruct the generative AI model to craft an engaging story based on the generated image:

```
query_text = "Providing an engaging story based on the generated image"
```

The code encapsulating the instructions is the same as in the previous steps. The CoT function now activates the `image_analysis` function as described previously in the *Image generation and analysis* section:

```
response = image_analysis(image_url, query_text)
```

The output is returned to the `response` variable and saved in the `image_text.txt` file for further use. This marks the completion of the CoT reasoning steps.

End

Upon completing all reasoning tasks, the CoT function signals the end of the process by clearing and updating the IPython display:

```
    # Clear output and notify completion
    with reasoning_output:
        reasoning_output.clear_output(wait=True)
        print("All steps completed!")
    return steps
```

The IPython interface takes over from here. Let's now run the CoT from a user perspective.

Running CoT reasoning from a user perspective

In this section, we'll seamlessly run the complex GenAISys we've been building since the beginning of the book. A single prompt will trigger the entire CoT process.

We'll simulate a user activating the reasoning capabilities of the GenAISys to obtain comprehensive ideation for an online travel agency. Specifically, we aim to predict customer-preferred activities, generate engaging images, and create storytelling narratives to evoke customers' episodic memories. These episodic memories might be real-world experiences or dreams of visiting a place and engaging in particular activities.

To run this scenario, make sure to check the **AI Agent** and **Files** checkboxes and enter the following prompt carefully:

```
Use reasoning to suggest customer activities.
```

The `Use`, `reasoning`, `customer`, and `activities` keywords will be recognized by the AI agent and trigger the CoT process we built in this chapter. Alternatively, we could have implemented a drop-down menu or performed a similarity search in the Pinecone index to retrieve specific instruction scenarios. STT input is also possible. In this chapter, however, we'll use typed prompts with keywords to clearly illustrate the CoT process.

In *Chapter 7*, we'll build a central keyword registry and an orchestrator to further optimize the AI agent's decision-making process.

Once the user presses *Enter*, all we have to do is sit back and watch just as we would with online ChatGPT-like copilots. The first process is to analyze the customer base to find the top-ranking activity based on daily data, as shown here.

```
Process: Performing machine learning analysis of the customer database.
```

Figure 5.10: Searching for activities

Once the whole process is complete, the decision tree classifier returns the results:

```
..Machine learning analysis result: The customers liked the Forum of Rome
because it reminded them of how…
```

The next stage involves searching for suitable activities matching customer preferences:

```
Process: Searching for activities that fit the customer needs.
```

Figure 5.11: Searching for activities matching customer needs

The creative output from GPT-4o provides structured steps to enhance the online offerings:

```
Activity suggestions: To enhance holiday trips with more activities,
especially focusing on cultural experiences, we can consider a variety
of options. Here's a structured approach to brainstorming and suggesting
activities:
…### Step 3: Suggest Activities
1. Historical Tours and Sites:
- Athens, Greece: Visit the Acropolis and the Agora, where democracy was
born. Include guided tours that explain the significance of these sites.
- Philadelphia, USA: Explore Independence Hall and the Liberty Bell,
focusing on the birth of modern democracy.
- Westminster, UK: Tour the Houses of Parliament and learn about the
evolution of the British democratic system…
```

Next, the CoT instructs DALL-E to generate an engaging image based on these suggested activities:

```
Process: Generating an image based on the ideation.
```

Figure 5.12: Image generation based on the output of the previous step

Because the **Files** checkbox is checked, the generated image is displayed. This image is a rather creative one and will vary with each run:

Figure 5.13: A cultural and historical image

In this case, the image contains text such as …`understanding of history and its impact on modern life.`, which perfectly fits our request.

Note that each run might produce a different output due to context variations and the stochastic (probabilistic) nature of generative AI models such as GPT-4o.

The next process involves asking GPT-4o to create a narrative for a storytelling promotion that leverages episodic memory of past real-life experiences or imagined trips:

```
Process: Providing an engaging story based on the generated image.
```

Figure 5.14: Creating an engaging story based on the image generated

The narrative output from GPT-4o, shown, is illustrative and will vary, as noted earlier:

```
…Story response: In the bustling town of New Haven, a place where history
and technology intertwined, a young historian named Clara discovered an
ancient artifact that would change everything. The artifact, a mysterious
tablet, was said to hold the secrets of the past, capable of bringing
historical figures to life through augmented reality…
```

Once the CoT sequence concludes, the GenAISys maintains its reasoning state, waiting for new standalone prompts or further CoT runs:

```
Reasoning activated
```

Figure 5.15: Reasoning is persistently activated in the GenAISys

The *Load and display the conversation history* and *Load and summarize the conversation history* sections in the notebook utilize the same functions detailed in *Chapter 4*.

We've now successfully built a small-scale ChatGPT-like GenAISys equipped with custom features, including multi-user support, domain-specific RAG, and tailored CoT capabilities. In the upcoming chapters, we'll apply this GenAISys framework across several practical business domains.

Summary

In this chapter, we have completed the basic framework of the GenAISys, consisting of three layers. The first layer is an IPython interactive interface that acts as an orchestrator. It now includes voice capability, file display, and CoT features, alongside user inputs, user selections, and the AI agent widget.

The second layer is the AI agent orchestrator, triggered by user prompts. This demonstrates that within the GenAISys, the boundaries between orchestration and control functions are somewhat blurred due to the interactive nature of these components. The AI agent distributes tasks between the Pinecone index for querying and the OpenAI API agent for generative tasks, such as content and image generation. The AI agent can also trigger the CoT process, and we will further enhance its capabilities in the following chapters.

The third and final layer contains the core functionality of the GenAISys, which involves AI workers powered by GPT-4o and DALL-E. In this chapter, we introduced DALL-E for image generation and utilized GPT-4o to provide insightful comments on these images. Additionally, we implemented a decision tree classifier to predict customer activities, incorporating machine learning capabilities into our GenAISys.

Introducing the CoT feature marked our initial step toward creating seamless reasoning capabilities from an end user perspective. Complex tasks require sophisticated AI systems that can emulate human reasoning. Therefore, we will expand upon the reasoning abilities of the GenAISys, among other features, in the next chapter.

Questions

1. The seamless interface of an online generative AI system shows that the system is easy to build. (True or False)
2. Selecting a **large language model (LLM)** is sufficient to build a GenAISys. (True or False)
3. A generative AI application requires an event-driven interactive interface. (True or False)
4. An AI system can mimic human reasoning. (True or False)
5. A **chain-of-thought (CoT)** process is just a sequence of classical functions. (True or False)
6. A CoT can process natural language but not computer vision. (True or False)
7. A CoT is a cognitive agent that can include non-AI or AI functions. (True or False)

8. Reasoning GenAISys can group a set of tasks for an end user. (True or False)
9. The continual acceleration of the economy requires more automation, including AI. (True or False)
10. A human-centric reasoning GenAISys can boost the productivity of a team. (True or False)

References

- Chan, Andy, Cassidy Ezell, Michael Kaufmann, Kevin Wei, Laurel Hammond, Hunter Bradley, Elliot Bluemke, Nandhini Rajkumar, David Krueger, Nikita Kolt, Lukas Heim, and Markus Anderljung. "Visibility into AI Agents." In Proceedings of the 2024 ACM Conference on Fairness, Accountability, and Transparency (FAccT '24), Rio de Janeiro, Brazil, June 3–6, 2024. New York: ACM, 2024. `https://arxiv.org/pdf/2401.13138`.
- Putta, Praveen, Eric Mills, Naman Garg, Soham Motwani, Chelsea Finn, Divyansh Garg, and Rohan Rafailov. "Agent Q: Advanced Reasoning and Learning for Autonomous AI Agents." Last modified 2024. `https://arxiv.org/abs/2408.07199`.
- Wiesinger, Jannis, Peter Marlow, and Vladimir Vuskovic. "Agents." Kaggle Whitepaper. Accessed July 8, 2025. `https://www.kaggle.com/whitepaper-agents`.
- OpenAI. OpenAI API Documentation. Accessed July 8, 2025. `https://platform.openai.com/docs/api-reference/introduction`.

6

Reasoning E-Marketing AI Agents

The foundational concept of a successful advertising campaign is *memory*. Think about the advertisements you saw yesterday. What about those from one year ago or even several years ago? The ads you remember most vividly are the ones most effective for you, but perhaps not for someone else. The primary challenge for any advertising agency is designing promotional content that triggers positive reactions in diverse individuals. More crucially, successful marketing campaigns strive to make consumers remember brands, products, and services.

The Nielsen Neuroscience team (Brandt & Nieuwenhuis, 2017) explains why memory is so important in advertising. They demonstrate that memory decays significantly after just 24 hours, making it difficult for advertisements to have lasting effects. Several factors, including repetition and the emotional or intellectual impact of the content, can enhance memory retention. The emergence of agentic systems such as the GenAISys has reshaped the marketing landscape because these systems can replicate human-like expert marketing reasoning.

In this chapter, we will enhance the GenAISys we've been building throughout previous chapters. First, we'll design a **consumer memory agent** tailored to a specific market segment. The goal of this agent is to analyze how consumers encode promotional messages. We'll begin by exploring why memory matters and how it is structured, examining key memory categories such as short-term, long-term, explicit, and implicit memory, as well as important dimensions such as intellectual and emotional encoding. Next, we'll expand the architecture of the GenAISys by integrating a deeper understanding of consumer memory into its knowledge base. We'll then develop a strategic consumer memory agent leveraging the multimodal capabilities introduced in earlier chapters. This agent will employ a neuroscience-inspired approach to craft customized marketing messages. By introducing **meta-cognition** through OpenAI's advanced **o3 reasoning model**, we will enable the agent to perform sophisticated, near-human self-reflection within its multistep CoT reasoning process.

Further, we will transform our generative AI model into a neuroscientific-like agent capable of analytic reasoning rather than mere content generation. Complex systems—like the human brain—are more than the sum of their parts, and the same applies to machine intelligence. The strategic consumer memory agent using OpenAI's o3 reasoning model will apply complex neuroscience-informed prompts to analyze consumer memory encoding patterns in hotel reviews. The resulting insights will feed into a multimodal **thread-of-reasoning pipeline**, building upon the CoT framework introduced in *Chapter 5, Adding Multimodal, Multifunctional Reasoning with Chain of Thought*. Ultimately, the GenAISys will leverage this detailed memory analysis to produce tailored marketing content using GPT-4o, accompanied by images generated by DALL-E.

Finally, we'll further enhance the IPython interactive interface by adding new features, including a widget capable of triggering agentic meta-cognition for memory analysis and customer service tasks. Users will have the option to analyze various types of content for memory-related insights or initiate customer-service-oriented CoT interactions.

By the end of this chapter, you will have learned how to build a customized, reasoning-driven GenAISys applicable to any domain based on the architecture of our consumer memory agent. We'll construct it step by step.

This chapter covers the following topics:

- The importance of consumer memory in marketing
- The high-level structure of human memory
- Building a strategic CoT consumer memory agent
- Analyzing hotel reviews with CoT
- Designing neuroscientific-like complex prompts
- Using o3, OpenAI's reasoning model that can analyze content in depth
- Using OpenAI's GPT-4o to generate content and DALL-E to generate images
- Assembling reasoning and generation in a thread-of-reasoning function
- Generalizing the consumer memory agent CoT to any content
- Enhancing the IPython interactive interface

Let's begin by designing the enhanced GenAISys interface and its AI-driven functionalities.

Designing the consumer GenAISys memory agent

Consumer neuroscience can significantly enhance brand memorability through emotionally resonant, personalized messaging. In this chapter, we begin by analyzing how consumers encode memories. Nicks and Carriou (2016) demonstrate that effective consumer neuroscience leverages storytelling through narrative transportation, where consumers become emotionally engaged and vividly remember promotional messages.

In our implementation, we'll deeply analyze how consumers encode memories, maintaining an authentic approach. If a consumer expresses dissatisfaction with a service, our system will tailor messages to emphasize improved offerings. Our goal is to create genuine connections through memorable, emotionally resonant messages.

In this section, we will describe how we enhance the GenAISys we have been building in the previous chapters:

- **Consumer-memory agent use case**: Shows how an AI-driven agent can apply memory principles—drawn from short-term and **long-term memory** (**LTM**) frameworks—to interpret consumer feedback.
- **Defining memory structures**: This extends beyond the basic categories of **short-term memory** (**STM**), LTM, episodic memory, and semantic memory that we previously examined and introduces new categories and dimensions, providing a deeper analysis of the input.
- **Enhancing the architecture of the GenAISys**: This adds new functionality to trigger the AI agent and integrates a new CoT scenario.

Let's first explore the consumer memory agent use case.

Consumer-memory agent use case

The use case in this chapter demonstrates how the GenAISys can be enhanced with a CoT consumer memory agent. Our goal is to understand a specific consumer's needs when selecting a hotel, using detailed analyses of hotel reviews. The system we develop performs comprehensive content analysis, evaluates sentiment, and generates personalized marketing content. The CoT agent initiates this process with a detailed memory and sentiment analysis of a hotel review. It identifies psychological features, emotional markers, and specific memory tags, assigning a sentiment score to each analyzed segment. These scores are combined to produce an overall sentiment score and a scaled rating (0–5). Through this approach, the agent effectively captures how particular hotel experiences are perceived, emotionally processed, and retained in a consumer's memory.

Based on these extracted insights and the resulting cognitive profile, the agent uses the scaled rating to determine the appropriate marketing strategy. Ultimately, the consumer memory agent produces tailored promotional content, including a customized message accompanied by a relevant image. For instance, if the analysis reveals a customer who generally dislikes hotels, the travel agency can emphasize alternative accommodations or improved services that the customer has previously mentioned positively. In such cases, the agent generates a personalized message with an engaging image, as illustrated in *Figure 6.1*.

Figure 6.1: Custom content-based image generated by DALL-E

The agent crafts a warm, personalized message by analyzing the cognitive profile identified from customer hotel reviews:

```
Customer message: Dear Customer,
We invite you to consider a unique vacation experience that goes beyond
the typical hotel stay. Imagine the warmth and comfort of staying with
family or friends, where every moment is filled with genuine connections
```

```
and cherished memories. From waking up to the aroma of freshly brewed
coffee made just the way you like it to sharing home-cooked meals and
laughter around a family table, this experience offers a heartfelt
alternative to the impersonal nature of hotels. Embrace the spontaneity
and freedom of living in the moment, surrounded by those who matter most.
Whether it's a cozy evening of board games or a relaxed afternoon in the
backyard, staying with loved ones provides a true home away from home.
This vacation, choose the warmth of a family home and create memories that
will last a lifetime. Book your stay with loved ones today and rediscover
the true meaning of home.
Warm regards,
[Your Company Name]
```

Generative AI is stochastic, so the same input will not necessarily generate the same output. The response may thus change from one run to another.

Let's now define the memory structure that the consumer memory agent will use.

Defining memory structures

Let's now expand upon the memory categories introduced in *Chapter 1* by outlining the human memory structures essential for building our consumer memory agent. Human memory is multifaceted: **STM** temporarily captures the information necessary for immediate tasks or emotional processing, quickly fading without reinforcement; **LTM** stores significant events, knowledge, and experiences over extended periods; **semantic memory** stores general knowledge and facts, independent of personal experience; **episodic memory** captures personally experienced events with context and detail; **procedural memory** enables unconscious retrieval of tasks, such as walking or driving; **emotional memory** categorizes experiences based on emotional intensity—positive or negative; and **explicit memory** involves conscious recall, whereas **implicit memory** operates unconsciously.

Our consumer memory agent will analyze consumer content using a flexible combination of these memory categories, as shown in *Figure 6.2*. The categorization provides the o3 OpenAI reasoning model sufficient freedom to interpret consumer data effectively.

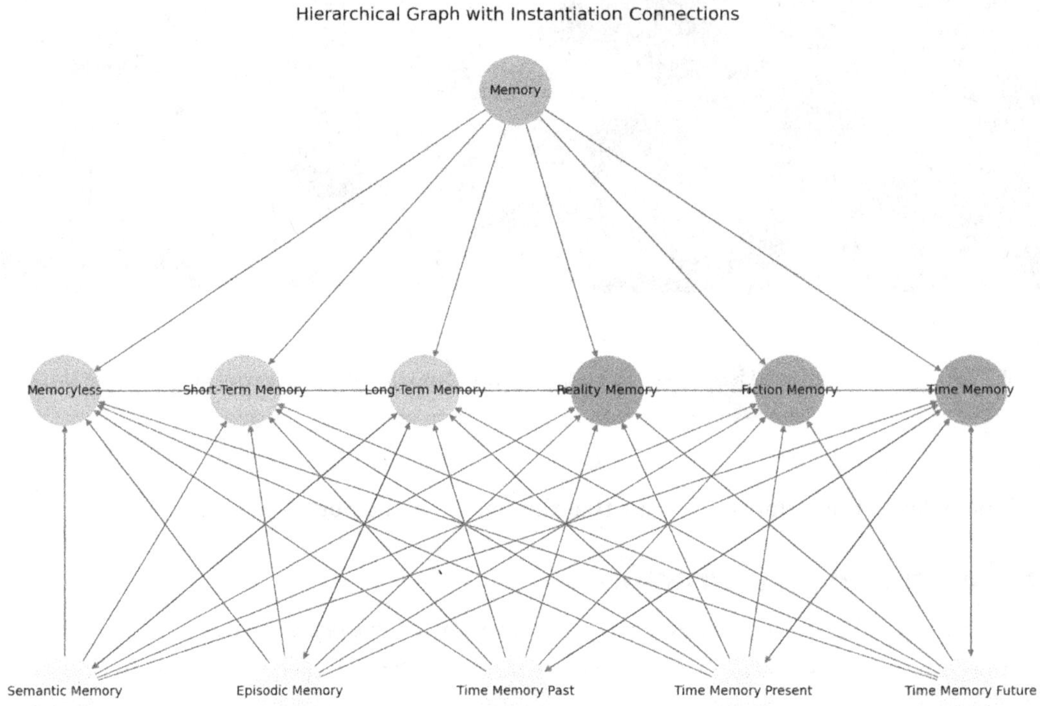

Figure 6.2: The memory categories of the memory agent

The main categories at the upper level are the following:

- **Memoryless** for systems or humans that do not remember information from a past event. These are events that we most forget, such as how many times we blinked yesterday.
- **Short-Term Memory** for the temporary storage of information to perform a task or the process of emotion. This memory decays rapidly if no event stimulates it again. It could be the working memory of reading a long text.

- **Long-Term Memory** for information we store over long periods, from days to years. This is a vital memory, such as knowing which country we are in, our age, and who our family members are.
- **Reality Memory** is what we know for sure related to actual events or facts, and the external world.
- **Fiction Memory** includes imagined or hypothetical internal events or narratives.
- **Time Memory** is critical to distinguish past, present, and future events. Otherwise, we would think that we had already eaten what we were going to eat for lunch tomorrow.

Notice how memoryless, short-term, and long-term memory form a subset (light green), and reality, fiction, and time memory (light orange) are all connected. These categories aren't isolated; they interconnect dynamically in real life. Our memory, in other words, doesn't function in subsets but with what we can call *tags* in AI. A memory can be a combination of multiple tags:

- A memoryless fiction, such as a dream
- A short-term reality, such as reading the news
- A long-term fiction, such as a novel we read a long time ago

When examining these memory subcategories, we quickly realize the vast number of possible tag combinations with the main memory categories—such as semantic STM or episodic LTM. Additionally, memories can seamlessly blend subcategories; for instance, the phrase "I visited Rome last year" combines episodic, semantic, and temporal memory tags simultaneously. Moreover, our memories range from implicit (subconsciously blinking our eyes all day) to explicit (intentionally blinking due to an irritation).

In our consumer memory agent, we will request a thorough analysis of content, assigning appropriate memory tags to each text segment. However, even this detailed tagging is not sufficient by itself. To effectively capture consumer experiences, we will enrich each memory tag with three analytical dimensions:

- **Intellectual dimension**: Identifies thoughts, logic, and reasoning within the text.
- **Emotional dimension**: Pinpoints emotions, feelings, and overall mood—critical for effective consumer engagement—and provides a quantifiable sentiment score ranging from 0 to 1, scalable to a familiar 1–5 rating used in customer satisfaction forms.
- **Physical dimension**: Highlights sensory experiences and physical sensations, such as "it was too cold to go swimming" or "my back hurt after sleeping in that hotel bed."

With these enhancements in mind, let's now explore how we'll integrate them into the architecture of our evolving GenAISys.

Enhancing the architecture of the GenAISys

In this chapter, we will build upon the existing three-layer architecture of the GenAISys, as illustrated previously in *Figure 5.3* and reproduced here:

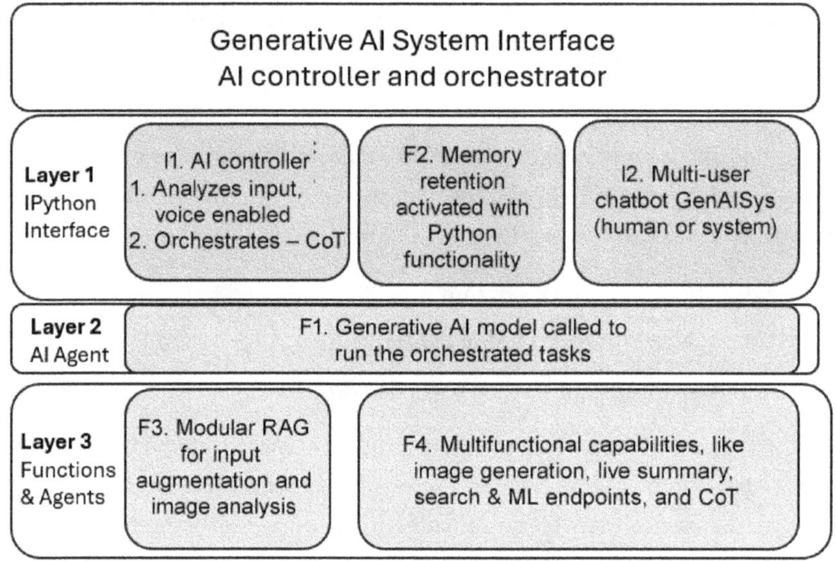

Figure 6.3: The three layers of the event-driven GenAISys

Our approach will be bottom-up, starting from the foundational functions and proceeding upward through the AI agent to the GenAISys interface:

- **Layer 3 (functions and agents)**: Here, we will introduce additional functionalities into our custom OpenAI library (`reason.py`), specifically tailored for the consumer memory agent and CoT reasoning. We will also develop a standalone memory analysis function that provides neuroscientific-like analyses applicable to any content.
- **Layer 2 (AI agent)**: This layer manages the behavior and decisions of our GenAISys. We will establish clear input triggers and naming conventions to activate and control the AI agent effectively.

- **Layer 1 (IPython interface)**: The interactive interface will be expanded to facilitate user interaction. We will add a new widget, allowing users to conveniently select how the consumer memory agent is invoked. Initially, we will focus on hotel reviews and subsequently generalize to any form of input.

Let's now begin building the consumer memory agent.

Building the consumer memory agent

In this section, we take our GenAISys to the next level by equipping it with neuroscientific capabilities for analyzing hotel reviews. The consumer memory agent will capture a user's cognitive, emotional, and physical mindset, decoding each review segment through a six-step CoT process, as illustrated in *Figure 6.4*. This CoT process relies on context chaining to pass insights from one analysis step to the next, building a comprehensive cognitive profile.

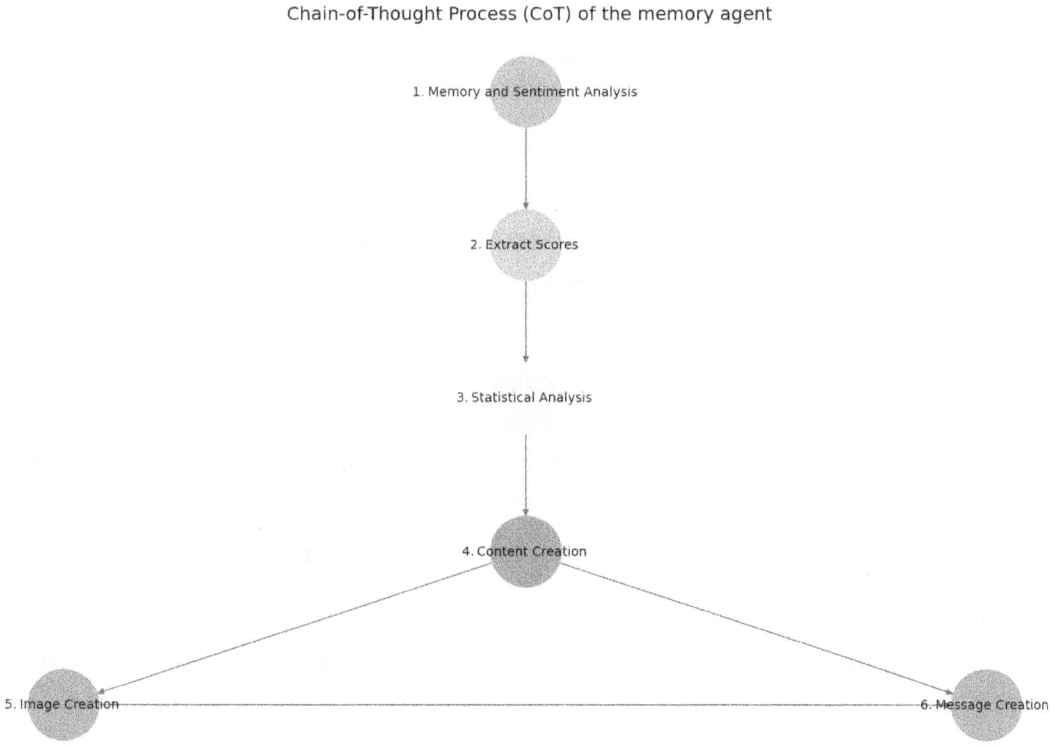

Figure 6.4: Chain-of-thought process of the memory agent

The consumer memory agent's CoT will use OpenAI's o3, GPT-4o, and DALL-E to run its six steps:

- **Step 1: Memory and Sentiment Analysis**: The agent will analyze the content of the hotel review with a complex memory structure system message. It will analyze and tag the content segment by segment.
- **Step 2: Extract Scores**: The agent processes the output of *Step 1* to extract the sentiment scores of each content segment.
- **Step 3: Statistical Analysis**: The agent uses the scores of all the tagged segments to produce an overall cognitive score for the content.
- **Step 4: Creating Content**: The agent now has a decision to make based on the score. If the score exceeds a positive threshold, it will generate a message encouraging the consumer to select hotels. However, if the score is negative, a guest house message will be created. Once the decision is made, the agent will use the consumer's memory tags to create a tailored promotional message.
- **Step 5: Image Creation**: The agent now uses the output of *Step 4* to create an image that will fit the consumer's mindset.
- **Step 6: Message Creation**: The agent now has all the information necessary to generate a custom message for the consumer.

After developing these steps individually, we'll integrate them fully in the upcoming section, *GenAISys interface: From complexity to simplicity*, aiming to generalize the CoT functionality beyond hotel reviews.

To begin our journey, open the `1_Building_the_Consumer_Memory_Agent.ipynb` notebook, which reuses previously built functionality, within the Chapter06 directory on GitHub (`https://github.com/Denis2054/Building-Business-Ready-Generative-AI-Systems/tree/main`). We will first download a dataset of hotel reviews to provide inputs to the AI agent.

The dataset: Hotel reviews

We will be using synthetic hotel reviews to build the memory agent. In this chapter, we will process hotel reviews but also generalize the memory structure of the agent to other content we wish to analyze. For copyright reasons, the dataset we are using is a synthetic dataset of reviews created manually and with a generative AI copilot.

If you wish to explore more datasets, you can use a similar dataset containing TripAdvisor hotel reviews available on Kaggle for non-commercial private implementations at `https://www.kaggle.com/datasets/andrewmvd/trip-advisor-hotel-reviews`.

Run the *Setting up the Environment* section on GitHub, identical to *Chapter 5*, and download the dataset directly from the GitHub repository:

```
download("Chapter06","hotel_reviews.csv")
```

Quick tip: Enhance your coding experience with the **AI Code Explainer** and **Quick Copy** features. Open this book in the next-gen Packt Reader. Click the **Copy** button

(1) to quickly copy code into your coding environment, or click the **Explain** button

(2) to get the AI assistant to explain a block of code to you.

```
function calculate(a, b) {
  return {sum: a + b};
};
```

Copy 1 Explain 2

The next-gen Packt Reader is included for free with the purchase of this book. Scan the QR code OR visit `https://packtpub.com/unlock`, then use the search bar to find this book by name. Double-check the edition shown to make sure you get the right one.

We will process the dataset with a pandas DataFrame. The program now loads the CSV file and displays the data:

```
import pandas as pd
# Load the CSV file into a Pandas DataFrame
dfta = pd.read_csv('/content/hotel_reviews.csv',sep=',')
# display the DataFrame
dfta
```

This dataset contains two primary columns: `Review` and `Rating`. For instance, record 0 has a relatively constructive rating of 3, while record 1 shows a clearly positive rating of 5:

index	Review	Rating
0	We got a nice hotel. The parking was rather expensive. However we got a good deal for my birthday. We arrived in during the night because of a late flight but the parking was open 24hours The check in was rapid. The room was a large size. Cool. But we didn't get the view expected. The bed was super comfortable and got a nice sleep for a few hours but then there was a raucous of a noisy crowd going to bed late. Anyway the price was acceptable and we made it to the shopping mall next to the hotel which reminded me of the one we had at home.	3
1	An absolutely flawless experience from start to finish. The valet was courteous check-in was a breeze and they even upgraded us to a suite with a breathtaking city view. The room was spotless the bed felt like sleeping on a cloud and the complimentary breakfast was restaurant-quality. It's expensive but you get what you pay for. We will definitely be back.	5

Figure 6.5: Excerpt of the hotel review dataset

Ratings alone, however, don't provide sufficient depth—we require a nuanced sentiment analysis to fully grasp why a customer was satisfied or dissatisfied. We will choose a challenging review to begin our analysis:

```
index_number = 0  # Specify the index number
```

The program now extracts the review and its rating:

```
        # Extract the desired fields
        review = row['Review']
        rating = row['Rating']
        # Display the results
        print(f"Review: {review}")
        print(f"Rating: {rating}")
    except IndexError:
        print(f"Error: Index {index_number} is out of bounds for the
DataFrame.")
    except KeyError as e:
        print(f"Error: Column '{e}' not found in the DataFrame.")
```

The output displays the review and its rating:

```
Review: We got a nice hotel. The parking was rather expensive. However, we
got a good deal for my birthday. We arrived in during the night because
of a late flight but the parking was open 24hours The check in was rapid.
The room was a large size. Cool. But we didn't get the view expected. The
bed was super comfortable and got a nice sleep for a few hours but then
there was a raucous of a noisy crowd going to bed late. Anyway the price
was acceptable and we made it to the shopping mall next to the hotel which
```

```
reminded me of the one we had at home.
Rating: 3
```

We have chosen a difficult review because it contains both negative and positive sentiment. The negative aspect of the review will challenge the agent to generate constructive solutions. Before continuing, analyze the memory tags, sentiment scores, and dimensions of each review segment yourself. This exercise clarifies memory category usage and provides a benchmark for comparing your insights to the agent's analysis. Set the extracted review as the initial input:

```
input1=review
```

We will now design a complex system message for *Step 1* for `input1`.

Step 1: Memory and sentiment analysis

This step introduces advanced reasoning to our consumer memory agent by incorporating meta-cognition and meta-reasoning through the OpenAI o3 reasoning model. In other words, the agent won't simply process text—it will actively reflect on its internal reasoning, performing a segment-by-segment analysis to categorize memory types and assign sentiment scores.

Specifically, the o3 model will operate within our carefully structured system message, which we will design in detail. This system message guides the model clearly, prompting deep reasoning and ensuring it assigns memory tags accurately based on human-like cognitive processes. We are definitely in the era of reasoning and self-reflecting AI!

In this section, we will do the following:

- Design a complex system message that incorporates the detailed memory structures defined earlier. This message, called `system_message_s1`, will be stored separately in a Python file for modularity.
- Rely on the reasoning abilities of OpenAI's o3 to do the heavy lifting and execute detailed segment-level analysis, relieving us from manual interpretation.

Note that we use o1 as an umbrella term to signal to the LLM its role as a reasoning model. Additionally, the LLM may refer to o1 itself in responses though we call o3 as much as possible in the API.

Let's now construct this detailed system message step by step.

Designing a complex system message for Step 1

We must design a system message comprehensive enough for the model to deeply understand and execute a neuroscience-inspired memory analysis. To achieve this, we carefully structure the message into clearly labeled sections, each guiding the agent through different aspects of the analysis.

0. Model introduction and role of the agent

The first line sets the tone for the agent at two levels. The first level provides the agent with the necessary concepts to understand advanced memory analysis for this task. The second level describes the role of the agent in detail:

```
You are a generative AI model, an advanced memory-analysis model. Your
role is to examine **each segment** of an incoming text and generate a set
of "memory encoding tags," similar to how the human brain encodes memories
in neuroscience. For every segment in the input, you will identify which
categories apply, discuss the rationale, and assign additional metadata
(dimension, sentiment, etc.).
```

Now, let's go through the instructions to grasp what the agent is learning through this first part of the message:

- `generative AI model, an advanced memory-analysis model`: We are setting the role of the system in a special way. We are asking the model to think, not just to generate text. For this task, we don't want the model to be created but to analyze and reason.
- `examine **each segment**`: We are teaching the model to replicate a neuroscience approach. Our brain encodes information in discrete packages. In this case, we are asking the model to mimic human memory processes. Each segment of text can be a sentence, a sentence piece, or a paragraph. This way, the model will analyze the text in a manner similar to how a human brain encodes information in independent packages.
- `generate a set of "memory encoding tags," similar to how the human brain encodes memories`: Human brains encode memories with *tags*, a term we can use at a high level without going into the biological process. Our brains apply tags to every bit of information that they encode to differentiate a past event from a future event, for example, from semantic data or personal emotional experiences. These tags represent the categories of memory we are looking for in human-generated text.

- `discuss the rationale, and assign additional metadata`: The model must explain the rationale behind the category of memory it tags. Each category, such as STM or LTM, must be explained. We need to know why a memory tag was attributed to the segment. The model is asked to add dimensions to its description, including intellectual and emotional reasons.

You might notice a Markdown divider (---) in the code. It shows the model that we are now moving to another topic. This may seem unimportant, but we need to emphasize topic changes as we do when giving instructions to humans. Now, we will give the model a purpose.

1. Purpose

Line 3 is a header that shows the model that we are entering the first significant section of the message:

```
### 1. Purpose
```

Line 4 defines the goal of o3, OpenAI's reasoning model:

```
The goal is for you, O1, to perform an **in-depth memory analysis** of
each segment of the text. In other words, you will classify and label each
segment you find using specific memory categories (also called "memory
encoding tags"). This process provides insight into how different parts of
the text might be encoded in human memory.
```

Note that the message contains "o1," which is used as an umbrella term for OpenAI's reasoning models here. The main idea is for the API to understand that we expect reasoning. This instruction will activate reasoning no matter which reasoning model you select. The key parts of this *Purpose* section insist on what we expect:

- `in-depth memory analysis`: We do not want a classical analysis but a reflection that goes into the details of each segment.
- `Classify and label each segment you find using specific memory categories`: This is a strong indicator of the memory categories the model is expected to tag. Once again, we remind the agent that we don't want to generate text but to classify and label segments.
- `provides insight into how different parts of the text might be encoded in human memory`: This is an explicit indication that we expect human-like thinking and replicates the way a brain encodes memories.

We now need to give the agent the heading it needs to learn the categories. The first lines provide clear instructions. Now, we have reached section 2 of the message.

2. Memory encoding tags

We now teach the agent how to recognize different categories of human encoding tags. We are getting to the core of human memory encoding. The memory categories are those discussed in the *Defining memory structures* section of this chapter:

```
### 2. Memory Encoding Tags (Categories)
```

This heading is vital as the agent will learn the tags we expect by taking a hint from this heading. Now, the agent has absorbed the heading. We then give the model a clear explanation of what actions we expect:

```
Below is a list of memory categories you must use. Consider them your
"tagging schema." A single segment may exhibit one or more categories. If
no category seems relevant, you may provide the special tag "memoryless"
to indicate no significant memory encoding.
```

Let's focus on the key parts of this message:

- `tagging schema`: Aligns the model with the way the human brain encodes different categories of memory, distinguishing the past from the present using *tags*
- `A single segment may exhibit one or more categories`: Explains to the model that a memory can be encoded in more than one category, just like in a human brain
- `If no category seems relevant … memoryless`: Tells the model that it should assign a memoryless tag if it cannot determine a category of a memory

We then clearly define the categories (e.g., STM, LTM, episodic memory, semantic memory, time memory, reality memory, fiction memory, memoryless), as previously discussed:

```
1. **Short Term Memory (STM)**
   - Used for information that seems fleeting, recently introduced, or
relevant only in the immediate context.
…
8. **Memoryless**
  - If a segment does not appear to connect to any memory encoding or
is purely functional text (e.g., disclaimers, random filler), label it
"memoryless."
```

The memory tags have been described but are insufficient to capture human memory, which relies on other dimensions to encode events.

3. Dimensions

The dimensions section adds intellectual, emotional, and physical features to the agent's investigation. The descriptions of these dimensions in the following message were described in the *Defining memory structures* section earlier:

```
Dimension Descriptions
  1. Intellectual
  2. Logical, analytical, or factual thought processes.
      - Explanation: If the text focuses on reasoned arguments, data,
        or factual details, it should be labeled "Intellectual."
  3. Emotional
  4. Feelings, mood, or affective elements.
      - Explanation: If the text displays happiness, sadness, or other
        strong emotional content, "Emotional" is assigned.
  5. Physical (with Sensations
```

With that, we have defined the memory categories and additional dimensions. However, we also need a more refined analysis of emotions.

4. Sentiment score

As defined in the *Defining memory structures* section, the sentiment score measures the emotional value of a segment. It provides a numerical score between 0 (negative) and 1 (positive), or 0.5 (neutral) if no sentiment can be detected:

```
### 4. Sentiment Score
Assign each segment a **sentiment score** between **0.0** and **1.0**,
where:
  - **0.0** = very negative
  - **0.5** = neutral
  - **1.0** = very positive
If a segment is purely factual with no emotional valence, use 0.5
(neutral).
```

Note that each section in the message begins and ends with clear Markdown indicators that show a change in topic.

Next, we are going to ask for a specific response format.

5. Response format

We need the response to clearly display each segment of the original text, provide memory tags for each segment, determine the dimension (intellectual, emotional, or physical) of each segment, provide a sentiment score, and provide a brief explanation to justify the analysis:

```
### 5. Format of the Response

For **each segment** in the incoming text:
  1. Show the segment excerpt or a short summary.
  2. **Memory Tags**: list any relevant categories.
  3. **Dimension**: choose intellectual, emotional, or physical.
  4. **Sentiment Score**: 0.0 → 1.0.
  5. **Brief Explanation**: why these tags/dimensions.
```

To make sure the model understands what we are asking for, we provide an example format:

```
Example format:
Segment 1: "Excerpt..."
  - Memory Tags: [Time Memory Past, Reality Memory]
  - Dimension: Emotional
  - Sentiment Score: 0.7
  - Explanation: The speaker refers to a past real event with positive
affect.
```

If we were writing a traditional generative AI model message, we could stop here. However, this is a complex message, so we need to add instructions to *insist on* what we expect.

6. Additional instructions

We avoided overloading the previous sections of the message. If we try to squeeze too many instructions in, the model might get confused. Let's remind the system that we always want a segment-by-segment analysis. We insist that if the model doesn't find a category, we want a "memoryless" tag and not a hallucination. Additionally, we only want short and clear explanations:

```
### 6. Additional Instructions
  - Always analyze segment-by-segment.
  - If no memory category applies, use "memoryless."
  - Use a short but clear explanation.
```

Now comes the tricky part. We told the model that if it didn't find a category at all, to use a "memoryless" tag. However, if the model has an idea but is not 100% sure, then it is allowed to pick the most probable memory tag along with a mandatory sentiment score:

```
    - If uncertain about the correct memory category, pick the most likely.
    - Always include a sentiment score.
```

At this point, we have provided the model with numerous instructions. Let's make sure it remembers its primary task.

7. Primary task recall

After all the instructions we have given the model, we will remind the model that its primary task is a memory tag analysis of text segments. We also expect the format of the output to be structured as defined:

```
### 7. Primary Task
When I provide multisegment text, you must do a thorough memory-tag
analysis for each segment. Return the results in the structured format
above.
[End of System Prompt]
```

Note that we added `[End of System Prompt]` to make sure that the model understands that the message part of the global prompt is now completely defined. We use the term `prompt` to make sure that it understands it as a set of instructions, not just a general message.

We are now ready to run the memory analysis.

Running the memory analysis

The complex system message we designed is stored in a variable named `system_message_s1` in `cot_message_c6.py` in the `commons` directory of the GitHub repository. The goal is to keep this message and those for other steps separate from the function calls so that the AI agent of the GenAISys can repurpose the function in this step or other steps for different tasks.

We first download the file that contains the messages:

```
download("commons","cot_messages_c6.py")
```

Then we import the `system_message_s1` message and the messages we will need for *Step 4*, which we will discuss later:

```
from cot_messages_c6 import (
    system_message_s1, generation,imcontent4,imcontent4b)
print(system_message_s1) # Print to verify
```

The `print` function is uncommented and will display the message we just created. It can be commented and used at any time to verify whether the message is correctly imported. We now prepare the messages for o3:

```
# Step 1 : Memory and sentiment analysis
mrole= system_message_s1
user_text=review
```

- `mrole` is `system_message_s1`, the system message we designed
- `user_text` is `review`, the review selected from the hotel reviews dataset

We now call o3 and store the result in a variable:

```
retres=reason.make_openai_reasoning_call(user_text, mrole)
```

`make_openai_reasoning_call` is located in `reason`, the AI library of the GenAISys. It takes the two arguments we just defined, creates an OpenAI client, makes the request, and returns the response:

```
# Implemented in Chapter06
def make_openai_reasoning_call(user_text, mrole):
    system_prompt=mrole
    client = OpenAI()
    rmodel = "o3-mini" # o1 or other models. model defined in this file
in /commons to make a global change to all the notebooks in the repo when
there is an OpenAI update
    response = client.chat.completions.create(
        model=rmodel,
        messages=[
            {"role": "system", "content": system_prompt},
            {"role": "user", "content": user_text}
        ],
    )
    return response.choices[0].message.content
```

For this call, we chose the `o3-mini` version of the o3 reasoning model series. Other versions and reasoning models can be chosen. The program displays the output received in `retres`:

```
# Print the generated output (memory analysis)
print(retres)
```

The output shows the depth of the system message and the o3 reasoning model. The AI model has broken the content down into segments and decoded the memory tags subconsciously used by the human reviewer, as shown in the first segment.

The model first provides the segment number and the content of that segment. Let's focus on segment 7, which requires our attention:

```
Segment 7: "But we didn't get the view expected."
```

It also provides the memory tags that encoded this segment:

```
- Memory Tags: [Episodic Memory, Reality Memory]
```

It continues by providing the dimension, which is as follows:

```
• Dimension: Emotional
```

It then gives a sentiment score, which is as follows:

```
• Sentiment Score: 0.4
```

Finally, it produces an explanation that sums up its analysis:

```
 - Explanation: The disappointment regarding the view introduces a
negative emotional element to this real-life account, impacting the
overall perception of the stay.
```

The model then continues its analysis for all the segments of the review. We have now performed a complex memory analysis that sets the stage for the subsequent steps. Let's proceed to extract the sentiment scores.

Step 2: Extract sentiment scores

From this point on, the original input stored in `review` is not used again. The CoT process uses context chaining, relying on the output of the previous step, which will continually vary depending on the initial input. The next step involves extracting the sentiment scores for all segments produced in *Step 1: Memory and sentiment analysis*. We will need this information to make decisions for *Step 4: Content creation*.

To extract the scores, we first create an `extraction` function and provide detailed instructions:

```
def extract(tasks_response):
    umessage = """
    1) Read the following text analysis that returns detailed memory tags 
for each part of the text
    2) Then return the list of memory tags with absolutely no other text
    3) Use no formatting, no hashtags, no markdown. Just answer in plain 
text
    4) Also provide the sentiment analysis score for each tag in this 
format(no brackets) : memory tag sentiment Score
    """
```

We have clearly instructed our GenAISys to provide the sentiment scores in a clean format only. We will now call GPT-4o with `reason.make_openai_api_call`, defined previously, and add `reason.py`, the AI library we began building in the previous chapters. The input to the API call is the output of the last step, `retres`, appended to the instruction message, `umessage`:

```
    umessage+=retres
```

The `system` role reminds the agent of its psychological marketing function:

```
    mrole = "system"
    mcontent = "You are a marketing expert specialized in the 
psychological analysis of content"
```

The `user` role introduces the user message, `umessage`, and the API call is made:

```
    user_role = "user"
    task_response = reason.make_openai_api_call(
        umessage,mrole,mcontent,user_role
    )
    return task_response
```

The agent returns `task_response`, from which we will extract the memory sentiment scores, process, and verify:

```
# Step 2: Extract scores
task_response=extract(retres)
print(task_response)
```

The output is the list of scores per segment we expected for each memory tag:

```
Reality Memory sentiment 0.8
Episodic Memory sentiment 0.8
Reality Memory sentiment 0.4
Episodic Memory sentiment 0.4
Episodic Memory sentiment 0.8
Reality Memory sentiment 0.8
Time Memory Past sentiment 0.8
Episodic Memory sentiment 0.5…
```

We now need to consolidate these scores to use them for decision-making.

Step 3: Statistics

We will use a simple non-AI **regular expressions (re)** module for this function for pattern matching and extraction. This shows that a GenAISys CoT can contain non-AI functions that expand its scope beyond generative AI models.

The text to analyze is the output of the previous step:

```
# Input text
text=task_response
```

We are looking for decimals:

```
# Regular expression to extract sentiment scores
pattern = r"(\d+\.\d+)"
scores = [float(match) for match in re.findall(pattern, text)]
```

We then display the scores:

```
# Output the extracted scores
print("Extracted sentiment scores:", scores)
```

The output contains the scores:

```
Extracted sentiment scores: [0.8, 0.8, 0.4, 0.4, 0.8, 0.8, 0.8, 0.5, 0.5,
0.5, 0.7,…
```

We first calculate an overall score if the function returned scores:

```
# Optional: calculate the overall score and scaled rating
if scores:
    overall_score = sum(scores) / len(scores)
```

Then we scale the score from 1 to 5:

```
scaled_rating = overall_score * 5
```

Finally, we display the overall score and the scaled score:

```
print("Overall score (0-1):", round(overall_score, 2))
print("Scaled rating (0-5):", round(scaled_rating, 2))
```

The output is what we expected:

```
Overall score (0-1): 0.63
Scaled rating (0-5): 3.14
```

The output requires some human analysis:

- In real-life projects, this process might not go so smoothly! Maybe the AI agent will not produce what we expect at all; perhaps it will for one step but not for the scores. When that occurs, we have to work on alternative steps. Building a GenAISys, as with any AI system, is an iterative process.
- The original rating in the hotel dataset for this review was 3, and we obtained 3.14, which is more refined. Online ratings are subjective and may not accurately reflect the content of the review. An AI agent will provide a more nuanced rating through advanced analysis processes similar to the one in this section. We could average the hotel review and ours. However, our goal is to generate a tailored message for the consumer. In a real-life project, we would reach out to consumers in marketing panels, utilizing the consumer memory agent, and obtain real-time feedback.

For now, however, we have the information we need to determine the content to create.

Step 4: Content creation

Before deciding on the content to create, the agent reads the information messages. The first message is `umessage4`:

```
from cot_messages_c6 import umessage4
```

The message contains instructions on how to create a promotional campaign. We are keeping the message in a variable so that the function can be called with different prompts depending on the task.

The agent must first use the memory tags analyzed to generate, not analyze, a text:

```
umessage4 = """

1) Your task is to generate an engaging text  for a customer based on a
memory analysis of a text
2) The analysis of the text is provided in the following format: text
segment, memory tags, dimension, sentiment score, and explanation
The text also contains the overall sentiment score and the list of memory
tags in the text
3) Use no other memory tags than those provided to generate your engaging
text
```

Then, the agent receives instructions on the sentiment analysis:

```
4) Use the overall sentiment score to give the tone of your response
If the overall sentiment score is positive write an engaging text
addressing each segment with its memory tag and sentiment score
If the overall sentiment score is negative analyze why and find ideas and
solutions to find a way to satisfy the customer
If the overall sentiment score is negative analyze make sure to show
empathy for this negative feeling and then make the transition from
negative to positive
4) Focus on the topic provided that begins with the term the topic which
focuses on the core topic of the text to make the customer happy
```

Then, the agent receives final instructions on the content to generate:

```
5) Use your training to suggest named entities for that topic to make
sure that the customer receives a message tailored to the memory tags and
sentiment score
```

We now create the input by adding the scaled rating we obtained and the memory tags the agent found:

```
ugeneration=generation + "The advanced memory analysis of each segment of
a text with a sentiment score:" + retres + " the scaled overall rating:
"+ str(scaled_rating)+ " and the list of memory tags of the text "+ task_
response
```

The agent now has a complete representation of the task expected. We explain the agent's role with `imcontent4`:

```
imcontent4 = "You are a marketing expert specialized in the psychological analysis of content"
```

The agent is now ready to run the generation with the `make_openai_api_call` call:

```
ugeneration=generation + …
mrole4 = "system"
mcontent4 = imcontent4
user_role = "user"
pre_creation_response = make_openai_api_call(
    ugeneration,mrole4,mcontent4,user_role
)
print(pre_creation_response)
```

The response is a `pre_creation_response` response that is empathetic if the sentiment is negative or adapts it to the tone of the review otherwise:

```
**Segment 7: "But we didn't get the view expected."**
Memory Tags: [Episodic Memory, Reality Memory]
Sentiment Score: 0.4
It's understandable to feel a bit let down when expectations aren't met.
For future stays, …
```

The output is in a cognitive format. We're going to run the same call but with a message to clean up and prepare the content for image generation:

```
umessage4b="Clean and simplify the following text for use as a DALL-E prompt. Focus on converting the detailed analysis into a concise visual description suitable for generating an engaging promotional image" + pre_creation_response
mrole4b = "system"
mcontent4b = imcontent4b
user_role4b = "user"
creation_response = make_openai_api_call(
    umessage4b,mrole4b,mcontent4b,user_role4b
)
print(creation_response)
```

The output is a clear instruction to create an image with an exciting *luxurious* offer that is always appreciated:

```
"Luxurious hotel stay with spacious rooms and swift check-in; enjoy a
comfortable bed and convenient 24-hour parking. Celebrate with special
deals and nearby shopping reminiscent of home. Despite minor noise and
view issues, the overall experience is positive and memorable."
```

The output of the message may vary each time we run the requests, but the tone should remain the same. Also, we can adapt the instructions to other content to generate. In this case, the agent is all set to use this instruction to create an image.

Step 5: Creating an image

At this stage, the consumer memory agent uses the instructions (`creation_response`) generated during *Step 4: Content creation* to create a tailored promotional image using OpenAI's DALL-E:

```
# Step 5: Creating an image
import requests
prompt=creation_response
image_url = reason.generate_image(prompt)
```

The `generate_image(prompt)` function is reused from the previous chapter. By consistently reusing functions, we reduce the development overhead and ensure code maintainability. As in *Chapter 5*, the image is generated and stored in a file as `c_image.png`:

```
save_path = "c_image.png"
image_data = requests.get(image_url).content
with open(save_path, "wb") as file:
    file.write(image_data)
```

The image is now ready to accompany our final personalized message. We will display the image at the end of the process.

Step 6: Creating a custom message

With the promotional image prepared, we now generate a concise and engaging customer message. First, we confirm that `creation_response` from *Step 5: Creating an image* is available:

```
if creation_response != "":
    umessage = """
    1) Read the following text carefully
```

```
    2) Then sum it up in a paragraphs without numbering the lines
    3) They output should be a text to send to a customer
    """
```

The output from the agent provides a polished message, suitable for customer communication:

```
Dear Customer,
Experience a luxurious hotel stay with spacious rooms and a swift check-in
process. Enjoy a comfortable bed and the convenience of 24-hour parking.
Take advantage of special deals and nearby shopping that feels like home.
While there may be minor noise and view issues, the overall experience
remains positive and memorable.

Best regards,
```

We can now display the output in another format if we wish to, with Python's `textwrap`:

```
import os
from IPython.display import Image, display
import textwrap
# Set the desired width for each line
line_width = 70
# Wrap the text to the specified width
wrapped_message = textwrap.fill(process_response, width=line_width)
print(wrapped_message)
```

The displayed message is clear, professional, and suitable for direct customer outreach:

```
Dear Customer,  Experience a luxurious hotel stay with spacious rooms
and a swift check-in process. …
```

To enhance this message, you might adjust the prompt to omit references to AI or any internal system details, focusing purely on customer-oriented language.

Finally, the system displays the generated image alongside the message to create an appealing, personalized promotional package:

```
# Define the image path
image_path = "/content/c_image.png"

# Check if the image file exists
if os.path.exists(image_path):
```

```
    # Display the image
    display(Image(filename=image_path))
else:
    print(f"Image file {image_path} not found.")
```

The resulting visual emphasizes the hotel's upgraded, luxurious offering, perfectly aligned with the customer's expectations based on their review analysis:

Figure 6.6: An upgrade for a stay in the hotel for a customer

You can now experiment with additional reviews, testing the depth and flexibility of the agent. We have successfully developed a sophisticated, neuroscience-inspired CoT consumer memory agent. In the next section, we'll integrate this full process into the `reason.py` AI library and further enhance our GenAISys framework.

GenAISys interface: From complexity to simplicity

Our journey in this chapter has taken us deeper into the era of self-reflecting, reasoning, and meta-cognitive agentic AI. In this final section, we shift from the intricate inner workings of our consumer-memory CoT to a clean, intuitive user experience. We'll add a CoT widget that lets any user trigger memory analysis or full content generation on arbitrary text. We'll then extend the AI agent so it reacts to that widget's options. Finally, we'll demonstrate the generalized workflow on a flight review to show how the same memory logic applies to new domains.

Open the `2_Running_the_Reasoning_GenAISys.ipynb` notebook on GitHub. Then run the *Setting up the Environment* section, which is identical to the notebook in *Chapter 5*. We will begin by adding a CoT widget to the IPython interface.

Adding the CoT widget

To make the memory agent simple and intuitive, we introduce a straightforward drop-down menu (*Figure 6.6*). Users can effortlessly select the task they wish the GenAISys agent to perform:

- **None** (default): No reasoning task is activated
- **Analysis**: Activates a standalone memory-analysis function
- **Generation**: Executes the full consumer memory agent workflow (*Steps 1–6*), including sentiment analysis, content generation, and custom message creation

This streamlined user interaction significantly reduces complexity for the end user, shifting the sophisticated internal operations into the background.

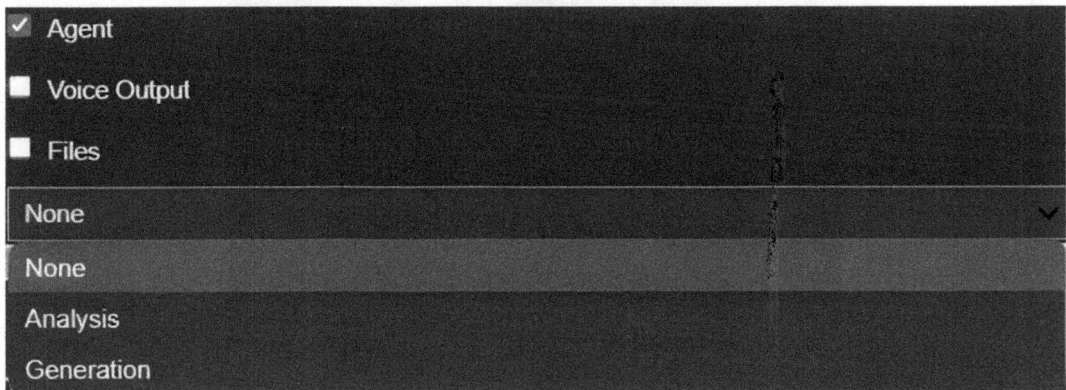

Figure 6.7: Choosing the reasoning task

The widget is implemented in three steps—adding the widget, adding an observer, and sending the options to the AI agent:

1. Adding the widget is done in a few lines. We define the drop-down menu (`instruct_selector`) within the IPython interface:

```
# Ensure 'Instructions' exists in the memory_selector options
instruct_selector = Dropdown(
    options=["None","Analysis", "Generation"],
```

```
    value="None",  # Ensure default active_memory is in the options
    description='Reasoning:',
    layout=Layout(width='50%')
)
```

The dropdown provides clear options, ensuring users easily understand their choices: **None**, **Analysis**, or **Generation**. Next, we incorporate `instruct_selector` into the existing interface layout (`VBox`):

```
VBox:
Box(
                [user_selector, input_box, agent_checkbox,
                    tts_checkbox, files_checkbox,instruct_selector],
                layout=Layout(display='flex', flex_flow='column',
                    align_items='flex-start', width='100%')
```

When the user submits their choice, a handler updates the output messages for the user to see that the choice has been taken into account using standard submission code:

```
def handle_submit(sender):
    user_message = sender.value
    if user_message.strip():
        sender.value = ""  # Clear the input box
        # Check if instruct_selector is "Analysis" or "Generation"
        if instruct_selector.value in ["Analysis", "Generation"]:
            with reasoning_output:
                reasoning_output.clear_output(wait=True)
                print("Thinking...")  # Display "Thinking..." only
when
            print("Reasoning activated")  # Restore default message…
```

We want "Thinking..." to be displayed to signal to the user that the system is working.

2. The second step is to insert an observer that will detect a change the user makes and update the display. The `instruct_selector` is called by `instruct_selector.observe`:

```
# Ensure 'Instructions' exists in the memory_selector options
instruct_selector = Dropdown(
    options=["None","Analysis", "Generation"],
```

```
        value="None",  # Ensure default active_memory is in the options
        description='Reasoning:',
        layout=Layout(width='50%')

    instruct_selector.observe(on_instruct_change, names='value')
```

3. Finally, we forward the user's selection seamlessly into the AI agent call. The chosen reasoning mode (`active_instruct`) is integrated into the agent's execution path:

```
    if agent_checkbox.value:
            pfiles = 1 if files_checkbox.value else 0
            active_instruct = instruct_selector.value  # Retrieve the
    selected instruction
            response = chat_with_gpt(user_histories[active_user],
                user_message, pfiles, active_instruct)
```

By clearly integrating these few lines, the AI agent dynamically activates the appropriate reasoning mode without additional user complexity. We can now enhance the AI agent.

Enhancing the AI agent

The AI agent will now receive the user's widget selection in a new argument named `active_instruct` that will trigger its decisions:

```
def chat_with_gpt(messages, user_message,files_status,active_instruct):
```

When the user selects **Analysis**, the AI agent triggers the previously built reasoning function, `make_openai_reasoning_call`, to perform memory analysis:

```
if active_instruct=="Analysis" and continue_functions==True:
        from cot_messages_c6 import (
            system_message_s1, generation,imcontent4,imcontent4b)
        mrole= system_message_s1
        reasoning_steps=reason.make_openai_reasoning_call(
            user_message, mrole
        )
        aug_output=reasoning_steps
        continue_functions=False
```

Similarly, when **Generation** is selected, the agent runs the complete memory agent workflow using the custom `memory_reasoning_thread` function:

```
if active_instruct=="Generation" and continue_functions==True:
        from cot_messages_c6 import (
            system_message_s1, generation,imcontent4,imcontent4b)
        reasoning_steps = reason.memory_reasoning_thread(
            user_message,system_message_s1,
            generation,imcontent4,imcontent4b
        )
        aug_output=reasoning_steps
        continue_functions=False
```

With these straightforward integrations, the GenAISys interface now provides powerful reasoning capabilities transparently.

Generalizing the GenAISys capabilities

With the enhanced interface, users can easily apply memory agent reasoning to diverse content beyond hotel reviews. Let's illustrate this flexibility with a general user comment regarding a flight. The user simply selects **Analysis**, then submits a flight review without additional instructions:

```
User01: The flight from Tokyo to Honolulu was nice in general but the food
was awful!
```

The GenAISys instantly returns a detailed, segment-by-segment memory analysis:

```
Agent: Segment 1: "The flight from Tokyo to Honolulu was nice in general"
- Memory Tags: [Reality Memory, Time Memory Past]
- Dimension: Emotional
- Sentiment Score: 0.7
- Explanation: This segment recounts a real past experience (a flight) and
expresses a positive emotional evaluation ("nice"). The reference to a
completed event classifies it under reality and past time memory…
```

The user then reenters the sentence, but this time with the **Generation** option and the **Files** option checked, so that the image generated with the text will be displayed. All the user has to do is select the option, enter the text, and submit it. Once again, the experience is seamless, no additional instructions are required on the part of the user, and the response is complete: the fully analysis, process, and final customer message as we designed in the previous section—beginning with memory analysis, proceeding to sentiment scoring, content generation, and concluding with a tailored customer message and an engaging image (see *Figure 6.8*) that will be displayed if `Files` is checked in the interface:

```
Process: Performing memory and sentiment analysis.
Memory analysis result: Segment 1: "The flight from Tokyo to Honolulu was nice in general but the food was awful!"
- Memory Tags: [Reality Memory, Episodic Memory, Time Memory Past]
- Dimension: Emotional
- Sentiment Score: 0.4
- Explanation: This segment recounts a specific past travel event (a flight between two cities), embedding personal experience and subjective evaluation (positive about the flight overall and strongly negative about the food). The use of experiential details and judgment indicates that the experience is encoded as a real, episodic, past memory with an emotional emphasis, resulting in an overall slightly negative sentiment.
…
Dear Customer,

Experience a serene flight from Tokyo to Honolulu while enjoying a gourmet meal inspired by the renowned Chef Nobu Matsuhisa. Indulge in diverse and vibrant dishes crafted to enhance your journey.

Best regards,
```

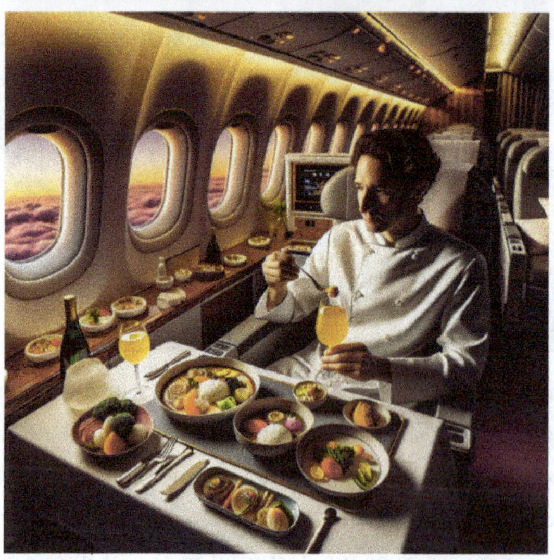

Figure 6.8: An engaging customer image to match the personalized message

We can see that from a user perspective, our GenAISys is running seamlessly. We are giving the user the illusion that everything in generative AI is simple. Of course, in a real-life project, we would have to spend resources trying all types of texts, finding the limitations, and solving the issues to cover edge cases and refine outputs. Let's now sum up our journey in this chapter and take the GenAISys to yet another level.

Summary

This chapter pushed our GenAISys far beyond classical AI, into the realm of meta-cognitive, self-reflective reasoning. We defined a pragmatic memory model combining primary categories (short-term, long-term, reality, fiction, and time) with semantic and episodic tags, then layered intellectual, emotional, and physical dimensions on top. Using this framework, we built a six-step CoT agent that decodes each review segment, tags memory categories, quantifies sentiment, and produces an overall cognitive score. Based on the cognitive profile and sentiment score, the agent generated personalized promotional text and created a matching DALL-E image—then wrapped everything into a polished customer message.

A new drop-down widget now lets users choose **None**, **Analysis**, or **Generation**, making sophisticated reasoning tasks a single-click experience. Behind the scenes, the AI agent routes requests to either a standalone memory analysis or the full consumer-memory workflow. We finally demonstrated the agent on a flight review, showing it can analyze, score, and respond to any text—extending GenAISys from hospitality into broader customer service scenarios.

With these advances, the GenAISys is ready for real-time, production-grade decision-making. The next chapter will focus on scaling the functionality of our GenAISys architecture for immediate, high-throughput AI operations.

Questions

1. Emotional memory is a key factor in e-marketing. (True or False)
2. OpenAI's o3 is a reasoning model that can perform complex tasks. (True or False)
3. Long-term memory does not include emotional factors. (True or False)
4. A generative AI model cannot analyze complex memory structures. (True or False)
5. A generative AI model can not only analyze sentiments but also provide numerical scores between 0 and 1. (True or False)
6. A Pinecone index can produce complex instructions based on system queries. (True or False)

7. A thread-of-reasoning agent can think through complex prompts and perform multiple coordinated tasks. (True or False)
8. A thread-of-reasoning scenario can be triggered with a user input. (True or False)
9. A reasoning agent can process reviews from sites such as TripAdvisor and generate custom messages. (True or False)
10. A generative AI system cannot process thread-of-reasoning agents. (True or False)

References

- Brandt, Denise, and Ilja Nieuwenhuis. 2017. Understanding Memory in Advertising. Nielsen. February. https://www.nielsen.com/insights/2017/understanding-memory-in-advertising/
- Nielsen Homepage: https://www.nielsen.com/
- Nicks, Guillaume, and Yannick Carriou. 2016. Emotion, Attention and Memory in Advertising. Ipsos Knowledge Centre. https://www.ipsos.com/sites/default/files/2017-07/Emotion-Attention-and-Memory-in-Ads.pdf
- Ipsos Homepage: https://www.ipsos.com/
- OpenAI. 2024. OpenAI o3 System Card: A Comprehensive Evaluation of the o3 Model Series, Outlining Advancements in Safety, Reasoning, and Robustness. OpenAI. https://openai.com/index/o3-mini-system-card/

Further reading

- Woodside, Arch G., Sanjay Sood, and Kimberly E. Miller. 2008. "When Consumers and Brands Talk: Storytelling Theory and Research in Psychology and Marketing." Psychology & Marketing 25 (2): 97–145. https://www.researchgate.net/publication/229889043_When_consumers_and_brands_talk_Storytelling_theory_and_research_in_psychology_and_marketing

Subscribe for a Free eBook

New frameworks, evolving architectures, research drops, production breakdowns—*AI_Distilled* filters the noise into a weekly briefing for engineers and researchers working hands-on with LLMs and GenAI systems. Subscribe now and receive a free eBook, along with weekly insights that help you stay focused and informed.

Subscribe at `https://packt.link/TRO5B` or scan the QR code below.

7

Enhancing the GenAISys with DeepSeek

The *DeepSeek-V3 Technical Report* arrived in December 2024, followed a month later by the **DeepSeek-R1** paper and a full set of open source resources. The release sent a shockwave through the AI community: download counts on Hugging Face exploded, DeepSeek apps topped store charts, and new API providers sprang up overnight. Governments debated moratoriums while the major generative AI players—OpenAI, X (with Grok 3), and others—stepped on the gas. Within weeks, we saw o3 versions improve OpenAI models, a clear signal that the AI race had entered a new phase. At the same time, real-world AI production teams watched these dizzying innovations pile up, disrupting existing AI systems. Teams that spent months adapting their systems to one generative AI model found themselves caught in a gray area between systems that work but could still be improved.

So, what should we do? Should we upgrade a stable GenAISys to follow the latest trend in an accelerating AI market with the cost and risks entailed? Or should we ignore the latest models if our system is stable? If we ignore evolutions, our system may become obsolete. If we keep following the trends, our system will become unstable!

This chapter shows how to strike a workable balance. Instead of rewriting entire environments for every model upgrade or new functionality, we introduce a **handler-selection mechanism** that routes user requests to the right tool at the right time. A **handler registry** stores every AI function we develop; the selection layer inspects each incoming message and triggers the appropriate handler. With this design, the GenAISys can evolve indefinitely without destabilizing the stack. We will begin the chapter by defining how a balanced approach can be found between model evolutions and real-world usage, illustrated through a product design and production use case. Next comes a concise look at DeepSeek-V3, DeepSeek-R1, and the distilled Llama model we'll implement. Then, we'll install **DeepSeek-R1-Distill-Llama-8B** locally with Hugging Face, wrap it in a reusable function, and then plug it into our GenAISys. At that point, we will develop the flexible, scalable environment of the handler-selection mechanism to allow us to activate the models and tasks we need for each project. By the end of the chapter, you will be able to have full control over the GenAISys and be ready for whatever the AI market throws at you.

This chapter covers the following topics:

- The balance between AI acceleration and usage
- An overview of DeepSeek-V3, R1, and distillation models
- Installing DeepSeek-R1-Distill-Llama-8B locally
- Creating a function to run DeepSeek-R1-Distill-Llama-8B
- Deploying DeepSeek-R1-Distill-Llama-8B in the GenAISys
- Building a handler registry for all the AI functions
- Building a handler-selection mechanism to select the handlers
- Upgrading the AI functions to be handler-compatible
- Running product design and production examples

Let's start by defining the balance between relentless AI evolution and day-to-day business usage.

Balancing model evolution with project needs

Before racing to adopt every new model, we must anchor our decisions on project needs. So far, our GenAISys has served mostly marketing functions for an online travel agency. Now, imagine that the agency has grown large enough to fund a line of branded merchandise—custom travel bags, booklets, and other goodies. To manage this new venture, the company hires a **product designer and production manager (PDPM)**. The PDPM studies customer feedback and designs personalized kits but quickly sees that AI could boost both creativity and throughput.

The examples in this chapter thus focus on product design and production workflows. Our goal is not to force DeepSeek (or any other model) into every task but to choose the model that best fits the need. To do that, we'll extend the GenAISys with a handler-selection mechanism that responds to user choices in the IPython interface and to keywords in each message. Depending on the situation, the operations team can configure the system to route requests to GPT-4o, DeepSeek, or any future model.

Before wiring DeepSeek into our GenAISys, let's review the DeepSeek model family.

DeepSeek-V3, DeepSeek-V1, and R1-Distill-Llama: Overview

DeepSeek's journey began with DeepSeek-V3, advanced to DeepSeek-R1—a reasoning-focused upgrade—and then branched into distilled variants built on Qwen and Llama architectures, as shown in *Figure 7.1*. V3 was responsible for putting the model on the map, and it was R1 that brought in robust reasoning.

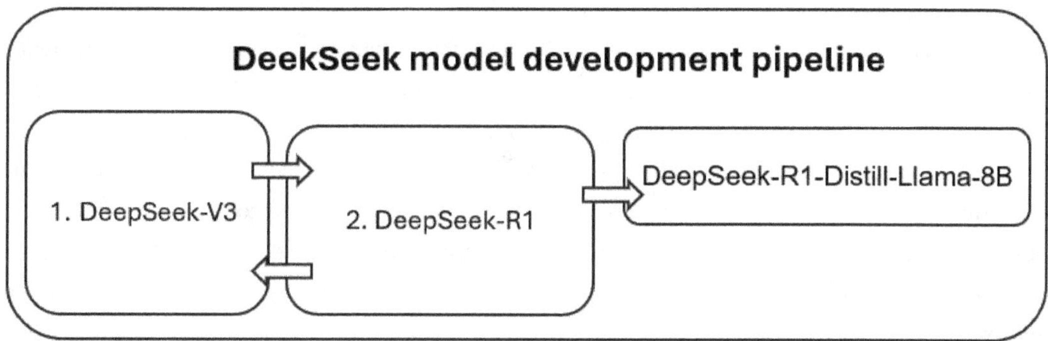

Figure 7.1: DeepSeek development cycle

According to DeepSeek-AI et al. (2024), V3 delivered striking efficiency gains. Its full training budget was only 2.788 million H800 GPU-hours (≈ USD 5.6 million at USD 2 per GPU-hour)—remarkably low for a modern frontier model. Even on a per-token basis, the cost is lean, needing just 180 K GPU-hours per trillion tokens. The cost is, therefore, very economical compared to what is typically reported for large-scale models.

When we examine the list of authors of the DeepSeek-V3 Technical Report (2024) on arXiv, `https://arxiv.org/abs/2412.19437`, we first notice that more than 150 specialists wrote the paper! In itself, this factor alone proves the efficiency of open source approaches that involve collective efforts to produce efficiency-driven architectures by opening ideas to every person willing to contribute. The list of *Contributions and Acknowledgements* in *Appendix A* is a tribute to open source developments.

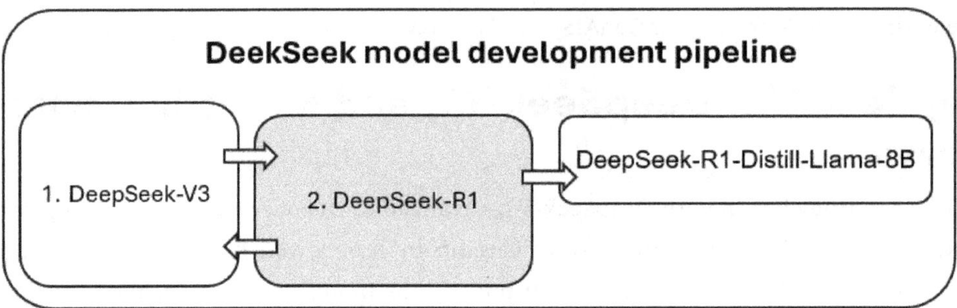

Figure 7.2: DeepSeek-R1 is derived from DeepSeek-V3

DeepSeek-R1 grew straight out of DeepSeek-V3. The team wanted V3's punch, but with feather-weight inference, so they wired the model to activate only a minimal subset of experts during inference, as shown in *Figure 7.2*. Furthermore, training stayed just as lean. R1 jumped directly into reinforcement learning with no supervised fine-tuning. The reasoning was high but faced limitations for classic NLP tasks. Rule-based rewards were introduced to avoid the neural network's training cycles. The training prompts were structured with neat `<think> … <answer>` tags, avoiding the smuggling of biases into the model's final answer. Moreover, the reinforcement learning process began with *cold-start* data containing **chain of thought** (**CoT**) examples focusing on reasoning. This approach reduced training time and costs.

DeepSeek evolved to R1 by refining MoE strategies and integrating multi-token prediction, significantly enhancing both accuracy and efficiency. Finally, DeepSeek-R1 was used to enhance DeepSeek-V3 with reasoning features. DeepSeek-R1 was also distilled into smaller models such as Llama and Qwen. The technique used was knowledge distillation, where a smaller "student" model (in this chapter, Llama) learns from a "teacher" model (in this chapter, DeepSeek-R1). This approach is effective in that it teaches the student model to achieve performance similar to that of the teacher while being more efficient and suitable for deployment on resource-constrained devices, which will be the case in this chapter, as you'll see.

Let's install and run DeepSeek-R1-Distill-Llama-8B and plug it into our GenAISys.

Getting started with DeepSeek-R1-Distill-Llama-8B

In this section, we will implement DeepSeek-RAI-Distill-Llama-8B, a distilled version of DeepSeek-R1, as shown in *Figure 7.3*. We will install Hugging Face's open-source `Transformers` library, an open framework for using and fine-tuning pre-trained transformer models.

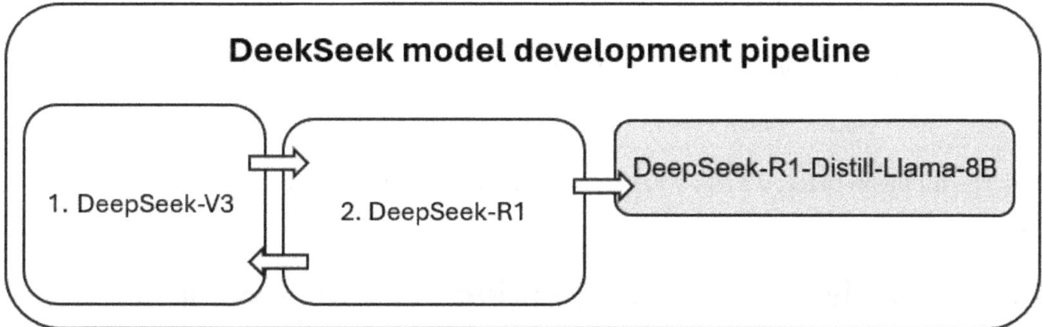

Figure 7.3: Installing DeepSeek-RAI-Distill-Llama-8B, a distilled version of DeepSeek-R1

We will be using the DeepSeek-RAI-Distill-Llama-8B documented by Hugging Face: `https://huggingface.co/deepseek-ai/DeepSeek-R1-Distill-Llama-8B`. Hugging Face also provides recommendations for this model: `https://huggingface.co/deepseek-ai/DeepSeek-R1-Distill-Llama-8B#usage-recommendations`.

The version we will download is an open source distilled version of DeepSeek-R1 provided by Unsloth, an LLM accelerator, on Hugging Face: `https://unsloth.ai/`. We will thus not use a DeepSeek API but only a locally installed open source version that does not interact with the web, leveraging Hugging Face's SOC 2 Type 2 certification that complies with privacy and security constraints: `https://huggingface.co/docs/inference-endpoints/en/security`.

To install deepseek-ai/DeepSeek-R1-Distill-Llama-8B locally on a recent machine, it is recommended to have about 20 GB of RAM. A bit less is possible, but it is best to avoid the risk. About 20 GB of disk space is also recommended.

To install DeepSeek-R1-Distill-Llama-8B on Google Colab, it is recommended to use Google Colab Pro to obtain GPU memory and power. For this section, the Hugging Face model is downloaded on Google Drive, which is mounted through Google Colab. The disk space required will exceed the free version of Google Drive, and a minimal subscription to Google Drive may be required. Check the costs before installing on Google Colab.

Open `Getting_started_with_DeepSeek_R1_Distill_Llama_8B.ipynb` within the Chapter07 directory on GitHub (`https://github.com/Denis2054/Building-Business-Ready-Generative-AI-Systems/tree/main`). We will follow the standard procedure of the Hugging Face framework:

- Run the notebook once to install DeepSeek-R1-Distill-Llama-8B locally:

  ```
  install_deepseek=True
  ```

- Run the notebook with no installation and interact with the model:

  ```
  install_deepseek=False
  ```

With the model in place, we can wrap it in a handler and plug it into our GenAISys in the next section.

Setting up the DeepSeek Hugging Face environment

We'll begin by installing DeepSeek-R1-Distill-Llama-8B (locally or in Colab) and then run a quick inference to confirm everything works.

We will first install DeepSeek in the first session:

```
# Set install_deepseek to True to download and install R1-Distill-Llama-8B
locally
# Set install_deepseek to False to run an R1 session
install_deepseek=True
```

The GPU needs to be activated, so let's check it:

```
Checking GPU activation
!nvidia-smi
```

If we are installing Google Colab, we can mount Google Drive:

```
from google.colab import drive
drive.mount('/content/drive')
```

We now set the cache directory in Google Drive and set the corresponding environment variables:

```
import os
# Define the cache directory in your Google Drive
cache_dir = '/content/drive/MyDrive/genaisys/HuggingFaceCache'
# Set environment variables to direct Hugging Face to use this cache
directory
os.environ['TRANSFORMERS_CACHE'] = cache_dir
#os.environ['HF_DATASETS_CACHE'] = os.path.join(cache_dir, 'datasets')
```

We can now install the Hugging Face `Transformers` library:

```
!pip install transformers==4.48.3
```

With that, we are ready to download the model.

Downloading DeepSeek

Let's now download the model from `unsloth/DeepSeek-R1-Distill-Llama-8B` within the Hugging Face framework with the tokenizer and the model:

```
from transformers import AutoTokenizer, AutoModelForCausalLM
import time
if install_deepseek==True:
    # Record the start time
    start_time = time.time()
    model_name = 'unsloth/DeepSeek-R1-Distill-Llama-8B'
    # Load the tokenizer and model
    tokenizer = AutoTokenizer.from_pretrained(model_name)
    model = AutoModelForCausalLM.from_pretrained(
        model_name, device_map='auto', torch_dtype='auto'
    )
    # Record the end time
    end_time = time.time()
    # Calculate the elapsed time
    elapsed_time = end_time - start_time
    print(f"Time taken to load the model: {elapsed_time:.2f} seconds")
```

The download time will be displayed and also depends on your internet connection and Hugging Face's download speed. Once installed, verify that everything is installed in your local directory. In this case, it is as follows:

```
if install_deepseek==True:
    !ls -R /content/drive/MyDrive/genaisys/HuggingFaceCache
```

The output should show the files downloaded:

```
/content/drive/MyDrive/genaisys/HuggingFaceCache:
models--unsloth--DeepSeek-R1-Distill-Llama-8B  version.txt

/content/drive/MyDrive/genaisys/HuggingFaceCache/models--unsloth--
DeepSeek-R1-Distill-Llama-8B:
```

```
blobs  refs  snapshots

/content/drive/MyDrive/genaisys/HuggingFaceCache/models--unsloth--
DeepSeek-R1-Distill-Llama-8B/blobs:
03910325923893259d090bfa92baa4088cd46573…
```

Now, let's run a DeepSeek session.

Running a DeepSeek-R1-Distill-Llama-8B session

To make sure the model is correctly installed and also to avoid overwriting the installation when starting a new session, go back to the top of the notebook and set the following:

```
install_deepseek=False
```

We will now load the `DeepSeek-R1-Distill-Llama-8B` tokenizer and model locally:

```
import time
from transformers import AutoTokenizer, AutoModelForCausalLM
if install_deepseek==False:
    # Define the path to the model directory
    model_path = '/content/drive/MyDrive/genaisys/
HuggingFaceCache/models--unsloth--DeepSeek-R1-Distill-Llama-8B/
snapshots/71f34f954141d22ccdad72a2e3927dddf702c9de'

    # Record the start time
    start_time = time.time()
    # Load the tokenizer and model from the specified path
    tokenizer = AutoTokenizer.from_pretrained(
        model_path, local_files_only=True
    )
    model = AutoModelForCausalLM.from_pretrained(
        model_path, device_map='auto', torch_dtype='auto',
        local_files_only=True
    )

    # Record the end time
    end_time = time.time()
    # Calculate the elapsed time
    elapsed_time = end_time - start_time
    print(f"Time taken to load the model: {elapsed_time:.2f} seconds")
```

The time it took to load the model is displayed and will depend on the configuration of your machine:

```
Time taken to load the model: 14.71 seconds
```

We can have a look at the configuration of the Llama model:

```
if install_deepseek==False:
    print(model.config)
```

The output shows interesting information. The `LlamaConfig` readout confirms we are running a compact, well-scoped model:

```
LlamaConfig {
  "_attn_implementation_autoset": true,
  "_name_or_path": "/content/drive/MyDrive/genaisys/
HuggingFaceCache/models--unsloth--DeepSeek-R1-Distill-Llama-8B/
snapshots/71f34f954141d22ccdad72a2e3927dddf702c9de",
  "architectures": [
    "LlamaForCausalLM"
  ],
 …
```

The distilled Llama model has 32 transformer layers and 32 attention heads per layer, totaling 1,024 attention heads. Also, it contains 8 billion parameters. By contrast, its teacher model, **DeepSeek-R1**, is an MoE giant with **61 layers** and a massive **671 billion parameters**, of which about **37 billion** are active on each forward pass. Let's now run an example with a prompt for a production issue:

```
if install_deepseek==False:
    prompt="""
    Explain how a product designer could transform customer requirements
for a traveling bag into a production plan.
    """
```

We first insert time measurement and tokenize the input using the GPU:

```
import time
if install_deepseek==False:
    # Record the start time
    start_time = time.time()
    # Tokenize the input
    inputs = tokenizer(prompt, return_tensors='pt').to('cuda')
```

Then, we run the generation:

```
# Generate output with enhanced anti-repetition settings
outputs = model.generate(
  **inputs,
    max_new_tokens=1200,
    repetition_penalty=1.5,       # Increase penalty to 1.5 or higher
    no_repeat_ngram_size=3,       # Prevent repeating n-grams of size 3
    temperature=0.6,              # Reduce randomness slightly
    top_p=0.9,                    # Nucleus sampling for diversity
    top_k=50        # Limits token selection to top-k probable tokens
)
```

The goal of our parameters is to limit the repetitions and remain focused:

- `max_new_tokens=1200`: To limit the number of output tokens
- `repetition_penalty=1.5`: To limit the repetitions (can be higher)
- `no_repeat_ngram_size=3`: To prevent repeating n-grams of a particular size
- `temperature=0.6`: To reduce randomness and stay focused
- `top_p=0.9`: Allows nucleus sampling for diversity
- `top_k=50`: Limits token selection to `top_k` to make the next token choice

This set of tokens tends to limit repetitions while allowing diversity. We can now decode the generated text with the tokenizer:

```
# Decode and display the output
generated_text = tokenizer.decode(
    outputs[0], skip_special_tokens=True
)
# Record the end time
end_time = time.time()
# Calculate the elapsed time
elapsed_time = end_time - start_time
print(f"Time taken to load the model: {elapsed_time:.2f} seconds")
```

The output shows the overall time it took the model to think and respond:

```
Time taken to load the model: 20.61 seconds
```

Let's wrap generated_text and display it:

```
import textwrap
if install_deepseek==False:
    wrapped_text = textwrap.fill(generated_text, width=80)
print(wrapped_text)
```

The output provides ideas as requested. It displays DeepSeek-R1's thinking abilities:

```
…Once goals & priorities become clearer, developing
prototypes becomes more focused since each iteration would aim at testing
one main feature rather than multiple changes simultaneously—which makes
refining individual elements easier before moving towards finalizing
designs, When prototyping starts: 1) Start with basic functional mockups
using simple tools -…
```

Integrating DeepSeek-R1-Distill-Llama-8B

In this section, we will add DeepSeek-R1-Distill-Llama-8B to our GenAISys in a few steps. Open GenAISys_DeepSeek.ipynb. You can decide to run the notebook with DeepSeek in the first cell, which will require a GPU:

```
# DeepSeek activation deepseek=True to activate. 20 Go (estimate) GPU
memory and 30-40 Go Disk Space
deepseek=True
```

You can also decide not to run DeepSeek in this notebook, in which case, you will not need a GPU and can change the runtime to CPU. If you decide on this option, OpenAI's API will take over, confirming that no GPU is required:

```
deepseek=False
```

Now, go to the *Setting up the DeepSeek Hugging Face environment* subsection of the notebook. We will simply transfer the following cells from `Getting_started_with_DeepSeek_R1_Distill_Llama_8B.ipynb` to this subsection. The following code will only be activated if `deepseek=True`:

- GPU activation check: `!nvidia-smi`
- Setting the local cache of the model: …`os.environ['TRANSFORMERS_CACHE'] =cache_dir`…
- Installing the Hugging Face library: `!pip install transformers==4.48.3`
- Loading the tokenizer and the model:

```
from transformers import AutoTokenizer, AutoModelForCausalLM
# Define the path to the model directory
model_path = …
```

The installation is now complete. The calls to the DeepSeek model will be made in the *AI Functions* section if `DeepSeek==True` with the parameters described in the *Running a DeepSeek-R1-Distill-Llama-8B session* section:

```
    if models == "DeepSeek":
      # Tokenize the input
      inputs = tokenizer(sc_input, return_tensors='pt').to('cuda')
….
      task_response =tokenizer.decode(outputs[0],skip_special_tokens=True)
```

With DeepSeek functioning, we're ready to build the handler selection mechanism, which will route every user request to GPT-4o, DeepSeek, or any future model—without touching the rest of the stack.

Implementing the handler selection mechanism as an orchestrator of the GenAISys

The PDPM at the online travel agency is experiencing increased demands, requiring the agency to design and produce large quantities of merchandise kits, including travel bags, booklets, and pens. The PDPM wants to be directly involved in the GenAISys development to explore how it can significantly boost productivity.

Given the growing complexity and variety of AI tasks in the system, the GenAISys development team has decided to organize these tasks using handlers, as illustrated in *Figure 7.4*:

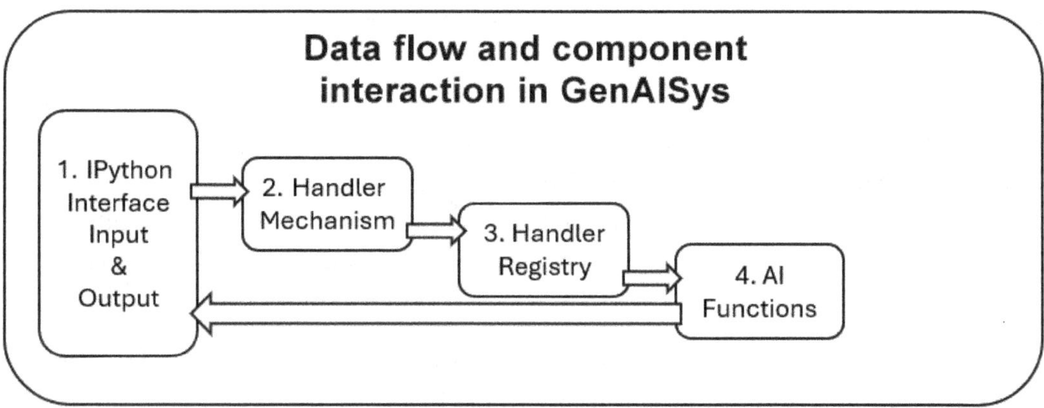

Figure 7.4: GenAISys data flow and component interaction

We'll, therefore, define, implement, and then invite the PDPM to run the enhanced GenAISys to evaluate functions aimed at improving productivity in merchandise design and production.

Figure 7.4 describes the behavior of the handler pipeline we are going to implement:

1. The **IPython interface** serves as the entry and exit point for user interactions, capturing user input, formatting it, and displaying responses returned by the handler mechanism.
2. The **handler mechanism** interprets user inputs, directing data among the IPython interface, the handler registry, and the AI functions. It ensures tasks triggered by user messages execute smoothly.
3. The **handler registry** maintains a list of all available handlers and their corresponding functions. It supports system modularity and scalability by clarifying handler registration and retrieval.
4. **AI functions** perform core tasks such as natural language understanding and data analysis, executing instructions received from the handler mechanism, and returning outputs to the IPython Interface.

In this setup, a user provides input through the IPython interface. This input is routed into a handler selection mechanism, which then evaluates the available handlers registered alongside specific conditions. Each entry in the registry is a (condition, handler) pair responsible for different operations such as reasoning, image generation, or data analysis. Once a matching condition is found, the corresponding AI function is activated. After processing, it returns the results to the interface. This structured pipeline—from user input through to the AI-generated response—is handled gracefully, with each handler clearly defined for readability and efficiency.

Before coding, let's clearly define what we mean by a "handler" in the GenAISys.

What is a handler?

A handler is essentially a specialized function responsible for addressing specific tasks or types of requests. Each handler is registered alongside a condition, typically a small function or lambda expression. When evaluated as `True`, this condition indicates that the associated handler should be invoked. This design neatly decouples the logic for deciding *which* handler should run from *how* the handler executes its task.

In our context, handlers are the orchestrator's building blocks—conditional functions designed to process specific input types. When a user provides input, the handler selection mechanism evaluates it against the handler registry, which consists of pairs of conditions and handlers. Upon finding a match, the corresponding handler is triggered, invoking specialized functions such as `handle_generation`, `handle_analysis`, or `handle_pinecone_rag`. These handlers execute sophisticated reasoning, data retrieval, or content generation tasks, providing precise and targeted outputs.

But why exactly is a handler better for our GenAISys than a traditional list of `if…then` conditions?

Why is a handler better than a traditional if…then list?

Using handlers improves maintainability and readability. Instead of scattering multiple `if...then` checks across the code, each handler is self-contained: it has its condition and a separate function that carries out the required action. This structure makes it easier to add, remove, or modify handlers without risking unintended interactions in a chain of lengthy conditionals. Additionally, since it separates the logic of "which handler do we need?" from "how does that handler actually work?" we're left with a more modular design that makes scaling seamless.

We will first go through the modifications to our IPython interface.

1. IPython interface

We'll start by reviewing the primary updates to our IPython interface, which remains the main interaction point, as shown in *Figure 7.5*. From a user perspective, the introduction of handlers doesn't alter the interface significantly, but some underlying code adjustments are necessary.

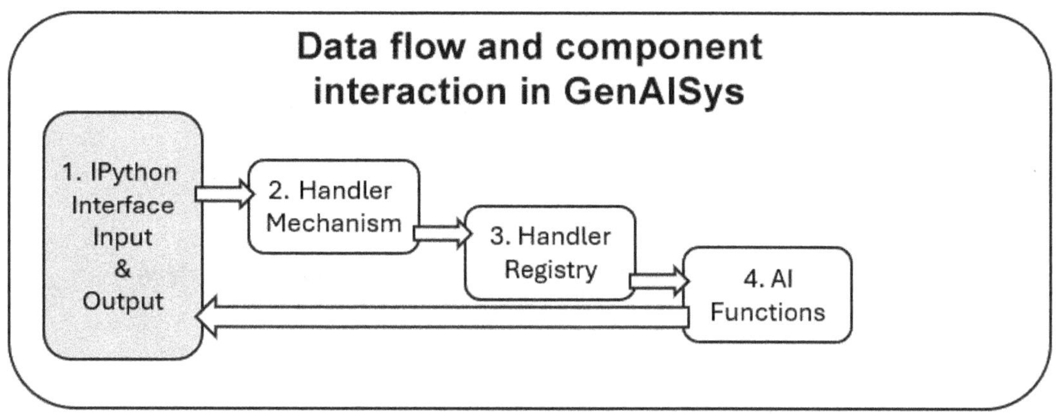

Figure 7.5: The IPython interface processes the user input and displays the output

The IPython interface calls `chat_with_gpt` as before:

```
response = chat_with_gpt(
    user_histories[active_user], user_message, pfiles,
    active_instruct, models=selected_model
)
```

Now, however, we can explicitly select either an OpenAI or a DeepSeek model with the following:

```
models=selected_model
```

To add the model to the `chat_with_gpt` call, we first add a drop-down model selector to the interface:

```
# Dropdown for model selection
model_selector = Dropdown(
    options=["OpenAI", "DeepSeek"],
    value="OpenAI",
    description="Model:",
    layout=Layout(width="50%")
)
```

The model selector is added to the `VBox` instances in the interface:

```
# Display interactive widgets
display(
    VBox(
        [user_selector, input_box, submit_button, agent_checkbox,
```

```
            tts_checkbox, files_checkbox, instruct_selector,
            model_selector],
        layout=Layout(display='flex', flex_flow='column',
            align_items='flex-start', width='100%')
    )
)
```

The user can now choose their preferred model directly from the interface, as shown here:

Figure 7.6: Selecting a model

🔍**Quick tip**: Need to see a high-resolution version of this image? Open this book in the next-gen Packt Reader or view it in the PDF/ePub copy.

🔒**The next-gen Packt Reader** is included for free with the purchase of this book. Scan the QR code OR go to `https://packtpub.com/unlock`, then use the search bar to find this book by name. Double-check the edition shown to make sure you get the right one.

An additional feature has been added to manage file displays.

File management

There are many ways to design file management. We will introduce a function here that can be expanded during a project's implementation phase as needed. Our file management code has three functions:

- Manage user-triggered file deletion
- Delete `c_image.png` when the checkbox is unchecked
- Use existence checks to prevent errors during deletion

We will build the code to handle user interactions directly by observing changes in the checkbox widget of our interface within the Jupyter Notebook environment. The code will then delete a specific image file (`c_image.png`) when the user unchecks the checkbox named `files_checkbox`. This ensures that files are removed cleanly when they are no longer needed, preventing clutter and saving storage space.

We first define the function:

```
def on_files_checkbox_change(change):
```

The event handler function defines a callback function named `on_files_checkbox_change` that will execute when the state of `files_checkbox` changes. `change` is provided by the observer, which contains information about the change event, including the following:

- `old`: The previous state of the checkbox
- `new`: The new state of the checkbox

```
    # Only remove images if the checkbox changed from True to False.
    if change['old'] == True and change['new'] == False:
```

The code verifies whether the checkbox was previously checked (`True`) and has now been unchecked (`False`). This guarantees that the file deletion only occurs when the user explicitly unchecks the checkbox, preventing accidental file removal. We now remove the file:

```
        if os.path.exists("c_image.png"):
            os.remove("c_image.png")
```

We also need to add an observer to inform the `on_files_checkbox_change` function when there is a file status change:

```
# Attach the observer to files_checkbox
files_checkbox.observe(on_files_checkbox_change, names='value')
```

The `files_checkbox.observe()` function links the `on_files_checkbox_change` function to the `files_checkbox` widget. `names='value'` specifies that the function should be triggered when the value of the checkbox changes (i.e., when it is checked or unchecked).

We will now move on to the next part of the pipeline and implement the handler selection mechanism.

2. Handler selection mechanism

The handler selection mechanism dynamically selects and executes the appropriate handler based on predefined conditions. It iterates through available handlers, evaluating conditions until it finds a match, ensuring efficient and structured processing of the user input. The handler selection mechanism is in the `chat_with_gpt` function we built in the previous chapters. However, it now contains an orchestration task, as shown in *Figure 7.7*:

- `chat_with_gpt` remains a pivotal function within the GenAISys and now contains the handler mechanism
- It checks conditions sequentially to decide which handler to invoke
- It falls back to a memory-based handler if no conditions match
- It ensures robustness with error handling for an uninterrupted user experience

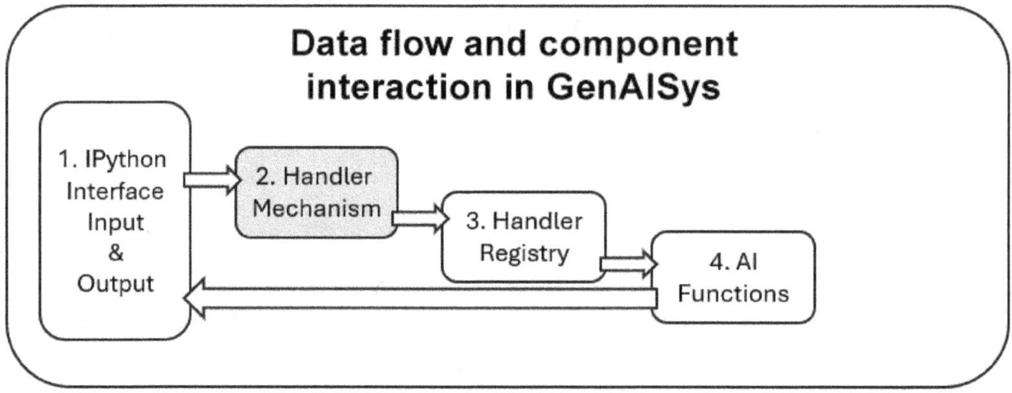

Figure 7.7: The orchestration role of the handler mechanism

In the broader GenAISys workflow, the handler mechanism acts as an orchestrator. It processes user inputs and identifies which AI functions to activate. When the IPython interface captures user messages, the handler mechanism evaluates these inputs to determine the appropriate handler from the handler registry. If no specific handler matches, it defaults to a memory-based response, which is then returned to the IPython interface.

The `chat_with_gpt` function encapsulates this logic. It iterates through a predefined list of handlers, each paired with a corresponding condition function. When a condition evaluates to true, the associated handler is executed. If none match, the fallback memory-based handler ensures a seamless response:

```
def chat_with_gpt(
    messages, user_message, files_status, active_instruct, models
):
     global memory_enabled  # Ensure memory is used if set globally
```

Let's go through the parameters of the function:

- `messages`: The conversation history between the user and the AI
- `user_message`: The latest message from the user
- `files_status`: Tracks the status of any files involved in the conversation
- `active_instruct`: Any instruction or mode that might influence how responses are generated
- `models`: Specifies the active AI model in use

The function uses `global memory_enabled` to access a global variable that determines whether memory should be applied to store/remember the full dialogue of a user. In this chapter, `global memory_enabled=True`.

The function attempts to execute the appropriate handler based on the provided conditions:

```
    try:
        # Iterate over handlers and execute the first matching one
        for condition, handler in handlers:
            if condition(messages, active_instruct, memory_enabled,
                models, user_message):
                return handler(messages, active_instruct, memory_enabled,
                    models, user_message, files_status=files_status)
```

As you can see, `for condition, handler in handlers` iterates over a list called `handlers`, where each item is a tuple containing the following items:

- A condition function to check whether a handler should be used
- A handler function to execute whether the condition is satisfied
- A generic `if` condition, `(...)`, to evaluate the condition function with the provided parameters

- The code returns the output of the corresponding handler if the condition is met, immediately exiting the function

Let's now add a fallback if no handlers match the input conditions:

```
        # If no handler matched, default to memory handling with full
conversation history
        return handle_with_memory(
            messages,  # Now passing full message history
            user_message,
            files_status=files_status,
            instruct=active_instruct,
            mem=memory_enabled,  # Ensuring memory usage
            models=models
        )
```

`handle_with_memory` is called as a default handler that does the following:

- Uses the full conversation history (`messages`)
- Considers memory if `memory_enabled` is true, which is the case in this chapter
- Returns the response directly if executed

Finally, let's add an exception to catch return errors:

```
    except Exception as e:
        return f"An error occurred in the handler selection mechanism:
{str(e)}"
```

With the handler selection mechanism defined, we can now proceed to build the handler registry that stores these handlers.

3. Handler registry

The **handler registry** is a structured collection of condition-handler pairs, where each condition is a lambda function that evaluates user messages and instructions to determine whether specific criteria are met. When a condition is satisfied, the corresponding handler is triggered and executed immediately, as illustrated:

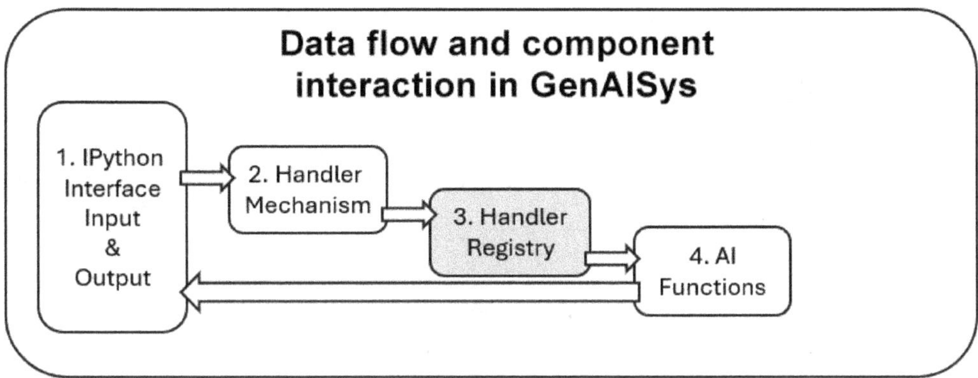

Figure 7.8: Creating the handler registry

All lambda functions have four parameters (`msg`, `instruct`, `mem`, and `models`). This ensures that the number of arguments matches when `chat_with_gpt()` calls a handler.

The handler registry has three main features:

- Is orchestrated by the handler mechanism and can be unlimited
- Routes inputs based on keywords, instructions, or model selection
- Guarantees a fallback response if no conditions match

We will design our handler registry with the following structure of four key properties:

- **Handler registration**: Creates a list of handlers, each with a condition function and a corresponding handler function
- **Specific handler conditions**: Sequentially checks whether an input meets any of the specific conditions
- **Fallback handler**: Adds a default memory-based handler if none of the conditions match
- **Execution**: When a condition is satisfied, the corresponding handler is executed immediately

The role of `**kwargs` in the code provides a flexible way to interact with the AI functions. `**kwargs` is short for *keyword arguments* and is used in Python functions to allow passing a variable number of arguments to a function. In the context of our handler registry code, `**kwargs` plays a crucial role by allowing handlers to accept additional, optional parameters without explicitly defining them in the function. It makes the handlers extensible for future updates or new parameters without requiring modifications to existing function signatures.

We will now begin to build the handler registry with the Pinecone/RAG handler.

Pinecone/RAG handler

The Pinecone/RAG handler manages the **retrieval-augmented generation** (**RAG**) functions previously defined. It activates when detecting the `Pinecone` or `RAG` keyword within the user message:

```
# Pinecone / RAG handler: check only the current user message
(
    lambda msg, instruct, mem, models, user_message,
    **kwargs: "Pinecone" in user_message or "RAG" in user_message,
    lambda msg, instruct, mem, models, user_message,
    **kwargs: handle_pinecone_rag(user_message, models=models)
),
```

This handler checks whether the user message contains "Pinecone" or "RAG," in which case `lambda`: returns `True`; otherwise, it returns `False`. We will now create the reasoning handler.

Reasoning handler

We have already built the reasoning function, but now we need a handler. The keywords that trigger the handler are `Use reasoning`, `customer`, and `activities`. Any additional text in the message provides context for the reasoning process. The handler uses `all()` to ensure all keywords are included in the message:

```
# Reasoning handler: check only the current user message
(
    lambda msg, instruct, mem, models, user_message, **kwargs: all(
        keyword in user_message for keyword in [
            "Use reasoning", "customer", "activities"
        ]
    ),
    lambda msg, instruct, mem, models, user_message, **kwargs:
        handle_reasoning_customer(user_message, models=models)
),
```

Let's move on and create the analysis handler.

Analysis handler

The analysis handler has been used for memory analysis up to now and is triggered by the `Analysis` instruction:

```
# Analysis handler: determined by the instruct flag
(
    lambda msg, instruct, mem, models, user_message,
        **kwargs: instruct == "Analysis",
    lambda msg, instruct, mem, models, user_message,
        **kwargs: handle_analysis(
            user_message, models=models)
),
```

Time to create the generation handler.

Generation handler

The generation handler takes memory analysis to another level by asking the generative AI model to generate an engaging text for a customer based on a memory analysis of the text. The `Generation` instruction triggers the generation handler:

```
# Generation handler: determined by the instruct flag
(
    lambda msg, instruct, mem, models, user_message,
           **kwargs: instruct == "Generation",
    lambda msg, instruct, mem, models, user_message,
            **kwargs: handle_generation(
                user_message, models=models)
),
```

Let's now build the image creation handler.

Image handler

The image creation handler is triggered by the `Create` and `image` keywords in the user message:

```
# Create image handler: check only the current user message
(
    lambda msg, instruct, mem, models, user_message,
    **kwargs: "Create" in user_message and "image" in user_message,
```

```
        lambda msg, instruct, mem, models, user_message,
            **kwargs: handle_image_creation(user_message, models=models)
    )
]
```

We will now create the freestyle handler for when there is no keyword or instructions.

Fallback memory handler

This handler is a general-purpose handler when there is no instruction or keyword to trigger a specific function. Let's append the fallback memory handler accordingly:

```
# Append the fallback memory handler for when instruct is "None"
handlers.append(
    (
        lambda msg, instruct, mem, models, user_message,
              **kwargs: instruct == "None",
        lambda msg, instruct, mem, models, user_message,
              **kwargs: handle_with_memory(
            msg,
            user_message,
            files_status=kwargs.get('files_status'),
            instruct=instruct,
            mem=memory_enabled,  # Replace user_memory with memory_enabled
            models=models
        )
    )
)
```

Note that we have replaced `user_memory` with `memory_enabled` to generalize memory management.

You can add as many handlers and AI functions as you wish to the handler registry. You can scale your GenAISys as much as you need to. You can also modify the keywords by replacing them with explicit instructions, as we did for the `Analysis` and `generation` functions. The handlers will then call all the AI functions you need.

Let's now go through the new organization of the AI functions.

4. AI functions

We will now run the AI functions that are activated by the handler registry. The functions build on those from earlier chapters but are now managed by the handler-selection mechanism introduced in this chapter. Additionally, the examples used in this section are based on typical prompts related to product design and production scenarios. Keep in mind that, due to the stochastic (probabilistic) nature of generative AI models, outputs can vary each time we run these tasks.

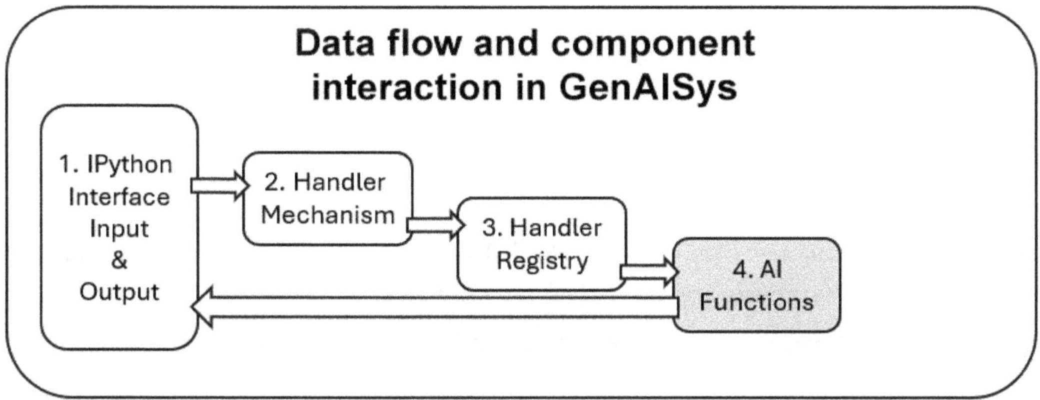

Figure 7.9: AI functions call by the handler selection mechanism and registry

We'll now execute all AI functions currently available in our GenAISys, incorporating DeepSeek model calls where applicable. Let's begin with the RAG functions.

Functions such as speech synthesis, file management, dialogue history, and summary generation remain unchanged from previous chapters.

RAG

This RAG function can run with OpenAI or DeepSeek with the `Pinecone` keyword in the user message. The RAG function's name has changed, but its process remains unchanged for the query:

```
# Define Handler Functions
def handle_pinecone_rag(user_message, **kwargs):
    if "Pinecone" in user_message:
      namespace = "genaisys"
```

```
    if "RAG" in user_message:
      namespace = "data01"
    print(namespace)

    query_text = user_message
    query_results = get_query_results(query_text, namespace)

    print("Processed query results:")
    qtext, target_id = display_results(query_results)
    print(qtext)

    # Run task
    sc_input = qtext + " " + user_message
```

However, the function now contains a DeepSeek distilled R1 call. The function first defaults to OpenAI if no model is provided or if DeepSeek is deactivated:

```
    models = kwargs.get("models", "OpenAI")  # Default to OpenAI if not
provided
    if models == "DeepSeek" and deepseek==False:
       models="OpenAI"


    if models == "OpenAI":
      task_response = reason.make_openai_api_call(
      sc_input, "system","You are an assistant who executes the tasks you
are asked to do.", "user")
```

If DeepSeek is activated, it will be called if chosen in the IPython interface for this task:

```
    if models == "DeepSeek":
      …
    return f"{namespace}:{models}: {task_response}"
```

We will first run a sentiment analysis.

Sentiment analysis (genaisys)

An example user input by the PDPM is the following:

```
Let's see what Pinecone thinks about this sentence: The customer did not
like the design of the traveling bag we gave out because it was not the
right color requested.
```

The output with OpenAI selected (default) and `Agent` checked will be as follows:

```
Agent: genaisys:OpenAI: Sentiment analysis score: 0.3

Explanation: The sentiment of the text is predominantly negative. The
customer expresses dissatisfaction with the traveling bag due to it not
being the right color as requested. The use of phrases like "did not
like" and "not the right color" indicates a negative sentiment, as the
customer's expectations were not met. The score of 0.3 reflects this
negative sentiment, though it is not extremely low because the issue is
specific to the color and not the overall quality or functionality of the
bag.
```

On the other hand, the output with DeepSeek selected (default) and `Agent` checked will be as follows:

```
Agent: genaisys:DeepSeek:
…The numerical value should be between 0 and 1, so let's say 0.2….

Wait, but the user didn't specify a particular method or model for
sentiment analysis. Maybe they want a simple rule-based approach? Let's
consider the presence of negative words like "did not like" and "not the
right color requested." These words indicate dissatisfaction. However,
the sentiment isn't extremely negative since it's just one aspect (color)
that's the issue, not the entire product or experience. So, a score of 0.2
seems appropriate because it reflects a moderate negative sentiment based
on the specific criticism given.
```

The sentiment score and explanation score are acceptable in both cases. Imagine receiving thousands of such customer feedback messages—the GenAISys filters the low scores and provides these outputs automatically, storing them in the customer database.

Now, the PDPM checks semantic analysis.

Semantic analysis (genaisys)

Consider another example input by the PDPM:

```
Let's see what meaning analysis Pinecone on this comment: The customer did
not accept the traveling bag we gave out for free because the size and
colors did not match the order processed.
```

This RAG function can run with OpenAI or DeepSeek with a "Pinecone" keyword in the user message.

OpenAI's output is acceptable and clearly outlines the semantic relationships within the message:

```
Agent: genaisys:OpenAI: Semantic analysis involves understanding the
meaning and interpretation of words, phrases, and sentences in context.
It is a crucial aspect of natural language processing (NLP) that helps
in extracting meaningful information from text. In the context of your
comment, semantic analysis would involve breaking down the sentence to
understand the underlying meaning and the relationships between different
components of the sentence.
Here's a breakdown of the semantic elements in the comment:
1. Subject: "The customer" - This refers to the individual who is the
focus of the action.
2. Action: "did not accept" - This indicates the customer's refusal or
rejection of something.
3. Object: "the traveling bag we gave out for free" - This is the item
that was offered to the customer without charge.
4. Reason: "because the size and colors did not match the order processed"
- This provides the explanation for the customer's refusal, indicating a
mismatch between what was expected (the order) and what was received (the
traveling bag).
…
```

DeepSeek's output is relevant as well:

```
Agent: genaisys:DeepSeek: 300,Semantic analysis…This is not an analysis
but a semantic search. Provide more information on the topic. …
```

```
The original problem was that the customer didn't accept the travel bag
due to mismatched size and color from the ordered one. So, my goal here
should be to understand why this happened and perhaps suggest solutions.

I think the first step would be to look into whether there was a
miscommunication in the ordering process. Maybe the website had incorrect
images or descriptions leading to wrong expectations. Or perhaps during
manufacturing, sizes got altered without notice. Another angle could be
customer feedback—have others reported similar issues? It might also help
to check if there were any system errors in processing orders which caused
discrepancies between what was ordered and what was delivered.
…
# Final Answer
After conducting thorough research and identifying potential contributing
factors, effective solutions have been implemented to address the
discrepancy issues, resulting in improved satisfaction among customers
receiving accurate products aligned with their expectations.
```

DeepSeek's answer is longer and more complex. However, what would a team prefer? A shorter answer like OpenAI's response or a longer one with more explanations? The decision can be reached through workshops and meetings.

Data retrieval (data01)

Both OpenAI and DeepSeek can be used for data retrieval. The user input for a product designer could be as follows:

```
I'm a product designer and am trying to analyze human memory types. Use
RAG to tell us what the CTO said about different memory types
```

The output is satisfactory:

```
Agent: data01:OpenAI: To analyze the CTO's statement about different
memory types using Retrieval-Augmented Generation (RAG), we can break down
the information into the relevant memory types and their application in
the hybrid travel marketing campaign:

1. Memoryless: This refers to systems or processes that do not retain any
information from past interactions. In the context of your campaign, this
might relate to interactions that do not require historical data, such as
one-time queries or transactions…
```

We thus have a flexible RAG system in our GenAISys that can run with the models we wish. However, we still have to evaluate the models for each set of tasks we want to perform.

Chain of thought

The CoT function operates with **Files** checked and defaults to **OpenAI** as the model provider. Its implementation remains consistent as it is built and run in the previous chapter. The key difference is that it is now integrated into the handler selection mechanism, which activates based on specific keywords in the input:

```
def handle_reasoning_customer(user_message, **kwargs):
    initial_query = user_message
    download("Chapter05", "customer_activities.csv")
    reasoning_steps = reason.chain_of_thought_reasoning(initial_query)
    return reasoning_steps
```

Consider an example user input from the PDPM:

```
Use reasoning to analyze customer activities so I can design custom travel
merchandise kits.
```

The output seems acceptable:

```
Agent: data01:OpenAI: To analyze the CTO's statement about different
memory types using Retrieval-Augmented Generation (RAG), we can break
down the memory types mentioned and their relevance to the hybrid travel
marketing campaign:

1. Memoryless: This refers to systems or processes that do not retain
any information from past interactions. In the context of a marketing
campaign, this would mean treating each customer interaction as a
standalone event without any historical context….
```

Let's now see how memory analysis will run with both model sources (OpenAI and DeepSeek).

Analysis (memory)

Both OpenAI and DeepSeek models handle memory-based customer profiles using neuroscientific-style categorizations. The function has been adapted to the handler selection process and contains a DeepSeek call:

```
def handle_analysis(user_message, **kwargs):
    from cot_messages_c6 import system_message_s1
```

```
    models = kwargs.get("models", "OpenAI")  # Default to OpenAI if not
provided
    if models == "DeepSeek" and deepseek==False:
      models="OpenAI"
    if models == "OpenAI":
      reasoning_steps = reason.make_openai_reasoning_call(
        user_message, system_message_s1)
    if models == "DeepSeek":
…
    return reasoning_steps
```

An example user input using the `Analysis` option in the `Reasoning` list could be:

```
The hotel was great because the view reminded me of when I came here to
Denver with my parents. I sure would have like to have received a custom
T-shirt as a souvenir. Anyway, it was warm and sunny so we had an amazing
time.
```

OpenAI's output contains a useful segment highlighting the emotional dimension related to the customer's wish for a personalized souvenir, which could help the product designer with their merchandise kit production endeavor:

```
…
Segment 2: "I sure would have like to have received a custom T-shirt as a
souvenir."
- Memory Tags: [Episodic Memory]
- Dimension: Emotional
- Sentiment Score: 0.4
- Explanation: Here the speaker expresses a personal wish or regret about
a missing souvenir from the event. Although it doesn't recount an actual
episode in detail, it still connects to the personal event and reflects a
feeling of slight disappointment, thereby engaging episodic memory and an
emotional dimension with a modestly negative sentiment.
```

DeepSeek's output, however, goes off track. It first finds the right task to do:

```
Okay let's see this through step by step now...

Alright, I need to tackle analyzing segments from the given user response
according to their detailed tagging system based on cognitive psychology
```

```
principles regarding STM vs LTM, semantic versus episodic memory, reality
vs fiction, among others plus dimensions such as Intellectual, Emotional,
Physical, along with assigning sentiments scores ranging from 0-1
reflecting positivity.

Let me start reading carefully paragraph-wise.

First sentence:"The hotel wasgreatbecauseviewremindedmeofwhencyamehere
todallas."
```

But it then gets lost and seems to struggle with formatting and coherence, introducing irregular spacing and even foreign characters:

```
…Butwait,theuser later talks about souvenirs
wantingcustomTshirtswhichmaybe indicatespositiveintent。…
但此处更多的是体验性的(Eating和Enjoying)=所以可能既有知识元素也有身体维度的食物
味道。但主要在这里是描述经历，因此属于Episode或语义吗？…
```

DeepSeek can certainly do better, but improving this result would require additional iterations of prompt refinement or selecting a more robust DeepSeek variant or API. Investing time in refining prompts carries some risk, as even then, the outcome may not meet your expectations. Whether to refine the prompt, switch to a DeepSeek API, explore another DeepSeek variant, or default to OpenAI should ultimately be decided collaboratively within the team and based on your project's needs.

Let's now move on to running the generation function.

Generation

The generation function (select `Generation` in the `Reasoning` list), active by default with **OpenAI**, **Agent**, and **Files** checked, supports the creation of engaging, memory-based customer messages:

```
def handle_generation(user_message, **kwargs):
    from cot_messages_c6 import (
    system_message_s1, generation, imcontent4, imcontent4b
)
    reasoning_steps = reason.memory_reasoning_thread(
        user_message, system_message_s1, generation,
        imcontent4, imcontent4b
    )
    return reasoning_steps
```

Let's consider a general user input as an example:

```
The hotel was great because the view reminded me of when I came here to
Denver with my parents. I sure would have like to have received a custom
T-shirt as a souvenir. Anyway, it was warm and sunny so we had an amazing
time.
```

OpenAI's output is an appealing customer-facing message, blending nostalgia and merchandising suggestions, accompanied by an appropriate custom T-shirt image:

```
Customer message: Dear Customer,
Experience the charm of Denver with a nostalgic hotel view and enjoy the
sunny weather. Explore the beautiful Denver Botanic Gardens and the iconic
Red Rocks Amphitheatre. Don't miss out on exclusive souvenirs from local
artists and a personalized T-shirt to remember your trip.
Best regards,
```

Figure 7.10: A personal image for a customer

Let's now create an image.

Creating an image

This functionality utilizes DALL-E to generate images, with the **Files** box checked. The function does not change beyond being adapted to the handler-selection mechanism, which activates this feature with the `Create` and `image` keywords in the user input:

```
def handle_image_creation(user_message, **kwargs):
    prompt = user_message
```

```
    image_url = reason.generate_image(
        prompt, model="dall-e-3", size="1024x1024",
        quality="standard", n=1
    )
    # Save the image locally
    save_path = "c_image.png"
    image_data = requests.get(image_url).content
    with open(save_path, "wb") as file:
        file.write(image_data)
    return "Image created"
```

The product designer could use it to ideate merchandising kits:

```
Create an image:  Create an image of a custom T-shirt with surfing in
Hawaii on big waves on it to look cool.
```

The output is a cool T-shirt that the production team could use and adapt for production:

Figure 7.11: Custom T-shirt design

We will now create freestyle prompts that are not triggered by any keywords or instructions.

Fallback handler (memory-based)

This general-purpose handler activates when no specific instruction or keyword matches the input. `handle_with_memory` runs with OpenAI and DeepSeek, depending on the model selected. The memory of a user dialogue is set with a global variable, `memory_enabled`, that is initialized at the beginning:

```
# Global variable to ensure memory is always used
memory_enabled = True  # Set to True to retain conversation memory
def handle_with_memory(messages, user_message, **kwargs):
    global memory_enabled  # Ensure global memory setting is used
```

The function will return a message and stop if `memory_enabled` is set to `False`:

```
    # If memory is disabled, respond with a message
    if not memory_enabled:
        return "Memory is disabled."
```

It will process the past messages of a user from the conversation history:

```
    # Extract all past messages (user + assistant) from the conversation
history
    conversation_history = [
        f"{msg['role'].capitalize()}: {msg['content']}"
        for msg in messages if "content" in msg
    ]
    # Combine all conversation history
    combined_history = "\n".join(conversation_history)

    # Append the latest user message to the history
    full_context = f"{combined_history}\nUser: {user_message}"
```

Then, the models are selected with OpenAI being the default model provider if no other model is selected in the IPython interface:

```
    models = kwargs.get("models", "OpenAI")  # Default to OpenAI if not
provided

    if models == "OpenAI":
        task_response = reason.make_openai_api_call(
            full_context, "system",
```

```
                "You are an assistant who executes the tasks you are asked to
do.", "user"
            )
        elif models == "DeepSeek":
            # Tokenize the full conversation history for DeepSeek
            …
            # Decode the output
            task_response = tokenizer.decode(outputs[0], skip_special_
tokens=True)
```

The response message is stored and returned:

```
        # Store bot response in memory
        messages.append({"role": "assistant", "content": task_response})

        return task_response
```

An example input by the PDPM could be the following:

```
We are new to production processes and experience issues with the
production of a customized travel bag by Supplier S100 for a trip to
Hawaii that should have an image of Hawaii on it with the name of the
customer. The production process contained operations 10,20,30,40,50 and
60. Operation 40 is the operation that prints the image and the name of
the customer on the bag. Supplier S100 lacks the special ink to perform
the operation. SupplierS200 has the ink but the production of operations
10 to 30 have already been performed. Instructions: think this problem
through and provide real-world production ideas to solve this issue.
```

OpenAI's answer is both acceptable and productive. Take your time to read the prompt and the response, which shows the transition of generative AI from NLP general tasks to zero-shot domain-specific tasks:

```
Agent: To address the issue of Supplier S100 lacking the special ink
needed for operation 40, while Supplier S200 has the ink, consider the
following real-world production solutions:

1. Outsource Operation 40 to Supplier S200:
- Coordinate with Supplier S200 to perform operation 40. This involves
transporting the semi-finished bags (after operations 10 to 30) to
Supplier S200 for printing.
- Ensure that Supplier S200 is capable of handling the volume and has the
```

```
necessary equipment to perform the operation to your quality standards.

2. Ink Transfer:
- If feasible, arrange for Supplier S200 to supply the special ink to
Supplier S100. This would allow Supplier S100 to continue with operation
40 without needing to transport the bags.
- Ensure that Supplier S100 has the necessary equipment and expertise to
use the ink provided by Supplier S200.

3. Collaborative Production:
- Establish a partnership between Supplier S100 and Supplier S200 where
both suppliers work together to complete the production process. This
could involve sharing resources, expertise, and logistics.

4. Logistics and Scheduling:
- Plan the logistics for transporting the bags between suppliers
efficiently to minimize delays.
- Adjust the production schedule to accommodate the additional time
required for transportation and coordination between suppliers.

5. Quality Control:
- Implement a quality control process to ensure that the printing meets
the required standards, regardless of which supplier performs the
operation.
- Conduct inspections before and after operation 40 to maintain quality
consistency.

6. Communication and Coordination:
- Maintain clear communication with both suppliers to ensure that
all parties are aligned on timelines, quality expectations, and
responsibilities.
- Use project management tools to track progress and address any issues
promptly.

7. Contingency Planning:
- Develop a contingency plan in case of further supply chain disruptions,
such as identifying alternative suppliers or inks that could be used in
the future.
```

DeepSeek's answer does not provide the same quality, although it contains some interesting points:

```
C.), Utilize Alternate Suppliers With Intermediaries Who Could Possibly
Acquire Required Inks From Other Sources Perhaps Via Importing Directly
Under Different Terms Agreements,

D ). Leverage Existing Inventory Wherever Feasible Reallocating Stockpiles
Smartly Across Production Lines Maybe Even Repurposing Materials When Not
Used For Primary Function Into Secondary Applications Which May Free Up
Costs And Resources While Mitigating Penalties Imposed By Delays,
```

Obtaining a better result would require further prompt design and output analysis cycles, evaluating DeepSeek models that are not distilled, such as DeepSeek-V3 or DeepSeek-R1. DeepSeek can surely do better, as demonstrated by using DeepSeek-R1 on `https://chat.deepseek.com/`, which produced the following output:

```
To resolve the production issue with the customized travel bags, consider
the following structured solutions:
### 1. **Outsource Operation 40 to Supplier S200**
   - **Process**:
     - S100 ships semi-finished bags (post-operation 30) to S200.
     - S200 performs operation 40 (printing) using their special ink.
     - Bags are returned to S100 to complete operations 50 and 60.
   - **Key Considerations**:
     - Use expedited shipping (e.g., air freight) to minimize delays.
     - Ensure packaging protects semi-finished goods during transit.
     - Negotiate cost-sharing for logistics between S100 and S200.

### 2. **Expedite Ink Procurement for S100**
   - **Action**:
     - Source the ink directly from S200 or an alternative supplier.
     - Prioritize urgent delivery (e.g., same-day courier).
   - **Feasibility Check**:
     - Verify if the ink is proprietary or available on the open market.
     - Address contractual barriers if S200 is a competitor.

### 3. **Supplier Collaboration for End-to-End Production**
   - **Option**:
```

```
        - S200 takes over all remaining steps (40, 50, 60) after S100
completes operations 10-30.
    - **Requirements**:
        - Confirm S200's capability to perform operations 50 and 60.
        - Align quality standards between suppliers to avoid inconsistencies.

### 4. **Process Re-engineering**
    - **Temporary Workaround**:
        - Explore alternative printing methods (e.g., heat transfer, patches)
if customer specs allow.
        - Assess feasibility of resequencing operations (e.g., printing
later), though this is unlikely.

### 5. **Proactive Communication and Risk Mitigation**
    - **Customer Transparency**:
        - Inform the customer of potential delays and offer solutions (e.g.,
partial deliveries, discounts).
    - **Prevent Future Issues**:
        - Implement dual sourcing for critical materials (e.g., ink).
        - Use inventory management tools to track material availability in
real-time.

### **Cost vs. Speed Tradeoff Analysis**
- **Outsourcing to S200**: Higher logistics costs but faster resolution if
suppliers are geographically close.
- **Expedited Ink**: Higher material costs but avoids production handoffs.

### **Recommended Approach**
1. **Prioritize Outsourcing Operation 40**:
    - This balances speed and feasibility if S200 can print without
disrupting the workflow.
2. **Backup Plan**:
    - If logistics are too slow, expedite ink procurement while negotiating
with S200 for a bulk discount.

By combining supplier collaboration, agile logistics, and proactive
planning, the production delay can be minimized while maintaining quality
and customer trust.
```

Read the prompts and outputs in this section carefully. For security and privacy reasons, we are using only a locally installed, distilled Hugging Face open source version of DeepSeek-R1. However, you could use the online version of DeepSeek for certain tasks, such as the production example in this section, if you have the necessary permissions, just as you would with ChatGPT or any other online platform. Depending on your project's specifications, you could also explore US-based DeepSeek APIs or alternative deployment approaches.

In any case, both OpenAI and DeepSeek are capable of delivering effective production solutions when correctly prompted and when the most appropriate model version is selected. Generative AI has clearly entered a new phase!

Summary

In this chapter, we moved further along our journey into generative AI systems. First, we took the time to digest the arrival of DeepSeek-R1, a powerful open source reasoning model known for innovative efficiency improvements in training. This development immediately raised a critical question for project managers: should we constantly follow real-time trends or prioritize maintaining a stable system?

To address this challenge, we developed a balanced solution by building a handler selection mechanism. This mechanism processes user messages, triggers handlers within a handler registry, and then activates the appropriate AI functions. To ensure flexibility and adaptability, we updated our IPython interface, allowing users to easily select between OpenAI and DeepSeek models before initiating a task.

This design allows the GenAISys administrator to introduce new experimental models or any other function(non-AI, ML, or DL) while maintaining access to proven results. For instance, when analyzing user comments, administrators can run tasks using the reliable OpenAI model while simultaneously evaluating the DeepSeek model. Administrators can also disable specific models when necessary, providing a practical balance between stability and innovation, which is crucial in today's fast-paced AI environment.

To achieve this balance practically, we began by installing and running DeepSeek-R1-Distill-Llama-8B in an independent notebook, demonstrating its capabilities through production-related examples. We then integrated this distilled model into our GenAISys, creating a need for enhanced flexibility and scalability.

The introduction of the handler selection mechanism and the structured handler registry ensures that our system can scale effectively and indefinitely. Each handler follows a unified, modular format, enabling easy management, activation, or deactivation by administrators. We demonstrated these handlers through a series of practical prompts related to product design and production.

We are now positioned to expand and scale our GenAISys, adding new features within this adaptable framework. In the next chapter, we'll continue this journey by connecting our GenAISys to the broader external world.

Questions

1. DeepSeek-V3 was trained with zero-shot examples. (True or False)
2. DeepSeek-R1 is a reasoning model. (True or False)
3. DeepSeek-R1 was first trained with RL-only. (True or False)
4. DeepSeek-R1-Distill-Llama-8B is the teacher of DeepSeek-R1. (True or False)
5. DeepSeek-V3 was enhanced with DeepSeek-R1, which was derived from DeepSeek. (True or False)
6. A handler registry that contains a list of handlers for all the AI functions is scalable. (True or False)
7. A handler selection mechanism that processes user messages makes the GenAISys highly flexible. (True or False)
8. Generative AI models such as OpenAI and DeepSeek reasoning models solve a wide range of problems with no additional training. (True or False)
9. A GenAISys with a solid architecture is sufficiently flexible to be expanded in terms of models and tasks to perform. (True or False)

References

- *DeepSeek-V3 Technical Report.* `https://arxiv.org/abs/2412.19437`
- Vaswani, A., Shazeer, N., Parmar, N., Uszkoreit, J., Jones, L., Gomez, A. N., Kaiser, Ł., & Polosukhin, I. (2017). *Attention Is All You Need.* Advances in Neural Information Processing Systems, 30, 5998–6008. Available at: `https://arxiv.org/abs/1706.03762`
- DeepSeekAI, Daya Guo, Dejian Yang, et al: `https://arxiv.org/abs/2501.12948`

- Touvron, H., Lavril, T., Izacard, G., Martinet, X., Lachaux, M.-A., Lacroix, T., Rozière, B., Goyal, N., Hambro, E., Azhar, F., Rodriguez, A., Joulin, A., Grave, E., & Lample, G. (2023). *LLaMA: Open and Efficient Foundation Language Models.* `https://arxiv.org/abs/2302.13971`)
- Grattafiori, A., Dubey, A., Jauhri, A., Pandey, A., Kadian, A., Al-Dahle, A., Letman, A., Mathur, A., Schelten, A., Vaughan, et al. (2024). *The Llama 3 Herd of Models.* `https://arxiv.org/abs/2407.21783`
- Hugging Face:
  - `https://huggingface.co/docs/transformers/index`
  - `https://huggingface.co/deepseek-ai/DeepSeek-R1-Distill-Llama-8B`
  - `https://huggingface.co/docs/inference-endpoints/en/security`
- Unsloth AI: `https://unsloth.ai/`

Further reading

- Frantar, E., Ashkboos, S., Hoefler, T., & Alistarh, D. (2023). *GPTQ: Accurate Post-Training Quantization for Generative Pre-trained Transformers* `https://arxiv.org/abs/2210.01774`)
- NVIDIA's data center GPUs: `https://www.nvidia.com/en-us/data-center/`

8

GenAISys for Trajectory Simulation and Prediction

As AI's role continues to expand, trajectory analysis has permeated all human activity, from pizza deliveries to genome sequencing. This chapter introduces city-scale mobility prediction, highlighting how missing or noisy coordinates can undermine real-world applications in deliveries, disaster management, urban planning, and epidemic forecasting. The architecture of our mobility system draws inspiration from the innovative work of Tang et al. (2024).

We will first build and integrate an advanced trajectory simulation and prediction pipeline into our GenAISys using the `1_Trajectory_simulation_and_prediction.ipynb` notebook. The main objective is to address the challenge of modeling human mobility, both short- and long-term, by leveraging synthetic data generation and **large language models** (**LLMs**). We then demonstrate how to build upon this idea using Python-based solutions, complete with a custom synthetic grid generator that simulates random trajectories through a two-dimensional city map, deliberately inserting missing data for testing. These random trajectories could represent deliveries or other sequences, such as travel packages (custom bags or booklets) for an online travel agency.

Next, we will build a multistep orchestrator function that merges user instructions, the synthetic dataset, and domain-specific messages before passing them to an LLM-driven reasoning thread. The model will detect and predict unknown positions marked by placeholder values (such as `999, 999`), filling these gaps through contextual interpolation. This approach demonstrates the interpretability of text-based predictions while maintaining a systematic chain of thought, including debugging steps such as logging missing points before producing the final JSON output.

To support robust user interaction, we will integrate the trajectory pipeline into the GenAISys multihandler environment we've built, allowing requests for "mobility" instructions to trigger the creation and analysis of trajectories. We will implement a trajectory simulation and prediction interface. Visualization components are incorporated, automatically producing and displaying the resulting path (including direction arrows, missing data markers, and coordinate fixes) as a static image. The synergy between data generation, LLM inference, and the user interface showcases the end-to-end viability of our method, empowering users to apply trajectory simulation and prediction across different domains as needed.

This chapter provides a blueprint for coupling synthetic trajectory datasets with a prompt-driven LLM approach in the GenAISys. By following the design patterns described by Tang et al., we will explore how purely text-oriented models can excel at spatial-temporal reasoning with minimal structural modifications. Bridging mobility simulation and user-friendly interfaces can provide highly interpretable, fine-grained predictions for a variety of mobility analytics scenarios.

This chapter covers the following topics:

- Trajectory simulations and predictions
- Building a trajectory simulation and prediction function
- Adding mobility intelligence to the GenAISys
- Running the mobility-enhanced GenAISys

Let's begin by defining the scope of the trajectory simulation and prediction framework.

Trajectory simulations and predictions

This section is inspired by *Instruction-Tuning Llama-3-8B Excels in City-Scale Mobility Prediction* by Tang et al. (2024). We will explore the essential background on the challenges of human mobility prediction, the paper's key contributions, and how these ideas can be translated into practical Python implementations.

Human mobility prediction focuses on forecasting where and when individuals (or groups) will travel, and it plays a critical role in an expanding set of domains, including the following:

- **Disaster response**, for predicting the paths of wildfires, population movements during crises, or the impacts of earthquakes
- **Urban planning**, for modeling short- and long-term mobility patterns to help city planners optimize public transport and infrastructure

- **Epidemic forecasting**, for simulating and predicting the spread of infectious diseases in a region

In our case, we will first apply mobility prediction to the delivery of customized products (e.g., bags, T-shirts, and booklets) for an online travel agency's customers.

Traditionally, these predictions relied on specialized machine learning models, such as **recurrent neural networks** (**RNNs**) with attention mechanisms or **graph neural networks** (**GNNs**). While these techniques can be effective, they often require labor-intensive feature engineering and are not easily generalizable across diverse locations or time horizons (e.g., short-term versus long-term predictions).

Let's now examine the key challenges motivating the use of LLMs to address these issues.

Challenges in large-scale mobility forecasting

Cutting-edge LLMs offer promising solutions to several challenges that have historically plagued traditional mobility analysis and prediction systems:

- **Long-term versus short-term forecasts:** Predicting the next few steps (short-term) often relies on temporal recurrences and immediate contextual information. However, extending this to multi-day, city-scale horizons introduces additional complexities, such as changes in user behavior, variations in daily routines, holidays, or unexpected events.
- **Generalization across cities:** A model trained on data from City A may fail when exposed to the unique geography, population density, or cultural travel habits of City B. True city-scale mobility solutions must be robust enough to handle these differences.
- **Computational constraints:** Real-world mobility datasets, especially those representing entire metropolitan areas, can be enormous. Training sophisticated deep learning models or LLMs can become computationally expensive.
- **Data quality and missing data:** Large-scale mobility datasets often have noise or missing coordinates. Handling "gaps" in user trajectories (e.g., from GPS dropout or anonymization processes) is a significant challenge.

While LLMs are not perfect, they provide an effective alternative to traditional models by addressing these key obstacles with minimal manual feature engineering. Let's see how.

From traditional models to LLMs

The journey from traditional approaches to LLMs can be traced through a few groundbreaking shifts. Traditional approaches consumed extensive human resources to design heuristics, engineer features, and implement complex domain-specific solutions. In contrast, recent breakthroughs in generative AI—such as Llama 3, GPT-4o, Grok 3, DeepSeek-V3, and DeepSeek-R1—have opened exciting new avenues in reasoning and multimodal machine intelligence. And make no mistake—this is just the beginning! Recent research highlights how these models can generalize well beyond text-based tasks, excelling in the following:

- Time-series prediction
- Zero-shot or few-shot adaptation to new tasks
- Data preprocessing

Recent research has shown that LLMs, when guided by carefully crafted prompts or lightweight fine-tuning, can even surpass specialized models in city-scale, long-horizon trajectory prediction. In this chapter, we'll demonstrate effective results with zero-shot prompting—without any additional fine-tuning—using GPT-4o.

To understand this promising direction clearly, however, let's first examine the key contributions of the paper that served as a basis for this chapter.

Key contributions of the paper

It took a team consisting of Tang, P., Yang, C., Xing, T., Xu, X., Jiang, R., and Sezaki, K. (2024) to take LLMs to the next level through three pivotal innovations.

Reformulating trajectory prediction as a Q&A

Instead of passing raw coordinate sequences into a standard regression or classification model, the authors transform the input into a question that includes the following:

- An **instruction block** clarifying the domain context (city grid, coordinate definitions, day/time indexing)

- A **question block** providing historical mobility data with placeholders for missing locations
- A request to generate the prediction results in a predefined, structured JSON format

This Q&A style leverages the LLM's inherent ability to read instructions and produce structured outputs.

Then, they fine-tuned the LLM.

Instruction tuning for domain adaptation

Instruction tuning is a technique where the LLM is fine-tuned with carefully designed prompts and answers, teaching it to produce domain-specific outputs while still retaining its general language reasoning capabilities. The authors showcase that even if you use only a fraction of the mobility dataset for fine-tuning, the model can still generalize to new users or new cities. In our case, we attained acceptable results without a dataset.

Surprisingly enough, as we will see when we build the Python program in the *Building the trajectory simulation and prediction function* section, we achieve strong results even with a zero-shot, no-fine-tuning approach, leveraging GPT-4o's exceptional reasoning capability without needing any domain-specific fine-tuning data.

The mobility research team then solved the issue of missing data.

Handling missing data

A common challenge in mobility datasets is the presence of missing coordinates, typically marked with placeholder values such as 999. The LLM-based system is tasked explicitly with filling in these gaps, effectively performing spatiotemporal imputation. Naturally, this approach is not without limitations, which we'll clearly illustrate through practical examples when we run our mobility simulation. But before exploring these boundaries, let's first dive into building our solution.

In the next section, we will develop a trajectory (mobility) simulation and analysis component using OpenAI models. We will then integrate this mobility function into **Layer 3** of our GenAISys, as illustrated in *Figure 8.1* with function **F4.1.** We will also update **Layer 2** to register the handler and ensure it can be activated at the IPython interface level in **Layer 1**.

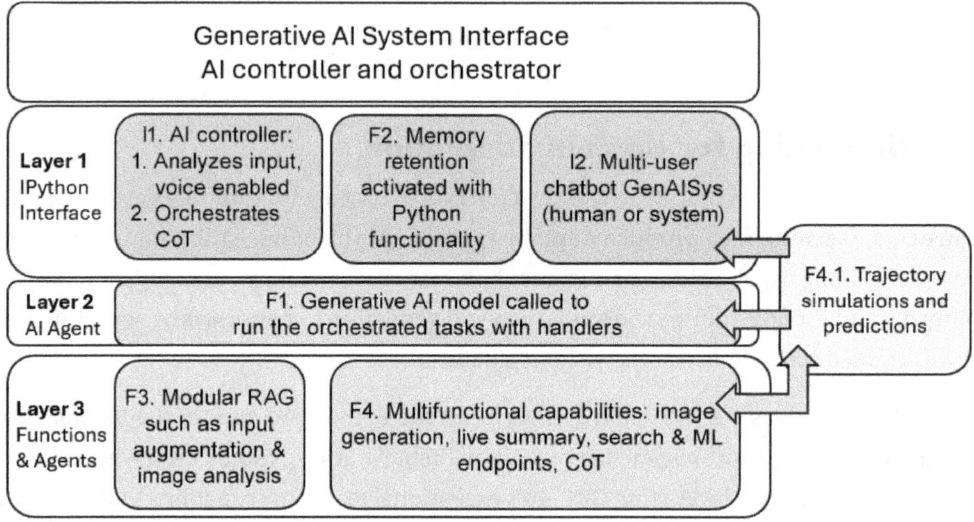

Figure 8.1: Integrating trajectory simulations and predictions

Once the trajectory simulation and prediction component is integrated into our GenAISys, it can be applied to deliveries and a wide range of mobility-related tasks. We will start by modeling the delivery of customized goodies—such as branded bags, T-shirts, and booklets—for customers of an online travel agency, and then explore other potential applications. For now, let's build our trajectory simulation!

Building the trajectory simulation and prediction function

The goal of this section is to create a robust trajectory simulation, prepare the predictive functions, and run an OpenAI LLM to analyze synthetic trajectory data and predict missing coordinates. Later, in the *Adding mobility intelligence to the GenAISys* section, we'll integrate this into our comprehensive GenAISys framework.

Open the `1_Trajectory_simulation_and_prediction.ipynb` notebook within the Chapter08 directory on GitHub (`https://github.com/Denis2054/Building-Business-Ready-Generative-AI-Systems/tree/main`). The initial setup mirrors the environment configuration in `Chapter07/GenAISys_DeepSeek.ipynb` and includes the following:

- File downloading script
- OpenAI setup
- Chain-of-thought environment setup

We will build the program in three main steps, as shown in *Figure 8.2*:

- Creating the grid and trajectory simulation to generate real-time synthetic data
- Creating a mobility orchestrator that will call the trajectory simulation, import the messages for the OpenAI model, and call the analysis and prediction messages for the OpenAI model
- Leveraging OpenAI's model for trajectory analysis and predictions, called via the mobility orchestrator

The mobility orchestrator will be added to the handlers registry in our GenAISys in the *Adding mobility intelligence to the GenAISys* section and managed by the handler selection mechanism when activated by the IPython interface. In this section, we will call the mobility orchestrator directly.

Figure 8.2 articulates the relationship between the mobility orchestrator, the trajectory simulator, and the generative AI predictor. This mixture of agents maintains close alignment with the framework of trajectory analysis and predictions.

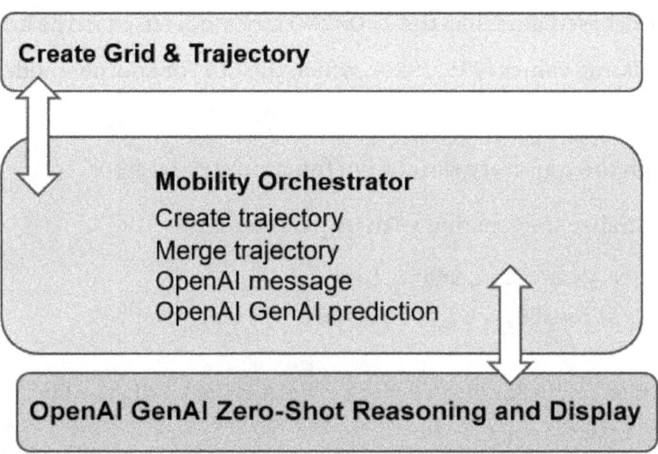

Figure 8.2: The functions of the mobility orchestrator

We will first begin by creating the trajectory simulation.

Creating the trajectory simulation

The reference paper by Tang et al. demonstrates how an LLM can be instruction-tuned to fill missing trajectory coordinates and predict future positions in a grid-based city map. Note that in our case, we will leverage the power of the OpenAI API message object to achieve an effective result with zero-shot prompts in real time, within the framework of the paper.

One important step in their methodology involves having *(day, timeslot, x, y)* records, with some coordinates possibly missing (e.g., `999, 999`) to indicate unknown positions.

The function that we will write, `create_grid_with_trajectory()`, essentially simulates a smaller-scale version of this scenario by doing the following:

1. Generate a two-dimensional grid representing a city (default: 200×200).
2. Create random agent trajectories within the grid over a certain number of points.
3. Intentionally insert missing coordinates (marked as `(999, 999)`) to simulate real-world data gaps.
4. Plot and save the trajectory visualization, highlighting direction with arrows and labels for missing data.

This kind of synthetic generation is useful for testing or proof-of-concept demos, echoing the spirit of the paper:

- You have grid-based data, like the 200×200 city model used in the article
- You inject missing values (`999, 999`), which the LLM or another model can later attempt to fill in

Let's now go through the trajectory simulation function step by step:

1. Let's first initialize the function with its parameters:

```
def create_grid_with_trajectory(
    grid_size=200, num_points=50, missing_count=5
):
    grid = np.zeros((grid_size, grid_size), dtype=int)
    trajectory = []
```

   The parameters are as follows:

   - `grid_size=200`: The size of the grid along one axis (so the grid is 200×200)
   - `num_points=50`: How many trajectory points (or steps) will be generated

- `missing_count=5`: How many of those points will be deliberately turned into missing coordinates `(999, 999)`

2. We now create the grid:

- `grid = np.zeros((grid_size, grid_size), dtype=int)` creates a two-dimensional array of zeros (of the `int` type). Think of `grid[x][y]` as the status of that cell, initially 0.
- `trajectory = []`: will hold tuples of the form *(day, timeslot, x, y)*.

This mirrors the discretized city concept in the paper, where each *(x, y)* cell might represent a zone within the city.

3. We can now set the initial state of the agent:

```
x = random.randint(0, grid_size - 1)
y = random.randint(0, grid_size - 1)
day = random.randint(1, 365)
timeslot = random.randint(0, 47)
```

- **Random start**: The agent's initial location *(x, y)* is chosen randomly anywhere on the grid.
- **Time setup**: A random day between 1 and 365 and a random timeslot between 0 and 47 is selected, aligning with the paper's time-slicing approach, where each day is divided into multiple discrete time slots.

4. We now determine the movement directions and turn probability:

```
directions = [(0, 1), (1, 0), (0, -1), (-1, 0)]
current_dir_index = random.randint(0, 3)
turn_weights = {-1: 0.15, 0: 0.70, 1: 0.15}
```

This structure is a classical mobility agent framework:

- `directions`: Represents four possible directions—up, right, down, and left.
- `current_dir_index`: Picks which of the four directions the agent faces initially.
- `turn_weights`: Probability distribution dictating how likely the agent is to turn left (`-1`), go straight (`0`), or turn right (`1`) at each step. In our case, there is a 15% chance of turning left, a 70% chance of continuing, and a 15% chance of turning right. This introduces randomness in how the agent moves and is a simple approximation of human or agent-like mobility patterns.

5. We are ready to generate the trajectory:

```
for _ in range(num_points):
    turn = random.choices(list(turn_weights.keys()),
        weights=list(turn_weights.values()))[0]
    current_dir_index = (current_dir_index + turn) % \
        len(directions)
    dx, dy = directions[current_dir_index]
    new_x = x + dx
    new_y = y + dy
    ...
    trajectory.append((day, timeslot, x, y))
    grid[x, y] = 1
    timeslot = (timeslot + random.randint(1, 3)) % 48
```

Let's go through the actions of our virtual agent:

- **Choosing a turn**: Based on `turn_weights`, the agent randomly decides whether to continue in the same direction, turn left, or turn right.
- **Updating the coordinates**:
    1. `dx`, `dy` are the increments along *x* and *y* for the chosen direction.
    2. The new location, `(new_x, new_y)`, is computed.
- **Checking the boundary conditions**: If `(new_x, new_y)` is outside `[0, grid_size-1]`, the code finds a valid direction or reverts to the old position to keep the agent inside the grid.
- **Recording the trajectory**:
    3. `(day, timeslot, x, y)` is appended to `trajectory`.
    4. Mark `grid[x, y]` as 1, signifying a visited cell.
- **Updating the time**: `timeslot = (timeslot + random.randint(1, 3)) % 48`: The timeslot jumps from 1 to 3 steps, staying in `[0, 47]`.

6. We now need to introduce the missing data, which will be the basis for the generative AI predictions:

```
missing_indices = random.sample(range(len(trajectory)),
                                min(missing_count,
                                len(trajectory)))
```

```
    for idx in missing_indices:
        d, t, _, _ = trajectory[idx]
        trajectory[idx] = (d, t, 999, 999)
```

The missing points are determined in two steps:

1. **Selecting the missing points**: Randomly choose `missing_count` points from the total `num_points` of the trajectory.
2. **Replacing the missing points with 999, 999**: For each chosen index, replace the valid `(x, y)` with `999, 999`.

In the paper, the authors define `999, 999` as the signal for unknown or missing coordinates that the LLM must later fill in. This code snippet simulates exactly that scenario—some coordinates go missing, requiring an imputation or prediction step.

We want to add a visualization function next that will help the user to see the trajectory and its missing points.

Visualizing the trajectory simulator

We will plot the grid and trajectory in Matplotlib:

```
x_coords = [x if x != 999 else np.nan for _, _, x, y in trajectory]
y_coords = [y if y != 999 else np.nan for _, _, x, y in trajectory]

plt.figure(figsize=(8, 8))
plt.plot(x_coords, y_coords, marker='o', linestyle='-',
         color='blue', label="Agent Trajectory")
...
plt.quiver(...)
...
plt.title("Agent Trajectory with Direction Arrows and Missing Data")
plt.xlabel("X coordinate")
plt.ylabel("Y coordinate")
plt.grid(True)
plt.legend()
plt.savefig("mobility.png")
plt.close()
```

Let's go through the visualization process:

- **Coordinates for plotting**: Converts missing `999, 999` values into `np.nan` so that Matplotlib will break the line and not connect them visually
- **Plotting with colors, arrows, and text:**
    1. The trajectory line is drawn in blue.
    2. Quiver arrows (`plt.quiver`) show the direction from each point to the next.
    3. Missing data points are highlighted with an `'X'` marker in magenta.
- **Titles and axes:** Labeling and legend for clarity
- **Save and close**: Saves the figure as `mobility.png`

Such plotting mirrors the style in the paper's *Case Study* section (*Section 4.4*), where the authors compare real versus predicted trajectories. Here, you're simply illustrating the synthetic path as well as the visual indications of missing data.

Output of the simulation function

The output of the function that we will process contains the grid and the trajectory:

```
    return grid, trajectory
```

These two variables will contain what our generative AI model needs to make a prediction:

- `grid`: A two-dimensional array marking the visited path
- `trajectory`: A list of *(day, timeslot, x, y)* tuples, with some replaced with `999, 999`

This final result will be fed into an LLM-based approach (such as the one described in the paper) with an OpenAI generative AI model that can produce an acceptable output in a zero-shot process. We will now begin to process the trajectory simulation.

Creating the mobility orchestrator

The trajectory simulation has generated the grid, the trajectory, and the missing coordinates in the trajectory. We will now develop the orchestrator function that integrates both the trajectory simulation and the predictive capabilities of the OpenAI model. We'll call this orchestrator `handle_mobility_orchestrator()`.

This orchestrator aligns with the method outlined by Tang et al. (2024) in their paper *Instruction-Tuning Llama-3-8B Excels in City-Scale Mobility Prediction*. It uses a form of context chaining to first generate synthetic data and then pass that data as context to the LLM for prediction. Its purpose is straightforward yet powerful, performing three critical functions:

- **Generating synthetic mobility data:** It invokes the `create_grid_with_trajectory ()` function to simulate a trajectory with possible missing points
- **Preparing data for an LLM call:** It formats the new trajectory data into a JSON string, appends it to the user's message, and then calls the reasoning function—presumably the LLM-based solution or orchestration logic (`reason.mobility_agent_reasoning_thread()`)
- **Returning a structured response:** It returns the final results clearly (`reasoning_steps`), to include both the newly generated trajectory data and the LLM reasoning steps

This approach remains true to the *Instruction-Tuning Llama-3-8B Excels in City-Scale Mobility Prediction* paper, where the authors emphasize creating structured input data—such as trajectories with missing points—and then passing it to an LLM for completion or prediction.

Let's now go through the orchestrator step by step:

1. First, the orchestrator function is initialized with the necessary parameters:

   ```
   def handle_mobility_orchestrator(
       muser_message1, msystem_message_s1, mgeneration,
       mimcontent4, mimcontent4b
   ):
   ```

   Immediately, it invokes the trajectory simulation function we built previously:

   ```
   grid, trajectory = create_grid_with_trajectory(
       grid_size=200, num_points=50, missing_count=5
   )
   ```

2. We now convert and process the trajectory in JSON:

   ```
   trajectory_json = json.dumps({"trajectory": trajectory}, indent=2)
       #print("Trajectory Data (JSON):\n", trajectory_json)
       muser_message = f"{muser_message1}\n\nHere is the trajectory
   data:\n{trajectory_json}"
   ```

   This code takes care of converting the trajectory and augmenting the user message:

   - Converting the trajectory to JSON:

     1. `trajectory_json` becomes a serialized version of the data so it can be embedded in text messages or API calls.

2. Under the hood, it's just `{"trajectory": [...list of (day, timeslot, x, y)...]}`.

- Augmenting the user message:
  1. The function takes the original user message, (`muser_message1`), and appends the newly generated trajectory data to it.
  2. This ensures the model (or reasoning thread) sees the complete context—*both the user's original query and the synthetic data*—when generating predictions or completions.

This step closely mirrors the Q&A-style interaction presented by Tang et al. (2024), where the trajectory data—marked clearly by placeholders (`999, 999`)—is delivered directly to the model.

3. With the context clearly defined, the orchestrator calls the mobility reasoning function (which we'll build next):

```
reasoning_steps = reason.mobility_agent_reasoning_thread(
    muser_message, msystem_message_s1, mgeneration,
    mimcontent4, mimcontent4b
)
```

Here's what happens behind the scenes:

- `reason.mobility_agent_reasoning_thread(...)` processes the mobility prediction logic through the selected LLM (such as GPT-4o)
- The provided arguments (`msystem_message_s1`, `mgeneration`, `mimcontent4`, and `mimcontent4b`) represent clear instructions and specific context for the generative AI model, guiding its reasoning and predictions

This mirrors the approach described in Tang et al.'s paper, where the model receives structured input data and is prompted to infer missing trajectories or forecast next movements.

4. Finally, the trajectory is added to the reasoning steps to provide a complete response:

```
reasoning_steps.insert(
    0, ("Generated Trajectory Data:", trajectory)
)
return reasoning_steps
```

Next, let's develop the AI reasoning function that the handler registry will call upon.

Preparing prediction instructions and the OpenAI function

In this section, we'll develop the function that allows our GenAISys to process mobility-related user messages. Specifically, we'll implement a function named `handle_mobility(user_message)` that integrates seamlessly into the AI functions of our GenAISys.

We'll approach this task in two main parts:

1. **Message preparation:** Clearly structuring the messages that guide the generative AI model
2. **Implementing these messages in the OpenAI API call:** Leveraging the structured messages in the AI reasoning thread

This aligns closely with the trajectory completion methodology described in *Instruction-Tuning Llama-3-8B Excels in City-Scale Mobility Prediction*, where structured prompts significantly enhance predictive accuracy.

Message preparation

We have four main message variables to send to the OpenAI function:

- `msystem_message_s1`: System message
- `mgeneration`: Generation message
- `mimcontent4`: Additional context
- `muser_message1`: User message

They each serve a distinct purpose in the final prompt that goes to the LLM (GPT-4o or similar) for the prediction task. The system message will set the stage for the task.

System message

The system message sets the overall context and constraints for the LLM, ensuring the model clearly understands its main objectives. The system message is stored in `msystem_message_s1`. We first specify the role of the model:

```
msystem_message_s1 = """
You are GPT-4o, an expert in grid-based mobility analysis. Your task
is to analyze the provided trajectory dataset and **identify missing
coordinates** flagged as `999,999`, then predict their correct values.
```

Now, we clearly detail the tasks expected in explicit natural language:

```
**Task:**
1. **Process only the dataset provided in the user input. Do not generate
or use your own sample data.**
2. Identify **every single** instance where `x` or `y` is `999`, including
consecutive and scattered occurrences.
3. Predict the missing coordinate values based on the trajectory pattern.
4. **Do not modify, reorder, or filter the data in any way**—your response
must reflect the dataset exactly as given except for replacing missing
values.
5. Before responding, **validate your output** against the original
dataset to confirm completeness and accuracy.
6. Maintain the exact order of missing values as they appear in the
dataset.
7. Include a debugging step: **first print the list of detected missing
values before structuring the final JSON output**.
```

The output format is specified:

````
**Output Format:**
```json
{"predicted_coordinates": [[day, timeslot, x, y], ...]}
```
````

These instructions mirror the approach of the paper we are implementing—the system message clarifies the *role* of the model and the *task instructions*, effectively reducing confusion or *hallucination*. The paper shows how a well-structured instruction block significantly boosts accuracy. Now, we can build the generation message.

Generation message

This secondary prompt provides generation instructions that will reinforce how the model should handle the data:

```
mgeneration = """
Scan the user-provided trajectory data and extract **every** point where
either `x` or `y` equals `999`.
You must process only the given dataset and not generate new data.
Ensure that all missing values are explicitly listed in the output without
skipping consecutive values, isolated values, or any part of the dataset.
**Before responding, verify that all occurrences match the input data
```

```
exactly.**

Then, predict the missing values based on detected trajectory movement
patterns. **Provide a corrected trajectory with inferred missing values.**

To assist debugging, **first print the detected missing values list as a
pre-response validation step**, then return the structured JSON output.
"""
```

This prompt focuses on scanning for missing values, ensuring none are skipped. Then, it addresses the next step: provide the corrected trajectory with inferred missing values.

Additional context

To make sure we obtain what we wish, we will now add additional context. The role of this additional context is to supplement the system/generation messages with domain-specific context:

```
mimcontent4 = """
This dataset contains spatial-temporal trajectories where some coordinate
values are missing and represented as `999,999`. Your goal is to
**identify these missing coordinates from the user-provided dataset
only**, then predict their correct values based on movement patterns.
Ensure that consecutive, isolated, and scattered missing values are not
omitted. **Before generating the final response, validate your results and
confirm that every missing value is properly predicted.**
"""
```

This additional context further guides the generative AI model toward precise predictions. We will now engineer a user message to further instruct the model.

User message

It's time to emphasize the instructions further to make sure we provide even more context to the input. The user message expresses the user's *explicit* request. It references the actual dataset with missing points. Realistically, in your code, you'll append or embed the actual trajectory data (with `999, 999` placeholders) *before* passing it to the generative AI model:

```
muser_message1 = """
Here is a dataset of trajectory points. Some entries have missing
coordinates represented by `999,999`.
You must process only this dataset and **strictly avoid generating your
own sample data**.
```

```
Please identify **all occurrences** of missing coordinates and return
their positions in JSON format, ensuring that no values are skipped,
omitted, or restructured. Then, **predict and replace** the missing values
using trajectory movement patterns.

Before returning the response, **first output the raw missing coordinates
detected** as a validation step, then structure them into the final JSON
output with predicted values.
"""
```

Let's fit the message together.

Fitting the messages together

The four messages converge to direct the generative AI model:

- The system message (`msystem_message_s1`) sets the stage and enforces top-level policies
- The generation message (`mgeneration`) clarifies the approach for scanning, verifying, and predicting
- The additional content (`mimcontent4`) ensures domain clarity
- Finally, the user's message (`muser_message1`) includes the data that needs to be processed (the partial or missing trajectory)

Together, they form the structure of a zero-shot advanced generative model's prediction.

Now, let's fit the message into the OpenAI API function. These messages are stored in `commons/cot_messages_c6.py` to be imported by the OpenAI API function.

Implementing the messages into the OpenAI API function

We will now create an AI mobility function for the *AI function* section in our GenAISys when we integrate it:

```
def handle_mobility(user_message):
```

We will now import the messages we stored in `cot_messages_c6.py`:

```
    from cot_messages_c6 import (
        msystem_message_s1, mgeneration, mimcontent4,muser_message1
    )
```

We'll now complete the function so that we can call it further in this program by plugging the messages in the generative AI call and return the reasoning steps:

```
    #call Generic Synthetic Trajectory Simulation and Predictive System
    reasoning_steps = handle_mobility_orchestrator(
        muser_message1, msystem_message_s1, mgeneration,
        mimcontent4, mimcontent4b
    )
    return reasoning_steps

    mimcontent4b=mimcontent4
```

We can now call the mobility orchestrator and return its reasoning steps:

```
    #call Generic Synthetic Trajectory Simulation and Predictive System
    reasoning_steps = handle_mobility_orchestrator(
        muser_message1, msystem_message_s1, mgeneration,
        mimcontent4, mimcontent4b)
    return reasoning_steps
```

We then create the `handle_mobility_orchestrator` function in the `reason.py` library we have been implementing in the previous chapters of this book. We first create the function:

```
# Implemented in Chapter08
def mobility_agent_reasoning_thread(
    input1,msystem_message_s1,mumessage4,mimcontent4,mimcontent4b
):
```

Then, we initialize the reasoning steps to display them in `VBox`:

```
    steps = []

    # Display the VBox in the interface
    display(reasoning_output)

    #Step 1: Mobility agent
    steps.append("Process: the mobility agent is thinking\n")
    with reasoning_output:
        reasoning_output.clear_output(wait=True)
        print(steps[-1])  # Print the current step
```

We then plug the messages received into the standard `make_openai_call` that we have been using in the previous chapters and return the steps:

```
    mugeneration=msystem_message_s1 + input1
    mrole4 = "system"
    mcontent4 = mimcontent4
    user_role = "user"
    create_response = make_openai_api_call(
        mugeneration,mrole4,mcontent4,user_role
    )
    steps.append(f"Customer message: {create_response}")
    return steps
```

We are now ready to run the trajectory simulation and prediction.

Trajectory simulation, analysis, and prediction

With our mobility functions built and clearly defined, we can now run the complete trajectory pipeline—generating synthetic trajectory data, identifying missing coordinates, and predicting them with a zero-shot LLM. This section will demonstrate the end-to-end execution and interpretation of results.

We'll use a simple, generic prompt to initiate the mobility analysis:

```
user_message="Check the delivery path"
output=handle_mobility(user_message)
```

This triggers the entire pipeline we set up previously, from synthetic data generation to coordinate predictions.

To clearly illustrate the trajectory and missing points, the system generates a visual plot (`mobility.png`). We can display this image directly:

```
# Display mobility.png if it exists and the "Mobility" instruction is
selected
if os.path.exists("mobility.png"):
    original_image = PILImage.open("mobility.png")
    display(original_image)
```

The output contains the grid, the trajectory, and the missing data, as shown in *Figure 8.3*:

Figure 8.3: Trajectory and missing data

The output is plotted with colors, arrows, and text as designs:

- Green is the starting point
- The trajectory line is drawn in blue
- Quiver arrows (`plt.quiver`) in red show the direction from each point to the next
- Missing data points are highlighted with an **x** marker in magenta

Then, we print the raw output:

```
print(output)
```

The output displayed is a single, unstructured line containing trajectory data and predictions:

```
[('Generated Trajectory Data:', [(50, 28, 999, 999), (50, ….
```

Clearly, we need to present this data more intuitively. Let's create a function to display a nice, formatted response:

```
def transform_openai_output(output):
    """
    Takes the 'output' (a list/tuple returned by OpenAI) and transforms
    it into a nicely formatted multiline string.
    """
```

The code breaks the output into well-presented lines:

```
    …
    lines = []
    …
    # Join all lines into one neatly formatted string
    return "\n".join(lines)
```

We then call the function to obtain the formatted output:

```
pretty_response = transform_openai_output(output)
print(pretty_response)
```

The output contains the three-step process we built:

1. Display the trajectory.
2. Isolate the missing data.
3. Make predictions for the missing data.

The output first contains the trajectory:

```
Generated Trajectory Data:
  (228, 6, 999, 999)
  (228, 7, 69, 79)
  (228, 9, 70, 79)
  (228, 11, 71, 79)
```

```
    (228, 13, 71, 78)
    (228, 16, 71, 77)
    (228, 18, 71, 76)
    (228, 21, 71, 75)
    (228, 24, 71, 74)
    (228, 26, 70, 74)
    (228, 27, 70, 73)
    (228, 29, 70, 72)
    (228, 32, 999, 999)
    …
```

Note the records with missing data containing 999 for *x,y* coordinates. Take the following example:

```
(228, 6, 999, 999)
```

The second step is the OpenAI GPT-4o thinking through the problem to isolate the missing data and display it:

```
Process: the mobility agent is thinking
Customer message: **Detected Missing Coordinates:**
1. [228, 6, 999, 999]
2. [228, 32, 999, 999]
3. [228, 9, 999, 999]
4. [228, 45, 999, 999]
5. [228, 47, 999, 999]
```

The third step is for the OpenAI generative AI to predict the missing data:

```
**Predicted Missing Coordinates:**
```

The output is displayed and the predictions with explanations:

```
1. [228, 6, 69, 79] - Based on the trajectory pattern, the missing values
at timeslot 6 are likely to be the same as the next known values at
timeslot 7.
2. [228, 32, 69, 72] - Interpolating between timeslot 29 (70, 72) and
timeslot 33 (68, 72), the missing values at timeslot 32 are predicted to
be (69, 72).
3. [228, 9, 64, 72] - The missing values at timeslot 9 are interpolated
between timeslot 7 (64, 71) and timeslot 10 (64, 73), resulting in (64,
72).
4. [228, 45, 58, 81] - Interpolating between timeslot 43 (58, 82) and
```

```
timeslot 46 (58, 80), the missing values at timeslot 45 are predicted to be (58, 81).
5. [228, 47, 58, 79] - The missing values at timeslot 47 are interpolated between timeslot 46 (58, 80) and timeslot 1 (58, 78), resulting in (58, 79).
```

The output also contains the predictions in JSON:

```
```json
{
  "predicted_coordinates": [
    [228, 6, 69, 79],
    [228, 32, 69, 72],
    [228, 9, 64, 72],
    [228, 45, 58, 81],
    [228, 47, 58, 79]
  ]
}
```
```

The results are acceptable and show that recent generative AI models have zero-shot capabilities to make predictions on missing data in sequences.

However, the real power lies in extending these predictions to a wide range of real-world applications. The next logical step is to integrate this functionality into our GenAISys interface, allowing users to customize prompts easily to suit diverse trajectory-related use cases.

Let's move forward to implement this user-friendly integration.

Adding mobility intelligence to the GenAISys

We will now integrate the trajectory simulation and prediction component into our GenAISys, allowing users to design domain-specific prompts. At the user interface level, we'll simplify the terminology from "trajectory simulation and prediction" to the user-friendly term "**mobility**." This shorter label is more intuitive for users, though technical documentation can maintain detailed terminology as required. Then it will be up to the users to decide what domain-specific terminology they wish to see in the interface.

We will add the mobility function we built in `1_Trajectory_simulation_and_prediction.ipynb` to the GenAISys at three levels, as shown in *Figure 8.4*:

1. **IPython interface:** Adding the mobility feature to the user interface
2. **Handler selection mechanism:** Adding the mobility handler to the handler registry
3. **AI functions:** Implementing the mobility feature in the AI functions library

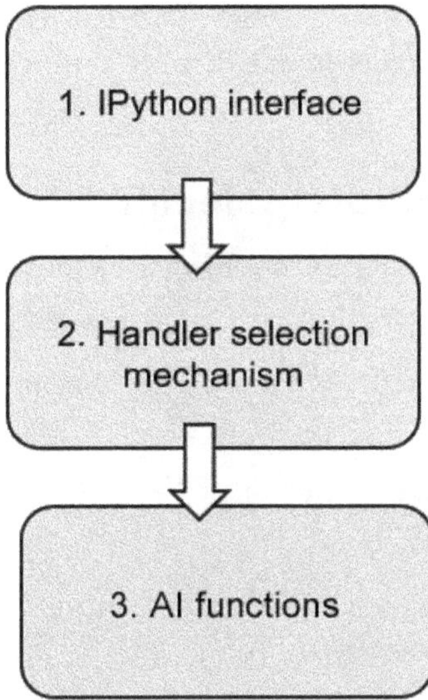

Figure 8.4: Integrating the trajectory simulation and prediction pipeline into the GenAISys

Open the `2_GenAISys_Mobility.ipynb` notebook. If needed, review the handler selection mechanism described in *Chapter 7* before continuing here. The notebook is not designed for voice outputs of lists of coordinates. As such, gTTS is best deactivated by default with `use_gtts = False` at the top of the notebook.

Let's first enhance the IPython interface.

IPython interface

The mobility option is primarily added to these parts of the IPython interface:

- To the `instruct_selector` dropdown with **Mobility** as one of its possible values
- To the display logic inside `update_display()`, which checks whether the user selected **Mobility** and, if so, displays the `mobility.png` file
- To the handling logic in `handle_submission()`, where the code prints `"Thinking..."` if `instruct_selector.value` is `"Analysis"`, `"Generation"`, or `"Mobility"`
- The mobility image (i.e., `mobility.png`) is only displayed when the **Files** widget is checked

We will begin by adding the option to the interface. We will create and add an option to `instruct_selector` and then handle the trajectory image display and submission code. Let's begin with the option in the interface.

Creating the option in instruct_selector

We will first add the **Mobility** option to the **Reasoning** drop-down list, as illustrated in *Figure 8.5*:

```
instruct_selector = Dropdown(
    options=["None", "Analysis", "Generation","Mobility"],
    value="None",
    description='Reasoning:',
    layout=Layout(width='50%')
)
instruct_selector.observe(on_instruct_change, names='value')
```

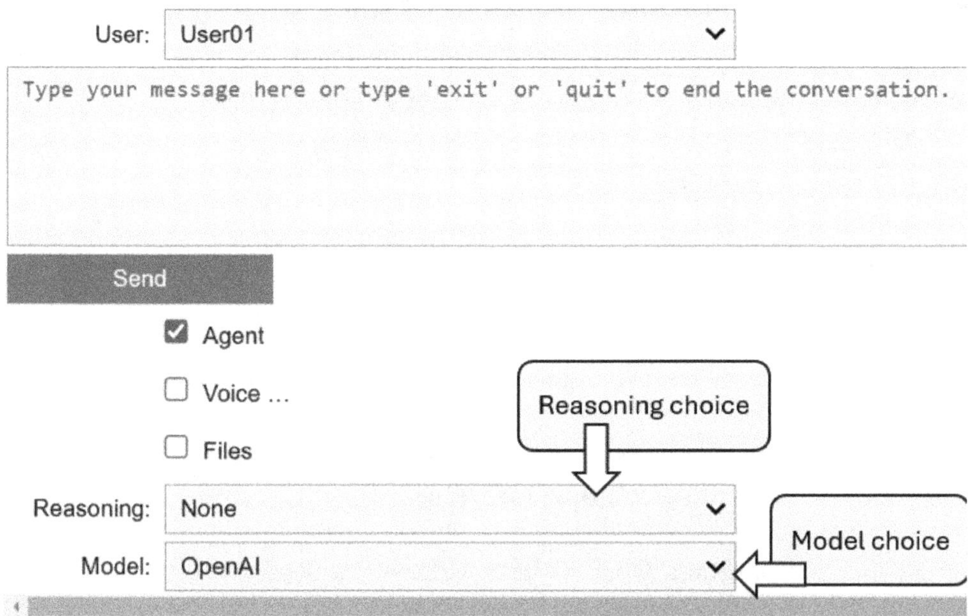

Figure 8.5: Adding Mobility to the dropdown

We can then select **Mobility**, as shown in *Figure 8.6*:

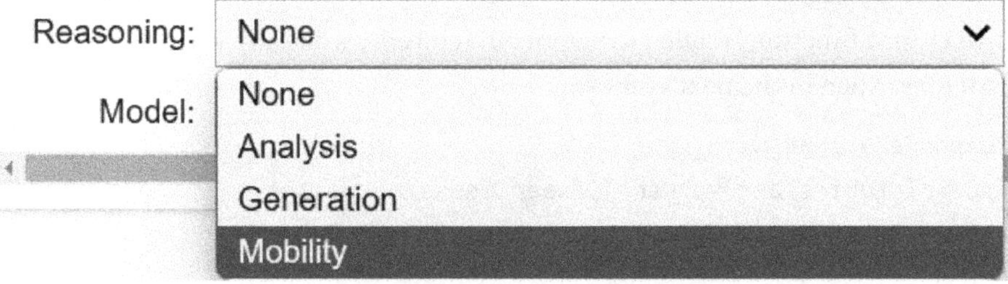

Figure 8.6: Selecting Mobility to activate the pipeline

Mobility is now selected. Notice the default model is set to **OpenAI**; however, you may extend this to other models, such as DeepSeek, during later phases, depending on your project needs.

Let's now handle the "mobility" value when we update the display.

Handling the "mobility" value in update_display()

We must ensure the generated trajectory visualization (`mobility.png`) is automatically shown when the **Mobility** option is selected and the **Files** checkbox is enabled:

```
def update_display():
    clear_output(wait=True)
    ...
    # Display c_image.png if it exists
    if files_checkbox.value == True:
    …
        # Display mobility.png if "Mobility" is selected
        if (
            os.path.exists("mobility.png")
            and instruct_selector.value == "Mobility"
        ):
            original_image = PILImage.open("mobility.png")
            display(original_image)
```

The image created by the trajectory simulation will be displayed. We now need to enhance the submission logic outputs to run the AI functions.

handle_submission() logic

The `chat_with_gpt` function is called as before but it interacts directly with the handler selection mechanism (described in the next section*)*:

```
The response = chat_with_gpt(
    user_histories[active_user], user_message, pfiles,
    active_instruct, models=selected_model
)
```

However, we will add the mobility functionality to the submission handling function:

```
def handle_submission():
    user_message = input_box.value.strip()
…
        if instruct_selector.value in [
            "Analysis", "Generation","Mobility"
```

```
        ]:
            with reasoning_output:
                reasoning_output.clear_output(wait=True)
                ...
```

We will now add the mobility function to the handler selection mechanism.

Handler selection mechanism

The handler selection mechanism contains two main parts. The first component, `chat_with_gpt`, remains unchanged from previous chapters and is directly called by the IPython interface:

```
def chat_with_gpt(
    messages, user_message, files_status, active_instruct, models
):
```

The second component is the handler registry, to which we'll now add the newly developed mobility handler:

```
handlers = [
...
# Mobility handler: determined by the instruct flag
    (
        lambda msg, instruct, mem, models, user_message, **kwargs:
            instruct == "Mobility",
        lambda msg, instruct, mem, models, user_message, **kwargs:
            handle_mobility(user_message, models=models)
    ),
...
```

This ensures that when users select **Mobility** from the **Reasoning** dropdown in the interface, the appropriate handler is activated automatically. We can see that the handler selection mechanism can be seamlessly scaled. Let's now add the functions we developed for this mobility function to the AI functions library.

AI functions

Next, we'll integrate the trajectory simulation and prediction functions—previously developed in the *Building the trajectory simulation and prediction* section—into the AI functions library within the notebook:

```
def create_grid_with_trajectory(
    grid_size=200, num_points=50, missing_count=5
):
...
```

This function is added just above the beginning of the functions called by the handler selection mechanism.

```
def handle_mobility_orchestrator(
    muser_message1, msystem_message_s1, mgeneration,
    mimcontent4, mimcontent4b
):
...
```

This function is also added just above the beginning of the functions called by the handler selection mechanism.

We now add the `handle_mobility` function we developed as well, and add `**kwargs` to process the arguments sent by the handler mechanism selection function:

```
def handle_mobility(user_message, **kwargs):
    from cot_messages_c6 import (
        msystem_message_s1, mgeneration, mimcontent4,muser_message1
    )
    mimcontent4b=mimcontent4
    #call Generic Synthetic Trajectory Simulation and Predictive System
    reasoning_steps = handle_mobility_orchestrator(
        muser_message1, msystem_message_s1, mgeneration,
        mimcontent4, mimcontent4b
    )
    return reasoning_steps
```

The code will run exactly like in the *Building the trajectory simulation and prediction function* section. With this setup, the mobility functionality is fully integrated into the GenAISys ecosystem, ready to be triggered via the intuitive IPython interface. Let's now get the user involved.

Running the mobility-enhanced GenAISys

In this section, we will demonstrate the mobility-enhanced GenAISys by running two distinct scenarios—a delivery use case and a fire disaster scenario—to illustrate the versatility of trajectory simulations and predictions, inspired by the work of Tang et al. (2024).

Open the `2_GenAISys_Mobility.ipynb` notebook. First, deactivate DeepSeek in the initial setup cell (you will only need a CPU):

```
deepseek=False
HF=False
Togetheragents=False
```

Then run the whole notebook. When it's finished, go to the *Running the interface* section in the notebook. We need to activate **Agent**, **Files**, and **Mobility**, and leave the default model as **OpenAI**.

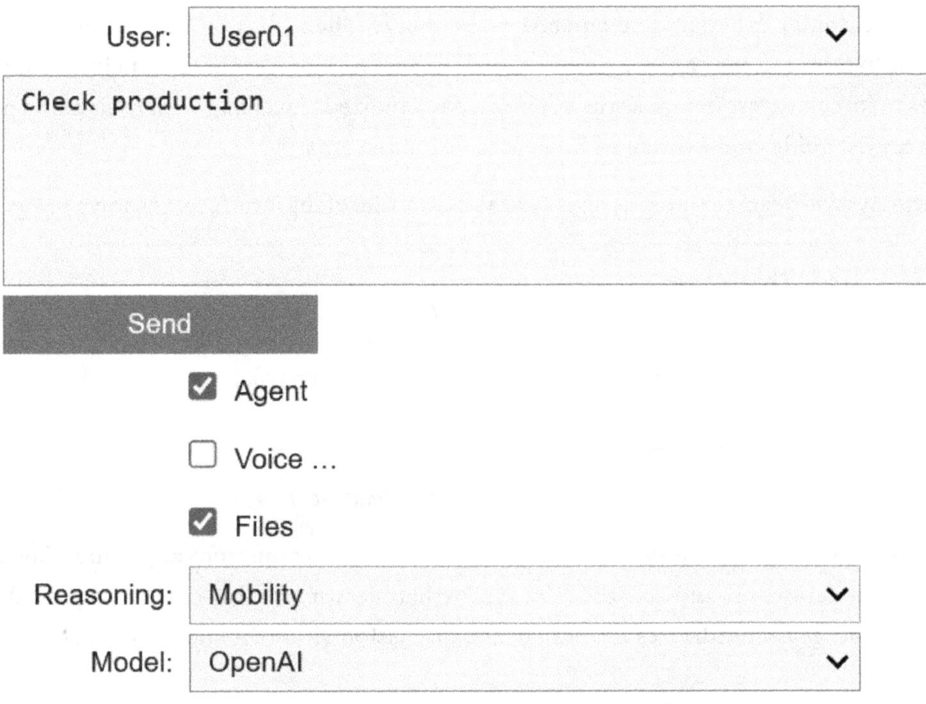

Figure 8.7: Running a delivery check with the mobility function

The synthetic trajectory simulates real-world input data and generates new data each time it runs. The explanations in this section reflect just one of these runs. When you execute the program, you'll obtain a new output every time, simulating real-time data.

Limit: Currently, the trajectory file is overwritten whenever a new trajectory is generated. If required, this functionality can be expanded during a project to save multiple files instead.

Let's first explore the mobility function with a delivery example.

Production-delivery verification scenario

To run the production-delivery verification, we simply need to activate **Agent** and **Files**, **Mobility** as the reasoning function, and **OpenAI** as the model. Then, click on **SEND** and let the AI do the work. In this case, we can imagine that an online travel agency wants to deliver customized goodies to its customers before a trip, such as a personalized travel bag with their name printed on it, a tourist guide, and a discount booklet to visit museums.

The GenAISys will begin to think, as displayed at the bottom of the interface, as shown in *Figure 8.8*:

```
Thinking...
```

```
Process: the mobility agent is thinking
```

Figure 8.8: The GenAISys has begun to think

The output first displays the synthetic trajectory, which can come from any source, such as real-time data, databases, and sensors. The synthetic trajectory, in this case, is displayed under the user message that indicates the title of the discussion since the mobility function itself is autonomous:

```
User01: Check deliveries
Agent: ('Generated Trajectory Data:', [(145, 20, 999, 999), (145, 22, 189, 125), (145, 25, 190, 125), (145, 28, 190, 124), (145, 29, 190, 123), (145, 31, 999, 999),
```

Then, the generative AI function takes over and indicates that it is processing the missing data in the trajectory:

```
Process: The mobility agent is thinking
```

When it is finished thinking, it provides the list of missing data:

```
Customer message: Detected Missing Coordinates:
1. [145, 20, 999, 999]
2. [145, 31, 999, 999]
3. [145, 34, 999, 999]
4. [145, 42, 999, 999]
5. [145, 3, 999, 999]
```

The GenAISys then provides its prediction:

```
Predicted Missing Coordinates: To predict the missing coordinates, we will
analyze the trajectory pattern before and after each missing value:
Predicted Coordinates:
1. [145, 20, 189, 125] - Based on the pattern before and after the missing
value.
2. [145, 31, 189, 122] - Interpolated from the surrounding points.
3. [145, 34, 189, 121] - Interpolated from the surrounding points.
4. [145, 42, 191, 128] - Based on the pattern before and after the missing
value.
5. [145, 3, 190, 124] - Interpolated from the surrounding points.
```

A structured JSON output is provided for integration with other systems or further processing:

```
json<br>{<br> "predicted_coordinates": [<br> [145, 20, 189, 125],<br>
[145, 31, 189, 122],<br> [145, 34, 189, 121],<br> [145, 42, 191, 128],<br>
[145, 3, 190, 124]<br> ]<br>}<br>
```

The original trajectory is also displayed for the user to make decisions, as shown in *Figure 8.9*:

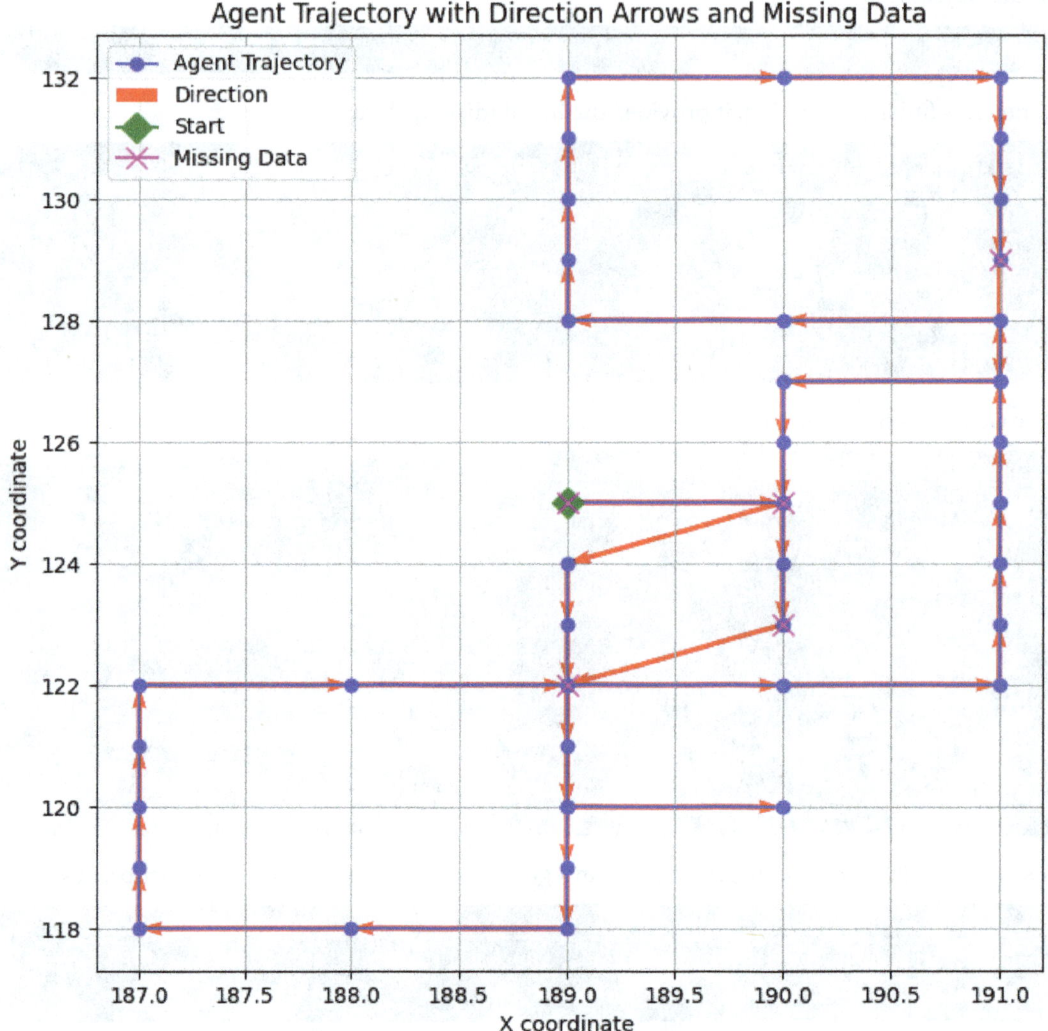

Figure 8.9: The original trajectory with missing data

Now, three **subject-matter expert (SME)** agents can intervene:

- **Human user**: A person evaluates the data, validates predictions, or provides feedback
- **AI agent**: An additional function can be added. The trajectory grid can be labeled, and instructions can be provided to an AI function. The instructions will rely on the coordinate sections of the grid to generate further functions or messages.

- **Hybrid (human-user + AI-agent)**: Once a human user has created a number of prompts successfully, it may be beneficial to automate some of them to alleviate the human user's workload if accepted within the scope of the project.

Let's now deactivate the **Mobility** function and **Files** display to put the cutting-edge generative AI to work, as shown in *Figure 8.10*:

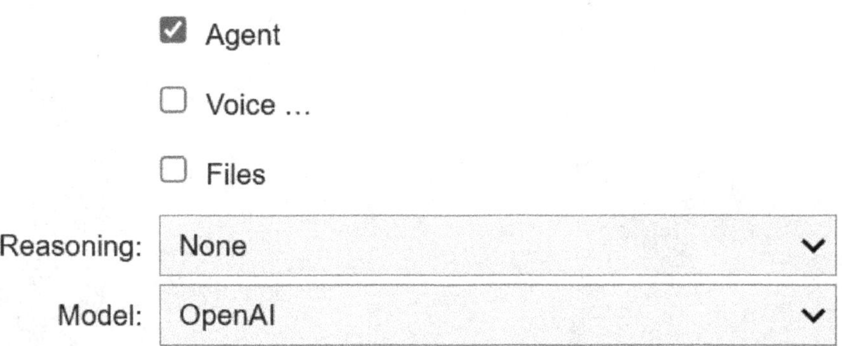

Figure 8.10: Generative AI takes over and generates messages

At this stage, a human user can interpret the AI's output and craft a tailored prompt. For instance, consider this scenario:

```
The missing x,y coordinates with 999,999 respectively in this dialog are
deliveries that were not made! The predicted x,y coordinates are inference
from the coordinates before and after.
```

Here's an example of a domain-specific prompt to address the issue:

```
Write an email to SUPPLIER COMPANY specifying that the missing coordinates
from x values 20 to 40 are located in the Midtown mall area. Provide the
missing coordinates and their predictions to inform the SUPPLIER COMPANY.
Be as precise and explicit as possible.
```

As demonstrated, such a prompt could easily be adapted to various domains. By simply adjusting the labeling of the coordinates and the intended recipient, you could generate a diverse array of messages. The exact scope and application depend entirely on the requirements of your specific project.

The generative AI's response to the preceding prompt was acceptable and detailed:

```
Dear [Supplier Company Contact Name],

I hope this message finds you well. I am writing to inform you about
certain discrepancies identified in the delivery data for the Midtown
Mall area. Specifically, there are missing coordinates in the trajectory
data, which indicate deliveries that were not completed. These missing
coordinates are represented by the values 999,999.

Below, I have detailed the missing coordinates along with their predicted
values, which have been inferred based on the surrounding data points:

1. Missing Coordinate: [145, 20, 999, 999]
- Predicted Coordinate: [145, 20, 189, 125]

2. Missing Coordinate: [145, 31, 999, 999]
- Predicted Coordinate: [145, 31, 189, 122]

3. Missing Coordinate: [145, 34, 999, 999]
- Predicted Coordinate: [145, 34, 189, 121]

4. Missing Coordinate: [145, 42, 999, 999]
- Predicted Coordinate: [145, 42, 191, 128]

5. Missing Coordinate: [145, 3, 999, 999]
- Predicted Coordinate: [145, 3, 190, 124]

These missing deliveries occurred within the x-value range of 20 to 40,
specifically in the Midtown Mall area. The predicted coordinates have been
calculated by analyzing the trajectory data pattern before and after the
missing points.

We kindly request that you review this information and take the necessary
actions to address these unfulfilled deliveries. Your prompt attention to
this matter will help ensure that our delivery operations continue to run
smoothly and efficiently.

Please feel free to reach out if you require further details or
```

```
clarification regarding this issue.

Thank you for your cooperation.

Best regards,

[Your Full Name]
[Your Position]
[Your Company Name]
[Your Contact Information]
```

Next, without restarting the session, let's ask our GenAISys how this methodology could be applied to a fire disaster scenario.

Fire disaster scenario

Now, let's explore how the concepts of trajectory analysis, missing coordinates, and predicted coordinates can be leveraged in disaster prediction scenarios, specifically in cases such as forest fires. We'll submit this scenario to the GenAISys and analyze its insights:

```
User01: How could the trajectory, missing coordinates, and predicted
coordinates be used for disaster prediction such as fires? Explain and
create an output for this use case.

Agent: Trajectory data, missing coordinates, and predicted coordinates
can be valuable tools in disaster prediction and management, such as for
fires, in the following ways:

1. Trajectory Analysis: By analyzing the movement patterns of vehicles,
drones, or other mobile units, authorities can identify areas that are
frequently visited or monitored. This can help in understanding the
normal patterns of activity and detecting anomalies that might indicate a
disaster, such as a sudden deviation from a typical route due to a fire.

2. Missing Coordinates: Missing data points can indicate areas where
monitoring or data collection was interrupted, possibly due to a disaster
like a fire. Identifying these gaps can help in pinpointing locations that
need immediate attention or further investigation.

3. Predicted Coordinates: By predicting the missing coordinates,
```

```
authorities can fill in the gaps in the data to maintain a continuous
monitoring system. This can help in reconstructing the events leading up
to a disaster and in planning the response by understanding the spread and
impact of the fire.

### Use Case Output for Disaster Prediction (Fires)

Scenario: A fire has been reported in a forest area, and trajectory
data from drones monitoring the region is being analyzed to assess the
situation.
Output:
json<br>{<br> "fire_prediction_analysis": {<br> "trajectory_analysis":
"Drones deviated from their usual path at coordinates [145, 20] to [145,
40], indicating potential fire activity.",<br> "missing_coordinates":
[<br> {"missing_coordinate": [145, 25, 999, 999], "predicted_coordinate":
[145, 25, 190, 130]},<br> {"missing_coordinate": [145, 30, 999, 999],
"predicted_coordinate": [145, 30, 191, 128]}<br> ],<br> "predicted_fire_
area": [<br> {"coordinate": [145, 25, 190, 130]},<br> {"coordinate": [145,
30, 191, 128]}<br> ],<br> "actionable_insights": "Immediate investigation
and firefighting efforts are recommended in the predicted fire area to
prevent further spread."<br> }<br>}<br>

This output provides a structured analysis of the situation, helping
disaster management teams to quickly identify and respond to potential
fire threats based on trajectory data and inferred information.
```

This structured output offers valuable insights, enabling disaster response teams to swiftly identify and respond to potential threats based on trajectory analysis, pinpointed data gaps, and predictive coordinates.

This methodology demonstrates that we can craft numerous specialized prompts across domains. Despite inevitable limitations, the era of GenAISys is just beginning, continually expanding into new, uncharted applications.

Summary

In this chapter, we began by recognizing that robust trajectory analysis is essential for applications ranging from deliveries and epidemic forecasting to city-scale planning. Guided by the innovative approach outlined in Tang, P., Yang, C., Xing, T., Xu, X., Jiang, R., and Sezaki, K. (2024), we emphasized the transformative potential of text-based LLMs for mobility prediction. Their framework directed our design of a method capable of intelligently filling gaps in real-time synthetic datasets through carefully structured prompts.

We then built a Python-based trajectory simulator that randomizes movement on a grid, mirroring typical user paths. It assigns day and timeslot indices, which enabled us to capture the temporal aspect of mobility. Critically, we inserted synthetic gaps marked as `999, 999`, approximating real-world data dropouts or missing logs. Next, we integrated an orchestrator function that adds instructions with this synthetic data before directing them to an LLM, in this case, an OpenAI GPT-4o model. The orchestrator composes prompts that accurately reflect the trajectory dataset, focusing the model's attention on flagged gaps. It employs a chain-of-thought routine, noting missing points for debugging prior to generating final JSON outputs.

We then merged this pipeline into the GenAISys environment by adding a dedicated mobility handler in the multihandler system. This handler streamlines the full process: trajectory generation, model inference, and visualization all in one place. Users can prompt the system to evaluate missing coordinates and instantly see the updated paths superimposed on a static city grid. Ultimately, we demonstrated that robust GenAISys forecasting need not remain abstract when grounded in purposeful, prompt design.

In the next chapter, we will open the GenAISys to the world with an external service that will lead us to enhance our system with security and moderation functionality.

Questions

1. A trajectory can only be a physical path in a city. (True or False)
2. Synthetic data can accelerate GenAISys simulation design (True or False)
3. Generative AI cannot go beyond natural language sequences. (True or False)
4. Only AI experts can run GenAISys. (True or False)
5. Generative AI can now help us with prompt design. (True or False)
6. Trajectory simulation and prediction cannot help with fire disasters. (True or False)
7. GenAISys's potential is expanding at full speed and can be applied to a growing number of domains and tasks. (True or False)

References

- P. Tang, C. Yang, T. Xing, X. Xu, R. Jiang, and K. Sezaki. 2024. "Instruction-Tuning Llama-3-8B Excels in City-Scale Mobility Prediction." *arXiv*, October 2024. `https://arxiv.org/abs/2410.23692`.
- Renhe Jiang, Xuan Song, Zipei Fan, Tianqi Xia, Quanjun Chen, Satoshi Miyazawa, and Ryosuke Shibasaki. 2018. "DeepUrbanMomentum: An Online Deep-Learning System for Short-Term Urban Mobility Prediction." *Proceedings of the AAAI Conference on Artificial Intelligence* 32, no. 1: 784–791. `https://ojs.aaai.org/index.php/AAAI/article/view/11338`.
- Jie Feng, Yong Li, Chao Zhang, Funing Sun, Fanchao Meng, Ang Guo, and Depeng Jin. 2018. "DeepMove: Predicting Human Mobility with Attentional Recurrent Networks." *Proceedings of the 2018 World Wide Web Conference*, Lyon, France, April 23–27, 2018, 1459–1468. `https://doi.org/10.1145/3178876.3186058`.

Further reading

- Haru Terashima, Naoki Tamura, Kazuyuki Shoji, Shin Katayama, Kenta Urano, Takuro Yonezawa, and Nobuo Kawaguchi. 2023. "Human Mobility Prediction Challenge: Next Location Prediction Using Spatiotemporal BERT." *Proceedings of the 1st International Workshop on the Human Mobility Prediction Challenge*, Tokyo, Japan, September 18–21, 2023, 1–6. `https://dl.acm.org/doi/10.1145/3615894.3628498`.

Upgrading the GenAISys with Data Security and Moderation for Customer Service

In this chapter, we will open up our GenAISys by integrating it with real-world online services—specifically, by connecting it to an online weather API. This will enable the fictional online travel agency that we've been supporting throughout the book to access real-time weather data. Weather reports for a specific location serve as an entry point for various tasks essential to the agency's operations, such as marketing initiatives, recommendations for tourist activities, and coordinating product deliveries.

Connecting our GenAISys to external online resources transitions our system from a controlled internal environment to the unpredictable realm of real-time data interactions. This transition, however, introduces critical security concerns. Opening a system without adequate protections can inadvertently expose sensitive data or cause security breaches, posing genuine risks both to users and the organization itself. As such, robust security measures are a prerequisite before fully integrating external services. Therefore, this chapter presents a threefold challenge: implementing the weather service using the OpenWeather API, building a moderation system leveraging OpenAI's moderation capabilities, and developing a RAG-based data security function with Pinecone, which will detect and prevent sensitive-topic breaches. We will rely heavily on the flexible and powerful handler selection mechanism of our existing GenAISys architecture to seamlessly integrate these new functionalities. Our objective remains clear—minimal code enhancements with maximum functional impact.

The chapter begins by detailing how these additional components—moderation, data security, and real-time weather—fit into the overall architecture of the GenAISys. Then, we'll dive under the surface to build the moderation function using OpenAI's moderation endpoint. We will then construct a RAG-powered data security module, using Pinecone, to proactively detect and filter out sensitive or inappropriate inputs. By integrating these two security layers directly into the GenAISys's handler selection mechanism, we ensure comprehensive protection against unwanted interactions.

With security firmly in place, we will then implement the OpenWeather API. This integration allows us to retrieve live, real-time weather information to power a range of engaging, user-centric tasks. Finally, we will demonstrate the capabilities of the enhanced GenAISys through practical, multimodal, multi-user scenarios—such as generating weather-based activity recommendations, crafting customized promotional images for travel merchandise, and dynamically creating personalized weather-aware messages for travelers. By the end of this chapter, you'll be fully equipped to adapt the core architecture and concepts we've explored in the GenAISys to real-world applications, confidently delivering a comprehensive, secure, and highly functional **proof of concept**.

This chapter covers the following topics:

- Enhancing the GenAISys
- Adding a security function to the handler selection mechanism
- Building the weather forecast component
- Running use cases in the GenAISys

Let's start by clearly mapping out how the new moderation, data security, and weather functions integrate into our GenAISys.

Enhancing the GenAISys

Integrating moderation, data security, and real-time weather functionalities into our GenAISys will affect all three architectural layers, as illustrated in *Figure 9.1*. We will rely on the framework built around our handler selection mechanism to make this three-level, three-function implementation seamless.

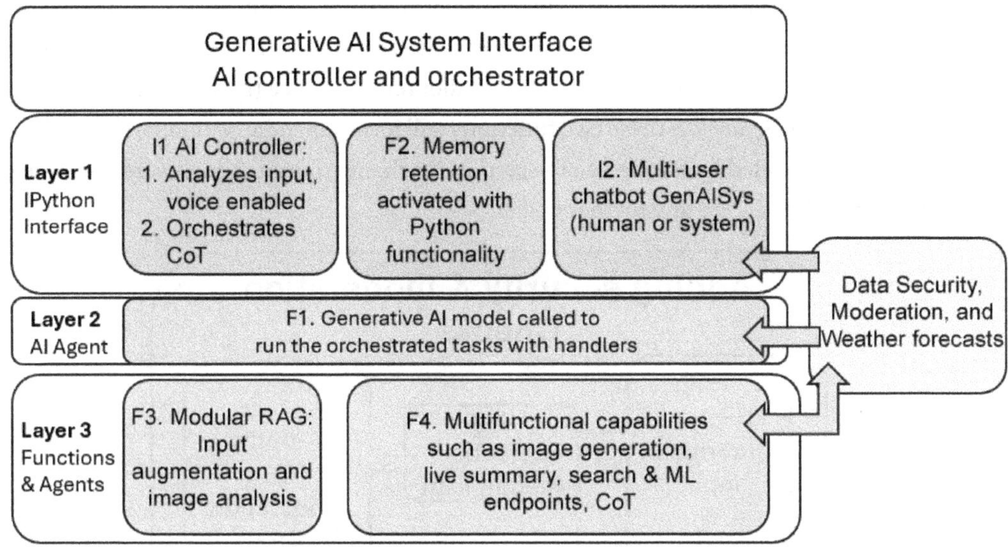

Figure 9.1: Moderation, data security, and weather report integration

The implementation will impact all three layers as follows:

- **Layer 1 (IPython interface)**: We'll introduce a new **Weather** option within the **Reasoning** drop-down menu of the IPython interface. Why place it here? Because in real-world project settings, the weather report function could easily be expanded into a more complex, multi-step **chain-of-thought** (**CoT**) pipeline, enabling the generation of sophisticated, context-aware outputs based on weather conditions.
- **Layer 2 (AI agent)**: The central orchestrator of our GenAISys, the handler selection mechanism, will now manage an additional component—the real-time weather forecasting capability. More critically, we'll enhance the handler with an integrated moderation and data security mechanism that will function as an immediate *kill switch*, instantly terminating the process and returning a security warning to the user when inappropriate or sensitive content is detected.

- **Layer 3 (functions and agents)**: Three new key functions—moderation, data security, and weather forecast retrieval—will be implemented at this level. The moderation and data security checks will reside outside the handler registry since their roles are not optional. As depicted in *Figure 9.2*, these two functions will form a mandatory global security barrier directly controlled by the handler selection mechanism, protecting users and company data before initiating any further tasks.

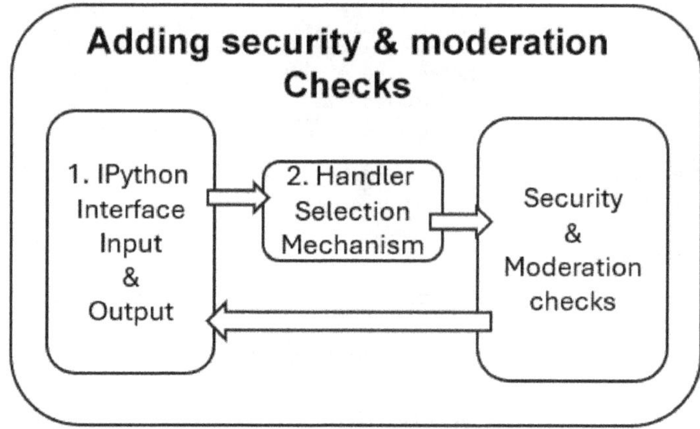

Figure 9.2: Security and moderation checks are directly managed within the handler selection mechanism

This security system will proactively intercept and evaluate each user input, blocking further processing and immediately alerting the user via the IPython interface if inappropriate or sensitive content is detected.

While building and showcasing your GenAISys as a flexible proof of concept, remember to demonstrate its scalability clearly. Highlight its potential for integration with additional AI functions, further security features, or alternative generative models. However, exercise caution: avoid prematurely adding functionalities without concrete project requirements, as this can lead to unnecessary complexity or overdeveloping your project. The goal is to clearly show the project's potential without investing resources in unrequested developments.

With this strategic clarity in mind, we will begin by constructing the security function.

Adding a security function to the handler selection mechanism

In this section, we will build the **security function**, the **moderation function**, and the **data security function**, as illustrated in *Figure 9.3*:

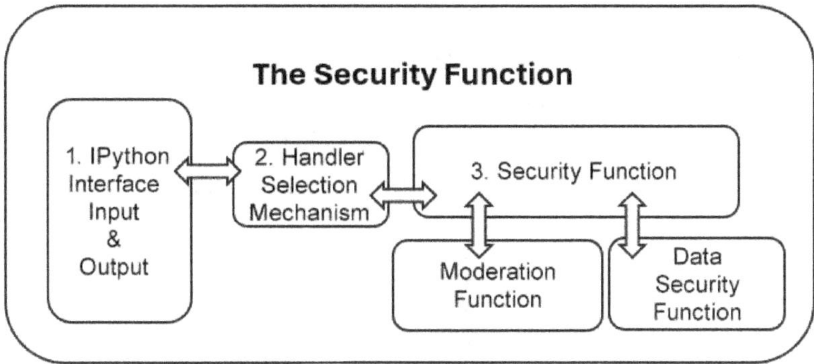

Figure 9.3: The handler selection mechanism directly calls the security function

We'll implement these functions as follows:

- The security function will call the moderation function
- The security function will also call the data security function
- The moderation function will contain subfunctions
- The data security function will ensure compliance with security standards

Let's first build the security function along with its calls.

Implementing the security function

The security function is directly integrated with the handler selection mechanism, as illustrated in *Figure 9.4*. It receives the user message directly, prior to any handler selection in the handler registry. If the user message violates the GenAISys content policy, the security function returns a `False` flag (indicating a content violation) to the IPython interface. This proactive filtering is a crucial defense against data poisoning as well, where malicious inputs could be used to manipulate the system's behavior.

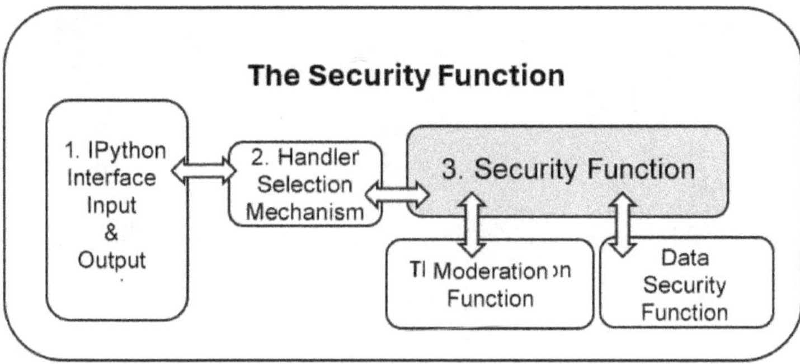

Figure 9.4: The security function is directly connected to the handler selection mechanism

Open `GenAISys_Customer_Service.ipynb` and go to the *Security* section within the Chapter09 directory on GitHub (`https://github.com/Denis2054/Building-Business-Ready-Generative-AI-Systems/tree/main`). We first create the function that sets the `securitym` flag to `True`, signifying the content does not violate content policies. Also in this function, `securityd=True` shows that no sensitive topic has been detected and that the message is data security-compliant:

```
def security(user_message):
  securitym=True # default value
  securityd=True # default value
```

The function begins by calling the moderation and content acceptability functions:

```
  response=moderation(user_message)
  # Moderation
  security = is_acceptable(user_message,response)
  #print(security)  # Outputs: True if acceptable, False otherwise
```

- `moderation(user_message)` invokes the OpenAI moderation API
- `is_acceptable(user_message, response)` processes the moderation response
- The debugging line (`print(securitym)`) can be uncommented during testing or troubleshooting

Next, the security function calls the data security function, which checks for sensitive topics within the user message:

```
# Data security
securityd = data_security(user_message)
#print securityd
```

The `securityd` variable will store either `True` or `False`, depending on whether the message meets the data security criteria.

The final part of the function evaluates both flags (`securitym` and `securityd`) and returns the security status accordingly:

```
if securitym==False or securityd==False:
  return False
else:
  return True
```

If you prefer not to activate the moderation or data security checks, you can comment out the relevant lines in the security function—specifically, the code between the assignments of `securitym=True` and `securityd=True`, and the `if securitym==False or securityd==False` conditional statement. This way, the function defaults to always returning `True`.

Let's now examine how the handler selection mechanism interacts with the IPython interface.

Handler selection mechanism interactions

The handler selection mechanism and IPython interface interact closely, as shown in *Figure 9.5*. When the IPython interface sends a user message to the handler selection mechanism, it determines whether the message complies with security policies.

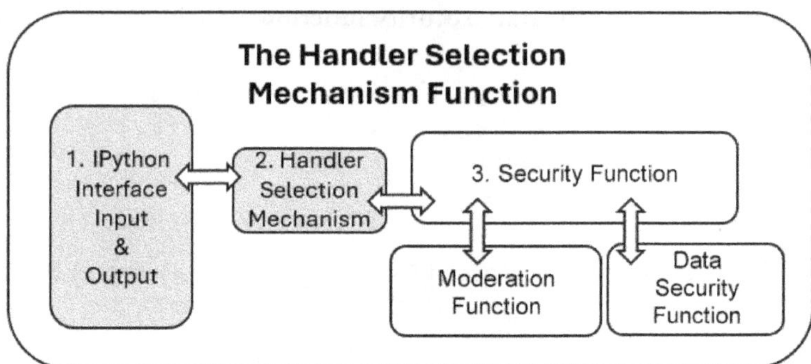

Figure 9.5: Handler selection mechanism and IPython interactions

The primary structure of the handler selection mechanism function remains unchanged from previous chapters:

```
def chat_with_gpt(
    messages, user_message, files_status, active_instruct, models
):
    global memory_enabled  # Ensure memory is used if set globally
```

However, at the start of this function, we now call the security function. It returns a security status (`True` for compliant or `False` for non-compliant):

```
    try:
        if not security(user_message):
            return "Your message could not be processed as it may violate
our security guidelines."d
```

If the message is flagged as non-compliant, a clear message will be immediately returned to the user interface. We will test security function examples thoroughly in the *Running security checks* section later. Before that, let's move forward by implementing the moderation function.

Implementing the moderation function

We will use OpenAI Omni, which has a comprehensive range of categorization options, as the moderation model: `https://platform.openai.com/docs/guides/moderation`.

Open the `GenAISys_Customer_Service.ipynb` notebook and navigate to the *Moderation* subsection within the *Security* section. The moderation function will be directly invoked by the handler selection mechanism, as illustrated in *Figure 9.6*. It classifies user messages and provides details about any inappropriate content flagged during processing.

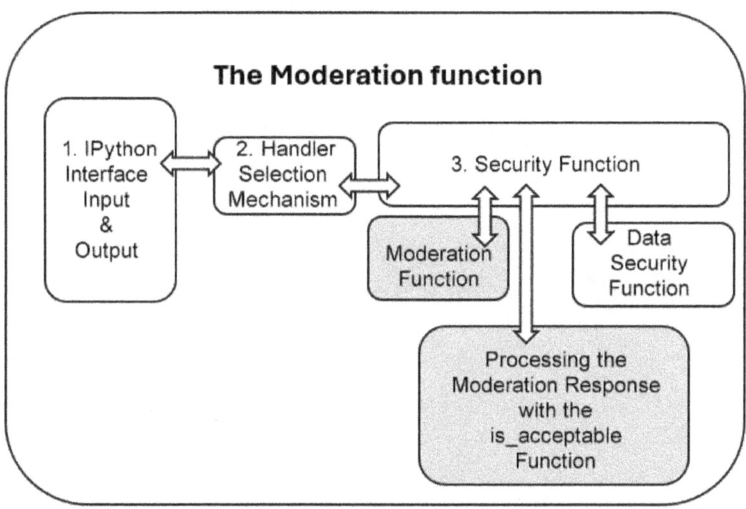

Figure 9.6: Calling the moderation function

We will implement the moderation function with OpenAI Omni as follows:

```
from openai import OpenAI
client = OpenAI()

def moderation(user_message):
  response = client.moderations.create(
      model="omni-moderation-latest",
      input=user_message,
  )
  return response
```

This function processes the user message and returns a moderation response. Once received by the security function, the response is sent to another function, `is_acceptable`, to evaluate whether the message is acceptable or not:

```
def is_acceptable(user_message, response):
    # Extract the 'flagged' status from the first result
    flagged = response.results[0].flagged
```

The response contains `response.results[0].flagged`, which provides a `True` value if the content is flagged or a `False` status if the response is acceptable. If the content is flagged, the details of the response will be saved in a timestamped JSON file:

```
    if flagged:
        # Generate filename based on current date and time
        timestamp = datetime.datetime.now().strftime("%Y%m%d%H%M%S")
        sanitized_message = ''.join(
            e for e in user_message if e.isalnum()
            or e in (' ', '_')
        ).strip()
        filename = \
            f"{sanitized_message[:50].replace(' ', '_')}_{timestamp}.json"

        # Ensure the 'logs' directory exists
        os.makedirs('logs', exist_ok=True)

        # Convert the response to a dictionary
        response_dict = response.model_dump()

        # Write the response to a JSON file in the 'logs' directory
        with open(os.path.join('logs', filename), 'w') as file:
            json.dump(response_dict, file, indent=4)
```

Finally, the function returns the flagged status as `not_flagged`:

```
    # Return True if content is acceptable, False otherwise
    return not_flagged
```

If the message is flagged, details are saved in a timestamped JSON file located within the `/logs` subdirectory. The resulting JSON file contains a unique ID, the model used, and the status of a wide range of categories:


```
{
    "id": "modr-bb021ae067c296c1985fca7ccfd9ccf9",
    "model": "omni-moderation-latest",
    "results": [
        {
            "categories": {
                "harassment": true,
                "harassment_threatening": false,
                "hate": false,
                "hate_threatening": false,
                "illicit": false,
                "illicit_violent": false,
                "self_harm": false,
                "self_harm_instructions": false,
                "self_harm_intent": false,
                "sexual": false,
                "sexual_minors": false,
                "violence": false,
                "violence_graphic": false,
                "harassment/threatening": false,
                "hate/threatening": false,
                "illicit/violent": false,
                "self-harm/intent": false,
                "self-harm/instructions": false,
                "self-harm": false,
                "sexual/minors": false,
                "violence/graphic": false
            },
```

```
        "category_applied_input_types": {
            "harassment": [
                "text"
            ],
            "harassment_threatening": [
                "text"
            ],…
```

In this case, the `harassment` category has been flagged, for example. The file also contains a score for each category, as shown in this excerpt from the file:

```
        "category_scores": {
            "harassment": 0.8075929522141405,
            …
        },
        "flagged": true
    …
```

To perform a quick quality control check or evaluate specific user messages manually, uncomment and use the following lines in your notebook:

```
# Uncomment to use as security user message evaluation
user_message="Your ideas are always foolish and contribute nothing to our discussions."
security(user_message)
```

The file containing the information can be processed further with other functions as required for your project. We could add the user to the file. We can also view the details of the dialogue to find which user entered the flagged message in the *Load and display the conversation history* section of the notebook. We will go through these features in the *Running security checks* section later.

Let's now build the data security function.

Building the data security function

We will now build a data security function designed to detect whether a user message contains sensitive topics. This function covers a wide spectrum of security-related concerns, from safeguarding confidential information to preventing inappropriate or sensitive discussions through the GenAISys interface. The data security function comprises two components, as shown in *Figure 9.7*: the first component populates a Pinecone index with sensitive topics, and the second component queries this index to detect sensitive topics within user messages.

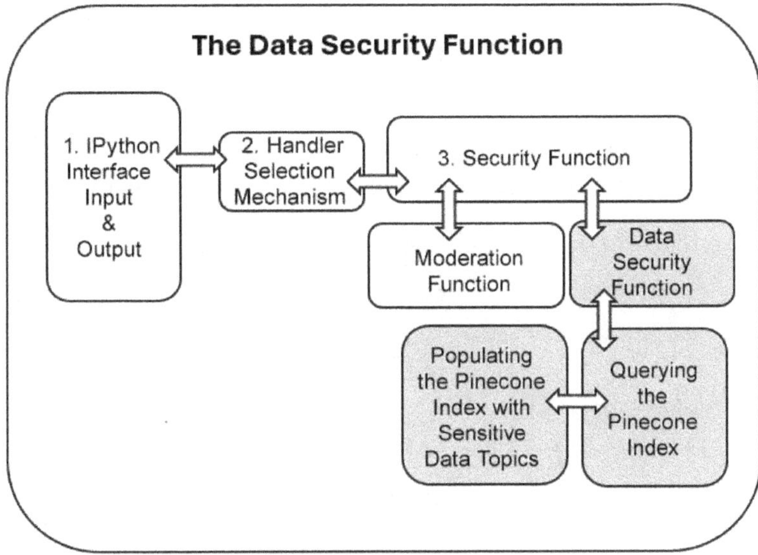

Figure 9.7: The two components of the data security function

Populating the Pinecone index

We will first populate the Pinecone index with sensitive topics. Open `Pinecone_Security.ipynb`. This notebook mirrors the structure of the earlier notebook from *Chapter 3* (`Chapter03/Pinecone_instruction_scenarios.ipynb`). We will simply adapt it here to accommodate sensitive topics. Feel free to revisit `Pinecone_instruction_scenarios.ipynb` for a detailed refresher if needed.

We'll briefly cover the code sections specifically adapted for this domain. The first step involves downloading our `sensitive_topics` dataset:

```
download("Chapter09","sensitive_topics.csv")
```

The file, `sensitive_topics.csv`, contains 100 sensitive topics structured as follows:

- `id`: A unique identifier ranging from `1000` to `1099`
- `values`: Text descriptions of sensitive topics

For example, the first topic is as follows:

```
Discussing the mismanagement of client funds that has resulted in
significant financial discrepancies which have raised concerns among
regulatory bodies and could lead to potential legal actions being taken
against the travel agency if corrective measures are not implemented
promptly and with strict accountability this issue demands immediate
thorough review.
```

If a user message contains content similar to this topic, for example, it will be flagged. Vector-based similarity searches provide more nuanced detection than traditional methods.

The next step is to chunk the data from the `sensitive_topics.csv` file:

```
import time
start_time = time.time()  # Start timing
# File path
file_path = 'sensitive_topics.csv'
# Read the file, skip the header, and clean the lines
chunks = []
…
```

We then create a namespace within Pinecone to hold this sensitive data. We will call this namespace `security`:

```
from pinecone import ServerlessSpec
index_name = 'genai-v1'
namespace="security"
cloud = os.environ.get('PINECONE_CLOUD') or 'aws'
region = os.environ.get('PINECONE_REGION') or 'us-east-1'
spec = ServerlessSpec(cloud=cloud, region=region)
```

The upserting function remains unchanged, retaining the same column names as in the `.csv` file:

```
def upsert_to_pinecone(batch, batch_size, namespace="security"):
    …
```

With the data now populated in Pinecone, we'll proceed to implement the query component.

Querying the Pinecone index

Open GenAISys_Customer_Service.ipynb and navigate to the *Data security* subsection in the *Security* section. The goal here is to determine whether a user message matches any sensitive topics stored in the Pinecone security index. If a match is found, the message will be flagged.

Initially, we set default security indicators:

```
# Define data security function
import datetime

def data_security(user_message):
    sec = True # Initialize to True (safe by default)
    target_id = 0
    score = None     # Initialize score
    security_level = 0.30 # Set your desired security threshold here (as a 
float, e.g., 0.0 to 1.0)
```

The default value, sec, is set to True, meaning that the user message is secure until found otherwise. Additionally, target_id is set to 0 to show that no record in the Pinecone index has been found yet. Additionally, we have a security level threshold, security_level, that can be set to the value deemed necessary in production. You can modify it here in the code or create a security interface once your strategy has been decided with your team. In this case, security_level is set to 0.30 to avoid filtering low-level security items in this educational example.

We then define our namespace for querying:

```
    namespace = "security"
    print(namespace)
```

We then query the Pinecone index to detect sensitive topics:

```
    query_text = user_message
    query_results = get_query_results(query_text, namespace)
```

We now extract the score, which we will use as our security level threshold:

```
# Extract score directly from query_results BEFORE calling display_results
    if (
        query_results
        and "matches" in query_results
        and query_results["matches"]
```

```
    ):
        score = query_results['matches'][0]['score']
        print(f"Extracted Score: {score}") # Optional: print to verify
```

Now, we can display the results:

```
    print("Processed query results:")
    qtext, target_id = display_results(query_results)
    print(qtext)
```

If a similar sensitive topic is found (indicated by a score that exceeds the threshold), we have a security logic process:

```
# --- LOGIC FOR SECURITY CHECK ---
    # Determine 'sec' based on score first, if a score is available.
    if score is not None:
        if score > security_level:
            sec = False # Breach detected: Score is above threshold
            print(f"Security flag triggered: Score ({score}) exceeds 
threshold ({security_level}).")
        else:
            sec = True # No breach: Score is below or equal to threshold
            print(f"Score ({score}) is below or equal to threshold 
({security_level}). Not a score-based breach.")
    else:
        # If no score is available (e.g., no match found), then use 
target_id as a fallback.
        if target_id is not None and int(target_id) > 0:
            sec = False # Breach detected: Target ID is positive 
(fallback)
            print(f"Security flag triggered: Target ID ({target_id}) is 
greater than 0 (fallback check).")
        else:
            sec = True # No breach detected by target_id fallback

    # --- END SECURITY CHECK  LOGIC -
```

The query results are tracked and the details are recorded:

```
 # Create a filename with the prefix "security_breach" and current
datetime (format: YYYYMMDD_HHMMSS)
    if not sec: # Only log if a security breach is detected
        now = datetime.datetime.now().strftime("%Y%m%d_%H%M%S")
        filename = f"security_breach_{now}.txt"

        # Include score and threshold in the log for context
        log_content = f"Security Breach Detected!\nUser Message: {user_
message}\nMatched Text: {qtext}\nTarget ID: {target_id}\nScore: {score}\
nThreshold: {security_level}"
        with open(filename, "w") as file:
            file.write(log_content)
        print(f"Security breach logged to {filename}")
    else:
        print("No security breach detected.")
```

The result, `True` (non-sensitive) or `False` (sensitive), is returned to the security function:

```
return sec
```

To manually evaluate messages, you can uncomment the following code snippet:

```
#Uncomment for security message evaluations
'''
user_message="Some client funds were mismanaged"
response=data_security(user_message)
print(response)
'''
```

Let's now run security checks for moderation and sensitive data classification.

Running security checks

We'll perform two security evaluations: one for moderation (harassment) and another for sensitive data. Let's first begin with a moderation user message.

Consider a user message that contains harassment content, as shown here:

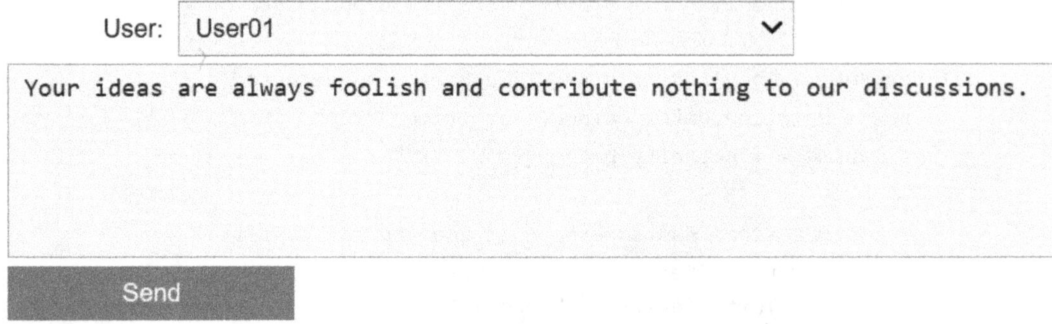

Figure 9.8: A harassing message

The moderation function detects the inappropriate content:

```
User01: Your ideas are always foolish and contribute nothing to our
discussions
Agent: Your message could not be processed as it may violate our security
guidelines.
```

A JSON file containing the flagged message is created in the `/logs` directory:

```
/content/logs/Your_ideas_are_always_foolish_and_contribute_
nothi_20250306213022.json
```

The file contains the moderation categories and has flagged the user message as `harassment`:

```
{
    "id": "modr-5ef1eeaaa3256a7c45feaa17396f11fd",
    "model": "omni-moderation-latest",
    "results": [
        {
            "categories": {
                "harassment": true,
```

The probability is also logged:

```
...
            "category_scores": {
                "harassment": 0.8054343118110135,
```

We now have a data security function. When we run a security check for a sensitive topic, such as referring to mismanaged client funds, the output confirms the detection of a sensitive topic:

```
User01: Some client funds were mismanaged and used in a case of
intellectual property violations.
Agent: Your message could not be processed as it may violate our security
guidelines.
```

A file containing the flagged message is created :

```
/content/security_breach_20250306_213240.txt
```

This file contains a record of the flagged message in the Pinecone index:

```
1063, Discussing disputes over intellectual property in travel brochure
designs which have led to accusations of copyright infringement potential
lawsuits and significant reputational damage to the travel agency thereby
necessitating an immediate review of all creative materials and strict
adherence to intellectual property laws with immediate legal consultation
and review now.
```

Both moderation and sensitive data checks have created detailed records. These logs can be further expanded with user profiles or additional data as your project needs evolve. Additionally, the GenAISys stores full dialogue histories (`conversation_history.json`) and summarizations for future reference and auditing. The conversation history is activated when we enter `exit`:

```
Conversation ended. Saving history...
Conversation history saved to conversation_history.json.
History saved. Proceed to the next cell.
```

The *Load and display the conversation history* outputs contain the log of the conversation with the usernames and security flags:

```
User01:..
Your ideas are always foolish and contribute nothing to our discussions…
assistant…
Your message could not be processed as it may violate our security
guidelines.
…
Some client funds were mismanaged and used in a case of intellectual
property violations.
```

```
assistant…
Your message could not be processed as it may violate our security
guidelines….
```

The raw log of the conversation is saved in `/content/conversation_history.json` for further use. The *Load and summarize the conversation history* section contains a summary of the dialogue, which includes the username, user messages, and assistant's responses:

```
List of Actions:
1. User01's First Message:
- Action: User01 criticized the quality of ideas in discussions.
- Assistant's Response: The message was blocked due to potential security
guideline violations.
2. User01's Second Message:
- Action: User01 reported an issue regarding the mismanagement of client
funds.
- Assistant's Response: The message was again blocked for potential
security guideline violations.
```

With these core functionalities in place, our system's moderation and security measures can be easily adapted or expanded to meet your project-specific requirements. We're now ready to move forward by integrating an external weather forecast component.

Building a weather forecast component

In this section, we'll cautiously open up the GenAISys by integrating a weather forecast library and building a dedicated weather forecast function using an external API. This integration allows our GenAISys to interact directly with real-time weather information, providing a pathway to more controlled interactions with external web services. After successfully implementing this functionality, the GenAISys could be authorized to interact with other websites as needed.

We'll implement real-time weather forecasts for specific locations (cities in this case) to support marketing, production planning, deliveries, and customer service, as illustrated in the upcoming *Running the GenAISys* section.

Open the `GenAISys_Customer_Service.ipynb` notebook. The notebook uses the OpenWeather API, available at `https://home.openweathermap.org/`. OpenWeather provides a wide range of weather forecasting services, but we will focus specifically on real-time forecasts suitable for our use case.

To use the OpenWeather API, sign up for an account, obtain your API key, and carefully review their pricing plans at `https://openweathermap.org/price`. At the time of writing, the API calls required for our examples are available under their free tier, subject to request limits. Please confirm the cost and limits before proceeding.

We will seamlessly integrate our weather forecast function into the GenAISys framework using the handler selection mechanism, as depicted in *Figure 9.9*:

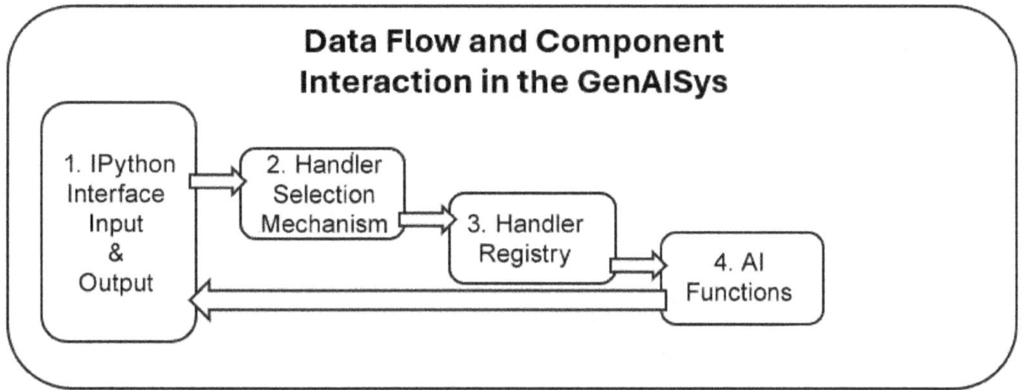

Figure 9.9: Integrating a weather forecast API in the GenAISys framework

Once we set the OpenWeather environment up, we will integrate the weather forecast function seamlessly with the handler selection mechanism framework:

1. **IPython interface**: Add a **Weather forecast** option to the interface.
2. **Handler selection mechanism**: No change.
3. **Handler registry**: Include a dedicated weather forecast handler.
4. **AI functions:** Develop a specific weather forecast function.

Let's first set up the OpenWeather environment.

Setting up the OpenWeather environment

In `GenAISys_Customer_Service.ipynb`, go to the *Weather* subsection under the *Setting up the environment* section.

First, download the script for retrieving your OpenWeather API key from Google Secrets:

```
download("commons","weather_setup.py")
```

The notebook then runs the API key initialization function:

```
google_secrets=True
if google_secrets==True:
  import weather_setup
  weather_setup.initialize_weather_api()
```

Then, initialize the API key using this function:

```
import requests
import os  # Make sure to import os to access environment variables
# Fetch the API key from environment variables
api_key = os.environ.get('Weather_Key')
if not api_key:
    raise ValueError("API Key is not set. Please check your 
initialization.")
```

You can also set the API key with another method, depending on the environment you are running the notebook in. We begin by double-checking that `api_key` is set:

```
def weather_location(city_name):
    # Fetch the API key from environment variables
    api_key = os.environ.get('Weather_Key')
```

This double-check is not obligatory; it just ensures that the code is robust if the session is interrupted by micro web interruptions, for example. Feel free to remove the redundant check if your environment is stable.

The OpenWeather call is remarkably simple. It requires only your API key and the city name:

```
# OpenWeatherMap API URL for city name
url = f"https://api.openweathermap.org/data/2.5/weather?q={city_
name}&units=metric&appid={api_key}"
```

Note that `metric` is a specific keyword defined by the OpenWeatherMap API, meaning degrees **Celsius (C)**, and also provides the wind speed in **meters per second (m/s)**. If you want US customary units, change metric to imperial, `&units=imperial`, and you will obtain degrees **Fahrenheit (°F)** and windspeed in **miles per hour (mph)**.

We now just have to make the request and retrieve the response:

```
    # Fetch real-time weather data
     response = requests.get(url)
     weather_data = response.json()
```

We will now extract and return the real-time weather information we need for our use case. We will use the current temperature, a brief weather description, and the wind speed:

```
# Extract relevant data
    current_temp = weather_data['main']['temp']
    current_weather_desc = weather_data['weather'][0]['description']
    wind_speed = weather_data['wind']['speed']
    return current_temp, current_weather_desc, wind_speed
```

Note that we only provided the name of the city, not the country. The OpenWeatherMap API has a smart system for handling requests for cities with the same name in different locations. When it searches for "Paris," it defaults to the most prominent and well-known location, which is Paris, France.

When the API receives a request with just a city name (`q={city_name}`), it uses an internal algorithm to determine the most likely intended location. This algorithm prioritizes several factors:

- **Population**: Larger, more populous cities are often ranked higher
- **Significance**: Capital cities and major cultural or economic hubs are given preference
- **Internal Ranking**: OpenWeatherMap maintains its own database and ranking system for locations

Because Paris, France, is a major global capital with a significantly larger population and international recognition than Paris, Texas, the API defaults to the French capital.

Since we are building an educational example using major tourist locations, such as Paris, the algorithm easily defaults to Paris, France. If needed in production, you could modify the function to include the country and country code, as in the following example:

```
def weather_location(city_name, state_code="", country_code=""):
...
query = city_name if state_code: query += f",{state_code}"
    if country_code: query += f",{country_code}"
```

The information that we are returning is sufficient for a generative AI model to interpret the real-time weather forecast and make decisions based on it. With that, we are now ready to add a weather forecast option to the IPython interface.

Adding a weather widget to the interface

We will now add a **Weather** option to the instruction drop-down list as shown in *Figure 9.10*. Why add a weather forecast option to a reasoning list? The motivation comes from the fact that we are building a GenAISys proof of concept. We could imagine several CoT scenarios based on the use cases we will run in our GenAISys. However, if we write these pipelines before having workshops with the end users, they might find the system too rigid.

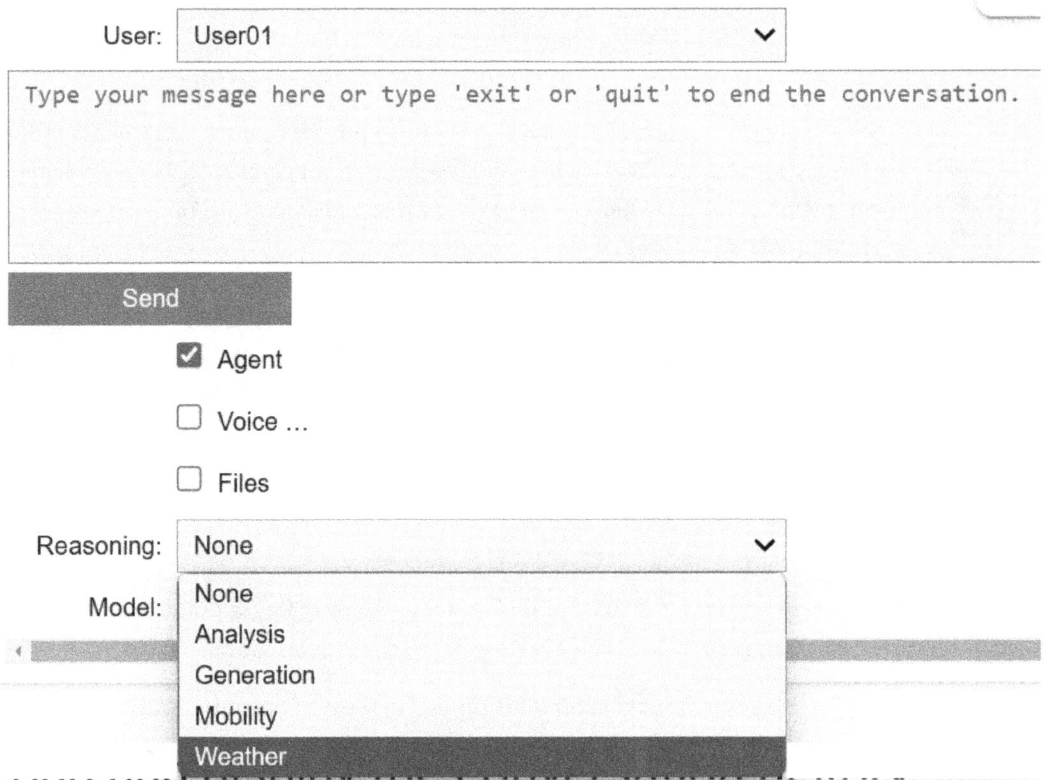

Figure 9.10: Add a weather forecast option to the IPython interface

The best approach is to have some CoT and pipeline scenarios to demonstrate the capabilities of the GenAISys, but leave room for flexibility until the users suggest that we automate some of the CoT scenarios they performed while running the GenAISys.

We will thus add the option to the IPython interface, in `instruct_selector`, leaving the way it is used open to discussion:

```
# Dropdown for reasoning type
instruct_selector = Dropdown(
    options=["None", "Analysis", "Generation","Mobility","Weather"],
    value="None",
    description='Reasoning:',
    layout=Layout(width='50%')
)
```

When running forecasts, users select **Weather** and simply enter a city name. Although adding a city selector or location autocomplete could improve the user experience, a simple text input is more practical for this real-time use case, especially since tourists typically know their exact destination names.

That is all we need at **Layer 1** in the IPython interface. The handler selection mechanism remains unchanged, so we move directly on to the handler registry.

Adding a handle to the handler registry

The weather handler in our registry only requires a one-word location in the user message and the `"Weather"` instruction in this implementation of OpenWeather:

```
# Weather handler: determined by the instruct flag
    (
        lambda msg, instruct, mem, models, user_message,
            **kwargs: instruct == "Weather",
        lambda msg, instruct, mem, models, user_message,
            **kwargs: handle_weather(
                user_message, models=models)
    ),
```

Why not use keywords? Using `"weather"` as a keyword instead of an instruction could be confused with a follow-up question in the following context, as follows:

- User 1 could ask: `What is the weather in Paris?` The assistant would answer: `20°C, clear skies, 5m/s wind speed.`
- User 2 could ask: `What can I visit in this weather?` In this case, `weather` could trigger an API weather call that would fail because no location is provided.

At the time of writing this chapter, even ChatGPT Plus has options before submitting a request, as follows:

- The choice between multiple models such as GPT-4o for general purpose tasks, DALL-E for images, o1-mini for reasoning, o3-mini for high reasoning, GPT-4.5, and more
- A button to activate web search and another one for **Deep Research**
- Manual file uploads

These multiple interactive choices make the interface flexible. However, as we work on user interfaces, we will see a progressive automation of many of these options, along with new generative AI models that encompass the functionality of several former models. It's an ongoing accelerated generative AI evolution!

We will now add the weather function to the AI functions library.

Adding the weather forecast function to AI functions

The real-time weather forecast function first checks whether `api_key` is still active during the session. This additional check ensures that the API call remains stable with no disconnection if there is a micro-interruption of the session in the VM we are using:

```
def handle_weather(user_message, **kwargs):
 # Fetch the API key from environment variables
  api_key = os.environ.get('Weather_Key')

  if not api_key:
      raise ValueError("API Key is not set. Please check your 
initialization.")
```

The city name is the user message; the weather forecast will be requested; and the temperature, description, and wind speed will be returned:

```
    city_name = user_message
    current_temp, current_weather_desc, wind_speed = \
        weather_location(city_name)
    return f"Current Temperature in {city_name}: {current_temp}°C\
nWeather: {current_weather_desc}\nWind Speed: {wind_speed} m/s"
```

We could add a keyword search to a longer user message.

At this stage, the constraint is that the user message must only contain the city's name. However, the functionality can be expanded in various ways based on user feedback gathered during project workshops, such as the following:

- Scan a user message for the location if the weather option is activated.

  Limitation: Only the name of the city is necessary.

- Have a drop-down list of all the possible city locations.

  Limitation: The list could contain hundreds of cities.

- Have a country selection list, then choose a region (or administrative name for a region), then choose the city.

  Limitation: A customer might not know the name of the region or state. Also, this takes longer than just entering the city.

- Automatically detect where the customer is.

  Limitation: The query might be for another city and not the one the customer is in.

- Just enter the city and nothing else, as we are doing now.

  Limitation: We might need to enhance the function with a city keyword search just in case the user enters more words than just the name of the city. Also, the customer might misspell the name of the location.

As you can see, there are several ways to implement the real-time weather forecast request at the user level. The best approach is to initially showcase the basic capabilities of the system clearly and then adapt and extend the interface based on the feedback you receive from workshops with a customer panel.

Now that we've built and integrated the weather forecast functionality into the handler selection mechanism framework, we're ready to demonstrate real-world use cases within the GenAISys.

Running the GenAISys

In this section, we will run multi-user, cross-domain, multimodal interactions using the GenAISys for real-time tourism-related services. Typically, when tourists visit a city, they check weather forecasts in one app and decide on their activities in another. Here, we're merging these domains into one seamless experience. Users will request real-time weather forecasts, and the GenAISys will suggest suitable activities—indoor, outdoor, or both—based on the current weather conditions. This integrated, cross-domain approach can be extended beyond tourism into areas such as construction planning, delivery scheduling, and nearly any scenario where multiple applications are traditionally needed to support user decisions.

The main objective of this section is to demonstrate the cross-domain capabilities of the GenAISys in a flexible manner, illustrating how the system can adapt to various domain-specific scenarios. Specifically, we will focus on the following:

- Activating and deactivating the weather forecast functionality
- Activating and deactivating the file display option

We aim to present a clear and practical proof of concept, highlighting potential workflows before automation decisions are finalized in user workshops. By not prematurely automating too many scenarios, the GenAISys maintains flexibility, allowing actual users first to experiment freely and determine which features should be automated. Remember, the more automation you introduce, the less flexibility you retain in user interactions—though, with careful planning, it is possible to balance both automation and interactivity.

As emphasized earlier, an effective strategy begins with supporting interactive, manual workflows, then progressively automating some or all tasks as users gain familiarity and provide feedback. *Time is not the objective; user maturity is.* Interactive scenarios can transition into fully automated pipelines or CoT processes at the users' pace. By letting user feedback shape development, adoption rates naturally improve.

Generative AI systems rely on probabilistic responses, meaning outputs can vary slightly with each run. This stochastic behavior ensures responses remain dynamic rather than rigid or repetitive.

At the beginning of the notebook (`GenAISys_Customer_Service.ipynb`), you have the option to activate or deactivate the DeepSeek model:

```
# DeepSeek activation deepseek=True to activate. 20 Go (estimate) GPU
memory and 30-40 Go Disk Space
deepseek=False
```

For this chapter, `deepseek` is set to `False`, and we will rely on OpenAI for running our use cases.

With all necessary functions ready for weather-based decision-making, let's now explore interactive use cases in the GenAISys.

A multi-user, cross-domain, and multimodal dialogue

We'll run an interactive scenario with the GenAISys, simulating a tourist seeking real-time, weather-based recommendations in a specific city—whether the weather is cloudy with possible rain or clear. The GenAISys must thus adapt its suggestions to real-time weather forecasts in that specific location.

The dialogue will be multi-user between the following users:

- `User01`: A tourist who expects weather-dependent suggestions on what to do
- `User02`: A customer service agent who provides support
- `User03`: A decision-maker who coordinates marketing, product design, production, deliveries, and customer service

Each interaction in this dialogue is clearly labeled, indicating the user initiating the request. If a new session starts, it will be prefixed by `New Session`.

The interface actions will be as follows:

- Writing and submitting a prompt
- Activating or deactivating the weather forecast function
- Activating or deactivating file displays

There are many enhancements, improvements, additional functions, CoT processes, and pipelines we could add to expand our GenAISys. However, this should be done in collaboration with a customer/user panel.

Let's start the dialogue in Paris.

```
User01: Paris
```

The customer activated the **Weather** function and entered **Paris**, knowing that in this proof of concept, those are the instructions to follow, as shown here:

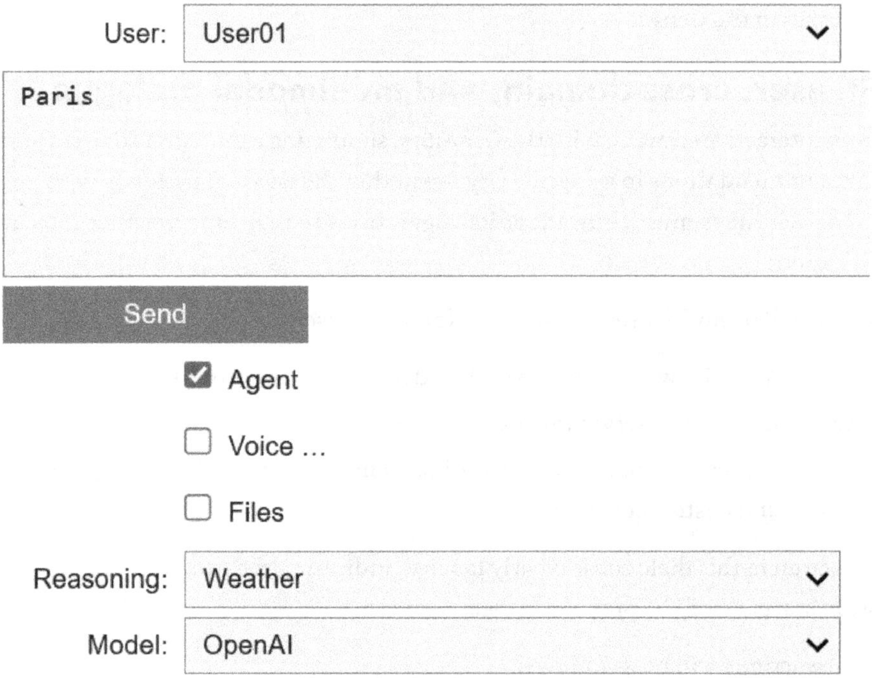

User: User01

Paris

Send

Agent

Voice …

Files

Reasoning: Weather

Model: OpenAI

Figure 9.11: Asking for a real-time weather forecast

The GenAISys agent provides the requested information:

```
Agent: Current Temperature in Paris: 10.68°C
Weather: clear sky
Wind Speed: 3.09 m/s
```

Note that the weather response will continually change, and thus, the whole dialogue that follows will take a different turn. Different locations, different weather conditions, and different points of interest for the customer will take this dialogue in unique directions. The customer now asks for recommendations:

```
User01: What should I first visit with this weather
```

With the **Reasoning** option set to **None**, the GenAI provides personalized suggestions:

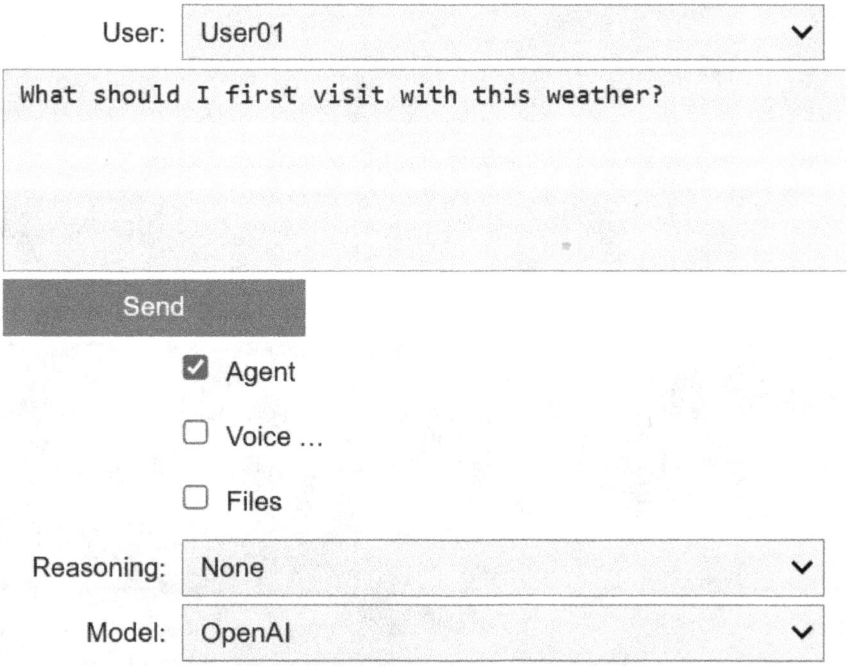

User: User01

What should I first visit with this weather?

Send

Agent

Voice ...

Files

Reasoning: None

Model: OpenAI

Figure 9.12: Asking what activities to do in this weather

The GenAI suggests some engaging activities based on the weather:

```
Agent: With clear skies and cool temperatures in Paris, it's a great
opportunity to explore outdoor attractions. Here are a few suggestions:

1. Eiffel Tower: Start your day with a visit to the iconic Eiffel Tower.
You can enjoy the view from the ground or take an elevator ride to the top
for a panoramic view of the city.
2. Champs-Élysées and Arc de Triomphe: Take a stroll down the famous
Champs-Élysées and visit the Arc de Triomphe. The clear weather will make
for great photos and a pleasant walk.
...
4. Seine River Cruise: Consider taking a Seine River cruise to see many
of Paris's landmarks from the water. The clear sky will provide beautiful
views of the city.
```

```
5. Luxembourg Gardens: Enjoy a leisurely walk in the Luxembourg Gardens.
It's a perfect spot to relax and take in the beauty of the well-maintained
gardens.
Make sure to dress warmly, as the temperature is quite cool. Enjoy your
visit!
```

The user finds the response useful but would also like to visit a museum:

```
User01:But what if I want to only visit one outdoor site and then go to a
museum?
```

Without changing the interface configuration, the agent provides a tailored itinerary:

```
1. Outdoor Site: Eiffel Tower: Start your day with a visit to the Eiffel
Tower. It's an iconic symbol of Paris and offers stunning views of the
city. You can choose to admire it from the ground or take a trip to the
top.

2. Museum: Musée d'Orsay: After visiting the Eiffel Tower, head to the
Musée d'Orsay. It's located along the Seine River and is housed in a
beautiful Beaux-Arts railway station. The museum features an extensive
collection of Impressionist and Post-Impressionist masterpieces by artists
like Monet, Van Gogh, and Degas.

This combination allows you to enjoy both the outdoor beauty of Paris and
its rich cultural heritage. Make sure to check the museum's opening hours
and any ticket requirements in advance. Enjoy your day!
```

Now, the customer service agent steps in to augment the personalized experience of the customer:

```
User02: Create an engaging image of the Eiffel tower and the Musée d'Orsay
museum in Paris for a tourist
```

This kind of multi-user interface could easily be built into a website with features such as text and image generation, customer profiles, and conversation history—your imagination is the only limit.

In this case, ensure **Files** is checked to display the image, as shown here:

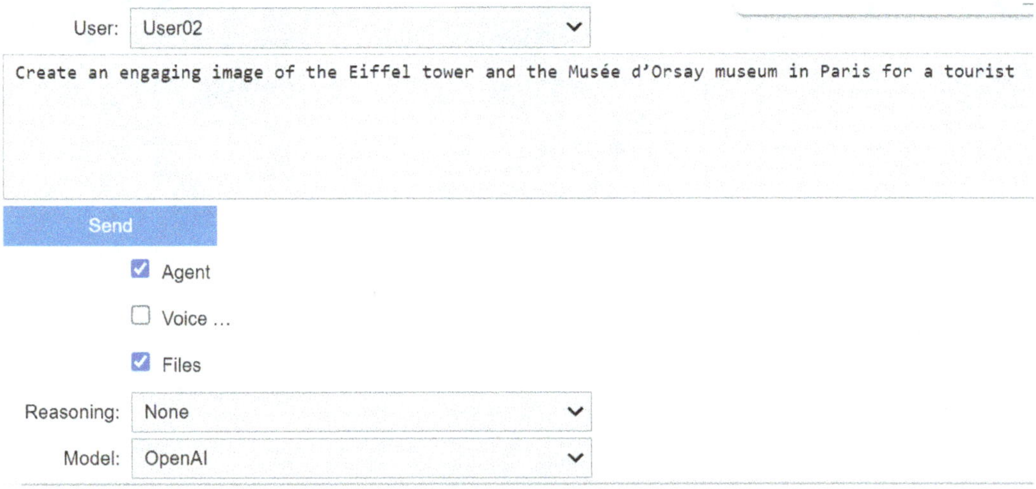

User: User02

Create an engaging image of the Eiffel tower and the Musée d'Orsay museum in Paris for a tourist

Send

☑ Agent

☐ Voice ...

☑ Files

Reasoning: None

Model: OpenAI

Figure 9.13: Creating a custom image for the visit

The GenAISys generates a captivating image representing the recommended custom visit, as displayed here:

Figure 9.14: A custom image for a customized visit

Our GenAISys could be expanded to send the image to the tourist's smartphone, or even better, the GenAISys could be developed as a smartphone app connecting with the GenAISys server.

The image is great and the weather fine. Pleased, `User02` requests a promotional offer:

```
User02: Create an offer with a 10% discount for a Paris Seine River Cruise
```

Files is unchecked so the GenAI can craft a text-based promotional message, as shown here:

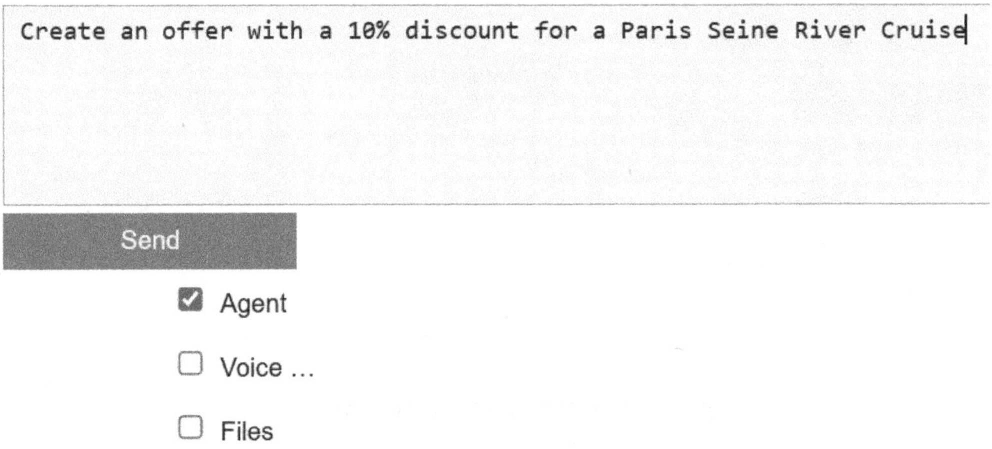

Create an offer with a 10% discount for a Paris Seine River Cruise

Send

☑ Agent

☐ Voice …

☐ Files

Figure 9.15: A custom image for a customized visit

The model generates an engaging message:

```
Agent: Exclusive Offer: 10% Discount on Paris Seine River Cruise!

Experience the magic of Paris from the water with our exclusive Seine
River Cruise. Glide past iconic landmarks like the Eiffel Tower, Notre-
Dame Cathedral, and the Louvre, all while enjoying the stunning views of
the City of Light.

Offer Details:
- Discount: 10% off your Seine River Cruise ticket
- Promo Code: SEINE10
- Validity: Book by [insert date] for travel until [insert date]

Highlights:
- Breathtaking views of Paris's most famous attractions
- Relaxing and scenic cruise experience
- Perfect for couples, families, and solo travelers
```

The river cruise offer is great, but now `User03` steps in with a cutting-edge idea to enhance customer engagement:

```
User03: Create an image of a cool travel bag with all the outdoor
locations a tourist can visit in Paris on a sunny day. Add the Eiffel
Tower and the Musée du Louvre to make it complete.
```

Files is checked again so the agent can produce the requested image, as shown:

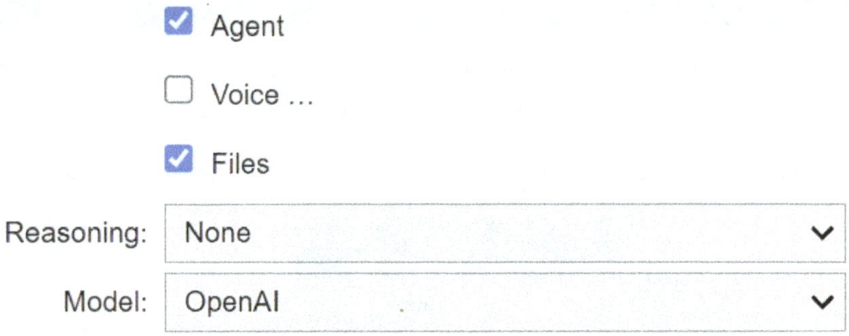

- [x] Agent
- [ ] Voice …
- [x] Files

Reasoning: None

Model: OpenAI

Figure 9.16: Checking files to display an image

The result is the eye-catching design displayed here:

Figure 9.17: A customized travel bag image

The manager forwards this custom design immediately to a local printing partner that maintains travel bag stock and specialized printing equipment. The travel bag will be printed and delivered to the customer's hotel within half an hour—a service as convenient and timely as ordering a pizza! This represents the future of real-time customer service.

For a final touch, the manager asks for an engaging email to be sent to the customer:

```
User03: Write an engaging email for the tourist that will receive this
custom-printed travel bag at the hotel desk within 30 minutes!
```

The message could be a notification in an app, a text message, and/or part of the online conversation on a cutting-edge, multi-user, cross-domain, multimodal website!

In this case, **Files** is unchecked to make room for the GenAI to generate the text, as shown here:

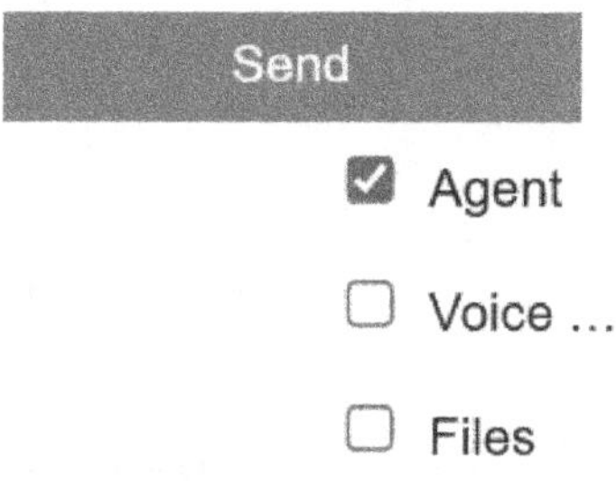

Figure 9.18: Unchecking files to send a message

If the manager wished, **Files** could have remained checked, and the image could have been processed to be attached to the message.

The message is engaging and highlights the exceptional level of customer service offered by the travel agency:

```
Dear [Tourist's Name],

Bonjour and welcome to the enchanting city of Paris!

We are thrilled to inform you that your custom-printed travel bag,
designed exclusively for your Parisian adventure, is ready and waiting for
you at the hotel desk. This isn't just any travel bag; it's your personal
guide to the most iconic outdoor locations in the City of Light, all
beautifully illustrated to inspire your journey.
```

```
….
We hope this bag not only serves as a practical companion but also as a
cherished keepsake of your time in Paris. Whether you're strolling along
the Seine or enjoying a café au lait at a local bistro, let this bag be
your guide to unforgettable memories.

Please visit the hotel desk within the next 30 minutes to collect your bag
and start your Parisian adventure in style.
…
P.S. Don't forget to share your adventures with us on social media using
#ParisWith[HotelName]! We can't wait to see where your travels take you.
```

The "P.S." message is important in a real-time delivery context, as it encourages customers to share their positive experiences on social media. This kind of user-generated content helps brands extend their reach and strengthen their market differentiation as competition intensifies across nearly every industry.

Now let's move to another location, using a different approach. Restart the session by entering `Exit` and rerunning the notebook's cell. The customer service agent will now explore options for customers visiting Tokyo:

```
New Session. User02: Tokyo
```

In this session, `User02`, the customer service agent, and `User03`, the manager, are trying to figure out what to suggest to their customers who have arrived in Tokyo. They want to customize the experience to make sure they keep ahead of the competition. They first want to know the weather for that day, so they select **Weather** (as in *Figure 9.19*) and enter `Tokyo`.

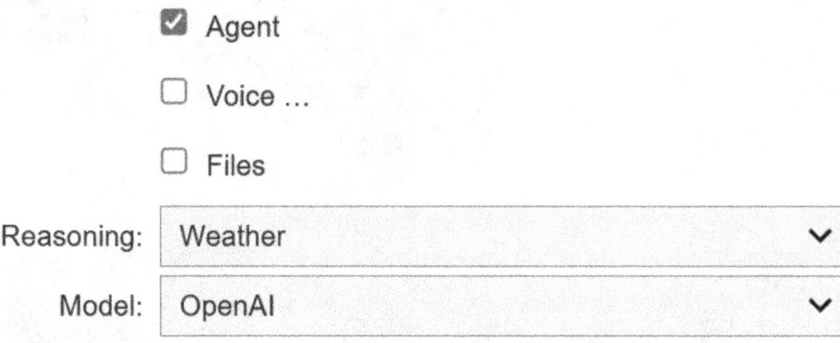

Figure 9.19: Files is unchecked, and Weather is selected

The GenAISys returns the weather in real time:

```
Agent: Current Temperature in Tokyo: 9.51°C
Weather: moderate rain
Wind Speed: 10.8 m/s
```

Now, `User02` would like to know what to suggest for this weather:

```
User02: What places would you suggest a tourist visit in this weather?
```

The **Weather** function is deactivated (see *Figure 9.11*). The GenAI must analyze the weather and then suggest activities.

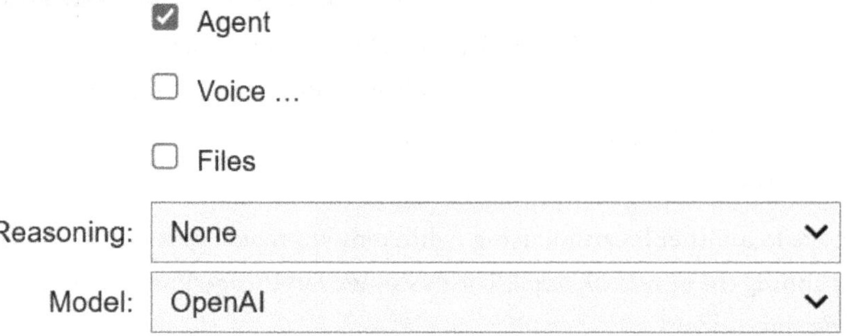

Figure 9.20: Files is unchecked, and Reasoning is set to None

The agent takes the weather into account and makes interesting suggestions:

```
Agent: In moderate rain and cooler temperatures, it's a good idea to
suggest indoor activities or places where tourists can enjoy themselves
without being too affected by the weather. Here are some suggestions for a
tourist visiting Tokyo in such conditions:

1. Museums and Art Galleries:
- Tokyo National Museum: Located in Ueno Park…
- Mori Art Museum: Situated in Roppongi Hills…

2. Shopping and Entertainment Complexes:
- Odaiba: This man-made island has shopping malls…
- Shibuya Scramble Square: Offers shopping, dining, …
```

```
3. Aquariums and Zoos:
- Sumida Aquarium: Located in…
- Ueno Zoo: While some parts…
…
```

We can see that the GenAISys is adapting to different scenarios in an acceptable way.

Now, `User02` is getting some ideas from the response. Since the weather is rainy, why not suggest a bus tour?

```
User02: Ok. Write an engaging email to offer a bus tour for some of these locations to visit in one day. Don't make it too dense. Make the bus tour nice and cool.
```

The configuration (**Files** unchecked, **Reasoning** set to **None**) does not change. The GenAISys makes an engaging offer for a bus tour:

```
Agent: Subject: Discover Tokyo's Hidden Gems with Our Exclusive Bus Tour!

Hi [Name],
Are you ready to explore Tokyo like never before? Join us for a one-of-a-kind bus tour that takes you to some of the city's most fascinating indoor attractions, perfect for a day of adventure, rain or shine!
Tour Highlights:
…
Our comfortable, climate-controlled bus ensures a relaxing journey between stops, with a friendly guide sharing insider tips and stories about each location. Plus, enjoy complimentary snacks and refreshments on board!

Tour Details:
- Date: [Insert Date]
- Time: [Insert Time]
- Pick-up Location: [Insert Location]
```

`User03` steps in and asks the GenAISys to write an engaging message for food and beverages:

```
User03: Write an engaging email for the food and beverages for tourists taking a bus tour of Tokyo with some stops. Do not mention the stops at all or the locations. Just focus on the super discount prices for food and beverages. Make sure to have food with and without meat. No alcohol on the bus is allowed. Also, think of some gluten-free food.
```

The configuration (**Files** unchecked, **Reasoning** set to **None**) does not change. The GenAISys makes an engaging offer for food and beverages during the bus tour. The idea is to show the customers that the travel agency is taking care of them in great detail.

The GenAISys understands the situation and provides an engaging message:

```
Agent: Subject: Delicious Deals Await on Your Tokyo Bus Tour!

Dear Traveler,

Get ready to embark on a delightful culinary journey as you explore Tokyo
with us! We're thrilled to offer you an array of mouth-watering food and
beverage options at unbeatable prices, ensuring your taste buds are as
entertained as your eyes.

Savor the Flavors:

- Meaty Marvels: …
- Veggie Delights: …
- Gluten-Free Goodies:…
Thirst-Quenching Beverages: Stay refreshed with our selection of non-
alcoholic beverages…
Exclusive Offer:…
…
P.S. Remember, no alcohol is allowed on the bus so that you can enjoy a
safe and family-friendly environment throughout your adventure!
```

We can see that the GenAISys is flexible and leverages the generative AI models efficiently.

Let's end our travelling journey experience in New York. Exit the session and rerun the IPython cell.

`User02` wants to anticipate activities in New York:

```
New session. User02: New York
```

In this session, `User02` (customer service agent) and `User03` (manager) are brainstorming personalized recommendations for tourists arriving in Tokyo. They want to tailor the experience to stay ahead of competitors. First, they check the current weather by selecting the **Weather** option (*Figure 9.21*) and entering `Tokyo`:

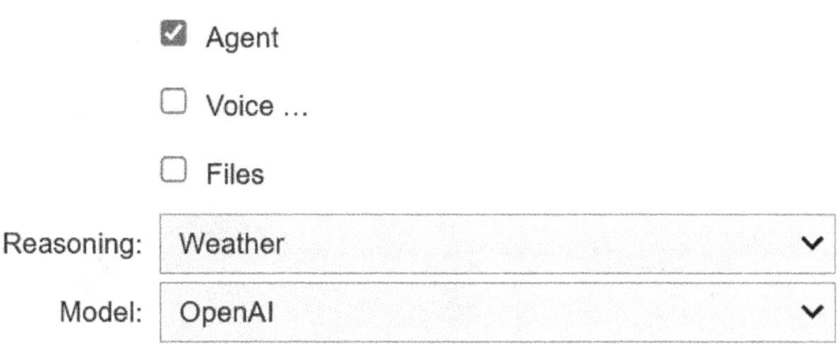

Figure 9.21: Files is unchecked and Weather is selected

The agent provides the weather in real time:

```
Agent: Current Temperature in New York: 10.06°C
Weather: mist
Wind Speed: 4.12 m/s
```

`User02` thinks that misty weather is not the best day to take customers on a tour of Central Park. What can the GenAISys come up with?

```
User02: I'm a tourist customer service agent. Considering the weather, I
was thinking of suggesting an exciting activity to my tourist group. Find
one activity for the morning that could be exciting.
```

`User02` runs the GenAISys with **Files** unchecked and no **Reasoning** instruction (*Figure 9.22*). The situation is not that easy because we want the customers to visit new places and possibly surprise them a bit:

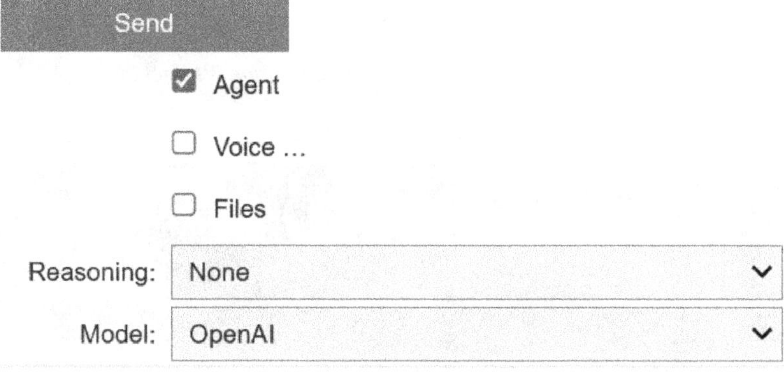

Figure 9.22: Files is unchecked and Weather is not selected

The GenAISys finds an interesting program that can add a cultural touch to the trip:

```
Agent: Considering the misty weather in New York, a great morning activity
for your tourist group could be visiting the American Museum of Natural
History. It's an exciting indoor experience where they can explore
fascinating exhibits, including dinosaur fossils, the Hall of Ocean Life,
and the Rose Center for Earth and Space. This activity will keep them
engaged and comfortable regardless of the weather outside.
```

At the end of the trip to New York, `User03`, the manager, steps in to offer the customers a souvenir T-shirt:

```
User03: The hotel was great for a tourist group in New York because the
view reminded some of them of when they first came to  New York with their
parents. I'm sure they would have like to receive a custom T-shirt as a
souvenir to remind them of that experience. Make sure that the image you
produce shows a T-shirt.
```

`User03` decides to activate a full-scale GenAISys CoT generation to analyze customer feedback and produce a nice T-shirt. **Files** is checked and **Generation** is activated, as shown:

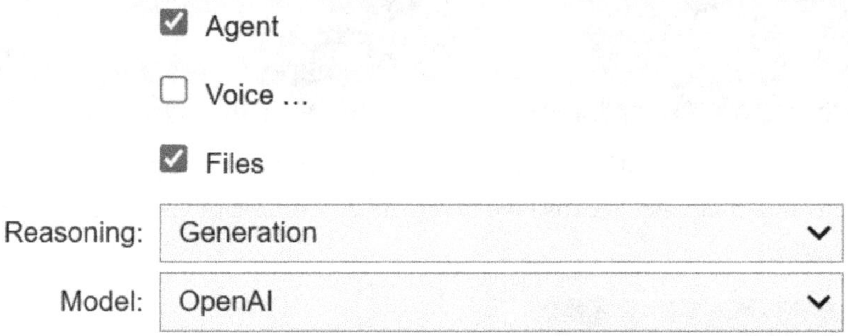

- [x] Agent
- [ ] Voice …
- [x] Files

Reasoning: Generation

Model: OpenAI

Figure 9.23: Files is checked and Generation is selected

Our GenAISys begins the generation process step by step:

```
"The hotel was great for a tourist group in New York because the view
reminded some of them of when they first came to New York with their
parents.I'm sure they would have like to receive a custom T-shirt as a
souvenir to remind them of that experience.Make sure that the image you
produce shows a T-shirt."
```

It first performs a memory and sentiment analysis of the feedback the tour guide detected:

```
Agent: Process: Performing memory and sentiment analysis.

Memory analysis result: Segment 1: "The hotel was great for a tourist
group in New York because the view reminded some of them of when they
first came to New York with their parents."
- Memory Tags: [Episodic Memory, Time Memory Past, Reality Memory]
- Dimension: Emotional
- Sentiment Score: 0.8
- Explanation: This segment references a real past event (their first
visit to New York with their parents) that triggers personal, nostalgic
recollections, placing it in episodic and time (past) memory while
affirming its reality and positive emotional tone.
```

It then extracts the scores from the content it analyzed:

```
Process: Extracting scores from response.

Memory analysis result: Episodic Memory sentiment 0.8
Time Memory Past sentiment 0.8
Reality Memory sentiment 0.8
Episodic Memory sentiment 0.75
Time Memory Past sentiment 0.75
Memoryless sentiment 0.5
```

It then performs a statistical analysis to provide an overall rating:

```
Process: Statistical analysis
Extracted sentiment scores: [0.8, 0.8, 0.8, 0.75, 0.75, 0.5]
Extracted sentiment scores: 0.73
Scaled rating (0-5): 3.65
Process: Creating content
```

The score is not 4 or 5 because they would like a souvenir, a T-shirt, and they don't have it yet! So, the GenAISys now creates a prompt for image generation:

```
Prompt created for image generation: "Design a custom T-shirt featuring
the iconic New York skyline or a memorable landmark like Times Square or
Central Park, capturing the essence of a nostalgic and cherished tourist
experience."
```

The GenAISys now creates an image for the prompt (*Figure 9.24*):

```
Process: Creating an image
Image created
```

Figure 9.24: A custom T-shirt

The GenAISys now creates a message for the customer:

```
Process: Creating a message.

Customer message: Dear Customer,

We invite you to design a custom T-shirt that beautifully showcases the
iconic New York skyline or a memorable landmark such as Times Square or
Central Park. This design aims to capture the essence of a nostalgic and
cherished tourist experience.

Best regards,
```

For the New York trip, the manager will now instruct the local printer to print the custom image onto a batch of T-shirts already stocked at the printer's site. The shirts will be produced and delivered to the tourist's hotel within the hour.

The travel agency is extremely satisfied with the GenAISys proof of concept and is now preparing to organize workshops with all team members to take the system to the next level.

Let's now summarize our journey—and get ready to explore what's next!

Summary

The acceleration of modern markets has led to the extinction of waiting times. *Real-time responsiveness is now the default.* The use case explored in this chapter demonstrates that a business-ready GenAISys proof of concept must be able to provide immediate, real-world services. We implemented a real-time travel agency scenario—one that can easily be adapted to other domains once you enter project mode.

To make our proof of concept suitable for live demonstration to a team, customer, or investor, we first integrated a moderation component into GenAISys. We implemented OpenAI's Omni model to detect and flag multiple categories of content policy violations. This moderation function is embedded at the handler selection mechanism level, ensuring that inappropriate user messages are flagged and rejected before any further processing. To further reassure stakeholders, we added a data security layer. Using vector similarity search, we built a function that detects whether a user message aligns with sensitive topics. We created a dedicated Pinecone namespace to store these topics and queried it against every incoming user message. If a match was found, the system flagged the content, displayed a warning, and rejected the request.

All flagged moderation and security violations were logged in separate files, and the full conversation history can be traced to individual users. With these essential safeguards in place, we proceeded to implement a real-time weather forecasting feature using OpenWeather—tightly integrated into GenAISys for weather-aware decision-making. Finally, we walked through a complete GenAISys dialogue, showcasing the system's ability to support real-time, multi-user, cross-domain, multimodal interactions—a powerful illustration of what generative AI can achieve in production settings.

The next step? Presenting your GenAISys proof of concept to secure the resources needed to expand and scale the system. But such a presentation must be carefully crafted—and that's exactly what we'll focus on in the next chapter.

Questions

1. Sensitive topics can be detected with vector similarity searches. (True or False)
2. A flagged user message containing a sensitive topic in the GenAISys does not stop the dialogue. (True or False)
3. The OpenAI moderation tool only has one category for foul language. (True or False)
4. A flagged user message containing content that violates the content policy of the GenAISys does not stop the dialogue. (True or False)
5. Weather forecasting in real time to automatically suggest activities can give a company a competitive advantage. (True or False)
6. A multi-user, cross-domain, multimodal GenAISys is the future of apps. (True or False)
7. Multi-user, cross-domain, multimodal apps don't exist yet. (True or False)
8. Adding GenAISys to apps will significantly boost productivity and engage users. (True or False)
9. A GenAISys should be developed with a customer panel. (True or False)

References

- OpenAI moderation: `https://platform.openai.com/docs/guides/moderation`
- The OpenAI Moderation object: `https://platform.openai.com/docs/api-reference/moderations/object`
- OpenWeather: `https://openweathermap.org/`

Further reading

- OpenWeather's solar radiance research: `https://home.openweathermap.org/solar_irradiance_history_bulks/new`

10

Presenting Your Business-Ready Generative AI System

Creating an innovative GenAISys alone will not get us customers, funding, or possibly even recognition. In today's fiercely competitive landscape, hundreds, if not thousands, of AI solutions, agentic systems, and similar offerings are vying simultaneously for market attention. Businesses, investors, and end users face an overwhelming flood of options—from industry giants such as Google Cloud, Microsoft Azure, **Amazon Web Services** (**AWS**), and IBM to an ever-expanding array of start-ups and mid-sized companies. Where does that leave us, and what should we do?

We need two things to penetrate the market's thickening wall: *implementation* and *profitability*. In this book, up to now, we have built a proof of concept of our ability to deliver an effective GenAISys capable of supporting core business functions across the supply chain—from marketing and production to delivery and customer service. But now, we must turn our technical achievements into a compelling story to effectively communicate and sell our ideas to stakeholders—be it internal teams, employers, potential customers, or investors. To successfully present our GenAISys, we need a well-crafted demonstration scenario that highlights our existing capabilities and clearly conveys our capacity for further innovation and expansion. Deciding exactly when to pause development and transition into presenting the proof of concept is critical.

In this chapter, the method described is only one of the many ways to achieve a successful presentation of a GenAISys amid relentless AI competition. We will focus on getting straight to the point. Given that the attention span of present-day AI-informed audiences is limited, capturing the attention of your audience will prove challenging. The chapter will thus focus on the first seven minutes of a presentation of our GenAISys. We will transition smoothly from the IPython-based interface developed throughout this book into a flexible, easily adaptable frontend web page. This strategy allows us to quickly customize our demonstration to address specific project requirements, supported by the credibility of our working GenAISys prototype. We will systematically present the key components: core functionality, vector store integration, essential **key performance indicators (KPIs)**, external database connectivity, and critical aspects of security and privacy. Additionally, we will emphasize how the system can be collaboratively customized through stakeholder workshops.

Finally, to showcase the ultimate flexibility and future potential of our framework, we will introduce a prospective evolution of our GenAISys: a swarm-based **multi-agent system (MAS)**, demonstrating a powerful new paradigm of human-AI co-worker collaboration. By the end of this chapter, you will know how to combine the actual GenAISys features with the potential evolutions that can be added to the system.

This chapter covers the following topics:

- Designing the presentation of the GenAISys
- Building a flexible frontend web interface
- Presenting the basic GenAISys functions
- Showing the possible evolutions in terms of integration, security, and customizations
- Introducing a MAS

Let's start by carefully designing the presentation of our GenAISys.

Designing the presentation of the GenAISys

Designing the presentation for the educational GenAISys we've built throughout this book goes beyond the technical dimension. Whether your audience is an internal team, an investor, or a client, the success of your GenAISys will largely depend on clearly communicating its *business value*.

A powerful GenAISys presentation must begin with a compelling introduction, especially during the first few critical minutes, as shown in *Figure 10.1*:

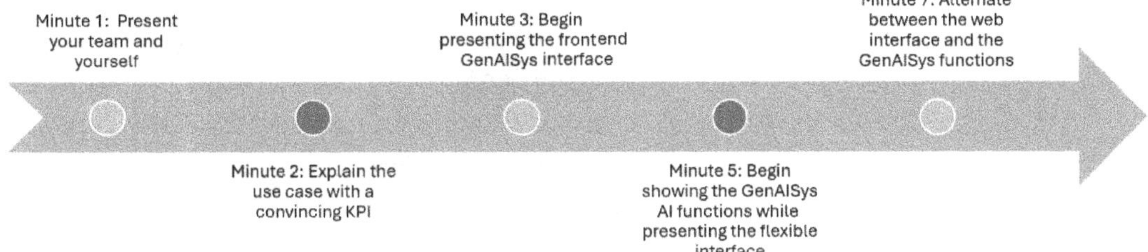

Figure 10.1: A fast-track presentation

The timing indicated (in minutes) is flexible and can be adjusted according to your needs. However, modern audiences expect concise, impactful, and effective presentations. Keep your audience engaged by clearly demonstrating expertise and efficiency. For simplicity, we'll refer to any audience—your internal team, investors, or clients—as the *customer*.

The customer will expect the following timeline:

- **Minute 1 – Presenting your team and yourself**: Present your team and yourself briefly. Highlight your strong points and then begin the presentation before they get bored. The length of this introduction is up to you, but don't overdo it.
- **Minute 2 – Explain the use case with a convincing KPI**: Make sure that the use case you present is profitable, whether it's marketing growth, production resource optimization, reduced delivery times, better customer service to increase sales, or any other activity booster. Also, explain that this KPI is achievable through precise context engineering, which ensures the AI performs its tasks efficiently and accurately. Very briefly introduce the generative AI foundation of the project. Then, rapidly explain that the KPI can be added to the interface and displayed in real time if requested, as shown:

Generative AI Chat Interface

Connected to Pinecone Index

Gross Margin

$577.78

Figure 10.2: Real-time KPI

- **Minute 3 – Begin presenting the frontend GenAISys interface**: No matter how hard you work, something will always be missing. Also, the GenAISys we built requires explanations at a higher level with a PowerPoint presentation and frontend web page that we will begin preparing in the *Building a flexible HTML interface* section of this chapter. The HTML page incorporates the functionality we built in the GenAISys and takes it further with ideas that we need for the presentation and ideas the customer wants. Show that you can rapidly adapt a flexible HTML page to meet customer needs interactively in workshops, as we will see in the *6. Customization* section.
- **Minute 5 – Begin showing the AI functions while presenting the flexible interface**: Continue the concept of working together to finalize the GenAISys interface and integration in a collaborative, human-centric AI approach. Emphasize the unique blend of next-generation technologies integrated in the GenAISys, such as OpenAI, DeepSeek, and Pinecone. Keep the presentation as human-centric as possible, showing that the goal is to leverage the power of the multi-user GenAISys you built to increase performance with AI as *copilots*, not replacements. Keep this in mind as much as possible to convey a collaborative spirit that will help build trust in your system. You can begin to show some of the functionalities of the educational GenAISys we built in this book. You can also show more if you went further to adapt to the use case.
- **Minute 7 – Alternate between the web interface and the GenAISys functions**: Now, it's time to show how your GenAISys can provide an efficient service by combining implementation and profitability. Navigate between the web interface and the GenAISys you built. Show your expertise with the GenAISys components. Demonstrate your flexibility and creativity with a web interface that can be rapidly adapted to meet the customer's needs.

By carefully structuring your presentation this way, you demonstrate clear expertise and practical flexibility. You've set the stage to convincingly showcase your business-ready GenAISys.

Next, let's build out the flexible web interface, preparing the foundation to seamlessly guide your audience through your fully integrated and user-friendly GenAISys.

Building a flexible HTML interface

The demonstration web interface needs to be flexible and adaptable. In some cases, we might even have to adapt the code on the spot during a coffee break! A customer might insist on adding a logo, changing the font, or changing the color. If we resist, this may become a stumbling block that casts a shadow over the whole presentation! We need to walk a fine line between adaptability and stability in this new era of generative AI-aware customers. Hence, in this section, we will first build an interface with GenAISys's core functionality, keeping flexibility in mind. You can have the web page open in your HTML editor and refresh it in real time if the customer insists on seeing a modification immediately. Remember, everybody has access to AI copilots and can overtake us in real time. We need to be faster than the competition, which could also be an end user.

The goal is to show that the IPython proof-of-concept interface is independent of the underlying AI orchestration functions. Thus, it can be adapted to any environment necessary for a project. In this case, we are choosing to show what a web page on a dedicated server would look like. Let's get started by opening `GenAISys_Presentation.ipynb` within the Chapter10 directory on GitHub (`https://github.com/Denis2054/Building-Business-Ready-Generative-AI-Systems/tree/main`).

First, we download the helper scripts and web pages required for the demonstration:

```
!curl -L https://raw.githubusercontent.com/Denis2054/Building-Business-Ready-Generative-AI-Systems/master/commons/grequests.py --output grequests.py

from grequests import download
download("Chapter10","01.html")
download("Chapter10","02.html")
download("Chapter10","03.html")
download("Chapter10","04.html")
download("Chapter10","05.html")
download("Chapter10","06.html")
```

Then, we define a reusable Python function to conveniently load and display the HTML interface within the notebook:

```
from IPython.display import HTML, display
def display_interface(filename):
    with open(filename, "r", encoding="utf-8") as file:
        html_content = file.read()
    display(HTML(html_content))
```

In this section, we'll carefully walk through the code for `01.html`. The goal is to ensure you clearly understand the interface's structure and logic. This clarity is critical for responding swiftly to potential customer requests during the presentation.

The code begins with a standard `<head>` section, which includes both metadata (such as `<meta charset="UTF-8">` and `<title>`). The CSS, visual styling, is embedded inside the `<style>` tag:

```
<!DOCTYPE html>
<html lang="en">
<head>
  <meta charset="UTF-8">
  <title>Generative AI Chat Interface</title>
  <style>
    body {
      font-family: Arial, sans-serif;
      margin: 20px;
      background: #f4f4f4;
    }
….
  </style>
</head>
```

Both the `<head>` and `<body>` sections are fully customizable based on the project's specific branding and design guidelines. In this case, the `<body>` container will wrap all the content and apply the CSS layout styling we just defined:

```
<body>
  <div class="container">
```

The header doesn't contain the term "Generative AI System." The choice, in this case, is to provide an accessible term that a broader audience can relate to:

```
<h1>Generative AI Chat Interface</h1>
```

The user selection block contains the three generic usernames we have been using throughout the book. It provides flexibility to adapt to any questions about who they are, depending on the project's specifications. You can choose to provide domain-specific names depending on the context of your presentation. You could also add more users. For the moment, let's keep it simple:

```
<div class="form-group">
  <label for="userSelector">User:</label>
  <select id="userSelector">
    <option>User01</option>
    <option>User02</option>
```

```
      <option>User03</option>
    </select>
  </div>
```

The user input message area contains standard information. We can modify it to suit an industry or task. It could even vary depending on the user and be domain-specific, such as "enter the customer's review here." In this case, we will display the message we have been using to build our GenAISys:

```
  <div class="form-group">
    <label for="messageInput">Your Message:</label>
    <textarea id="messageInput" placeholder="Type your message here or type
  'exit' or 'quit' to end the conversation." rows="4"></textarea>
  </div>
```

A standard send button is then implemented, but keep in mind that it can be visually customized according to specific branding requests during your presentation:

```
  <div class="form-group">
    <button id="sendButton">Send</button>
  </div>
```

Let's now add the checkbox options/widgets for the AI agent to enable voice or activate file management:

```
  <div class="checkbox-group">
    <label>
      <input type="checkbox" id="agentCheckbox" checked>
      Agent
    </label>
    <label>
      <input type="checkbox" id="voiceCheckbox">
      Voice Output
    </label>
    <label>
      <input type="checkbox" id="filesCheckbox">
      Files
    </label>
  </div>
```

Study the code, regardless of how it was designed, to be prepared to modify it in real time if the situation becomes tense over how something is displayed. This will show that you are flexible and can easily adapt to customer needs. It can be tricky, but it might be necessary. Just make sure to modify something only if the customer insists so as to avoid taking unnecessary risks.

The **Reasoning** dropdown highlights the powerful reasoning capabilities within our GenAISys, as implemented throughout the previous chapters. You can readily adjust or add reasoning options to match your project specifications:

```
<div class="form-group">
  <label for="reasoningSelector">Reasoning:</label>
  <select id="reasoningSelector">
    <option value="None" selected>None</option>
    <option value="Analysis">Analysis</option>
    <option value="Generation">Generation</option>
    <option value="Mobility">Mobility</option>
  </select>
</div>
```

The model selection block contains strategic information. The customer will want to know whether the model is secure and open source, or which country it originates from. We can add other models or explain that we can add more models with the flexibility of the drop-down list:

```
<div class="form-group">
  <label for="modelSelector">Model:</label>
  <select id="modelSelector">
    <option value="OpenAI" selected>OpenAI</option>
    <option value="DeepSeek">DeepSeek</option>
  </select>
</div>
```

Finally, the output area is a standardized section for displaying conversation responses clearly and legibly:

```
<div class="output-area">
  <p><em>Conversation output will appear here...</em></p>
</div>
```

We are now ready to write the closing tags and open the file:

```
    </div>
</body>
</html>
```

The `01.html` file was created with a basic text editor, uploaded to our GitHub repository, and then downloaded automatically by our notebook. You can directly open the file in a browser or within your presentation environment. In this case, we will now open it in the notebook.

The choice of whether to use HTML for the presentation is yours. For this particular strategy, a PowerPoint presentation may be too static, while directly running Python code risks distracting top executives from core concepts. An HTML page strikes an effective balance—more dynamic than static slides, yet clearer and more flexible than live code.

I would advise, however, that you keep your focus on the message, not the medium. Articulate the human-centric and automated gains that will take your audience to the next level.

Let's now outline the seven-step roadmap we'll follow to effectively showcase our GenAISys before exploring each component in greater depth in the upcoming sections:

1. **Present the core GenAISys** by focusing on the practical features of the GenAISys. Don't get lost in low-level technical terms about Python, CSS, HTML, or anything else. There is only one goal: to show the audience how this interface can help them build a human-centered, next-generation GenAISys.
2. **Present the vector store** to show how an organization's data can be safely stored in an innovative manner. The concept to convey is that the data is not static and stored in a single location, but rather dynamic. We can access the data directly, just like a GPS in our car.
3. **A human-centric approach to KPIs** to show that although we can automate many tasks, a human-centered system can be a profit center. If humans have real-time KPIs to prove productivity gains in real time, a powerful collaborative human-AI team will take a company to the next level.

4. **Integrating platforms and frameworks** can be an option. It will be up to the company to decide. We demonstrate our expertise in building a GenAISys from scratch. We are thus valuable assets to deploy the complex functions that any GenAISys requires, whether through a platform or not. We will also demonstrate that we can integrate powerful MASs to enhance the company's productivity.
5. **Security and privacy** are key components of a project. We will demonstrate that we can implement standard practices in this field based on the best practices and regulations currently available on the market.
6. **Customization** is our brand image! We have demonstrated our ability to build a cutting-edge GenAISys from scratch. We are thus able to customize any AI features that are required.
7. **GenAISys resources (RACI)** are our realistic touch. We know how hard it is to build a GenAISys. We recognize that transitioning from a model to a system is a complex process. We recognize that the one-to-one, user-to-copilot experience alone is insufficient. We have a one-to-many vision, an architecture that scales to groups of users, which requires careful resource planning.

With that roadmap in place, let's begin by presenting the core GenAISys.

1. Presenting the core GenAISys

Navigate to the *1. The Generative AI Chat Interface* section of the notebook, and run the following cell to display the `01.html` page:

```
display_interface("/content/01.html")
```

This output mirrors exactly the IPython interface we have carefully developed throughout the book. However, by displaying it as a web page, we emphasize that our GenAISys interface is platform-independent—it can be hosted anywhere, embedded in presentations, or accessed via web browsers, as illustrated in *Figure 10.3*:

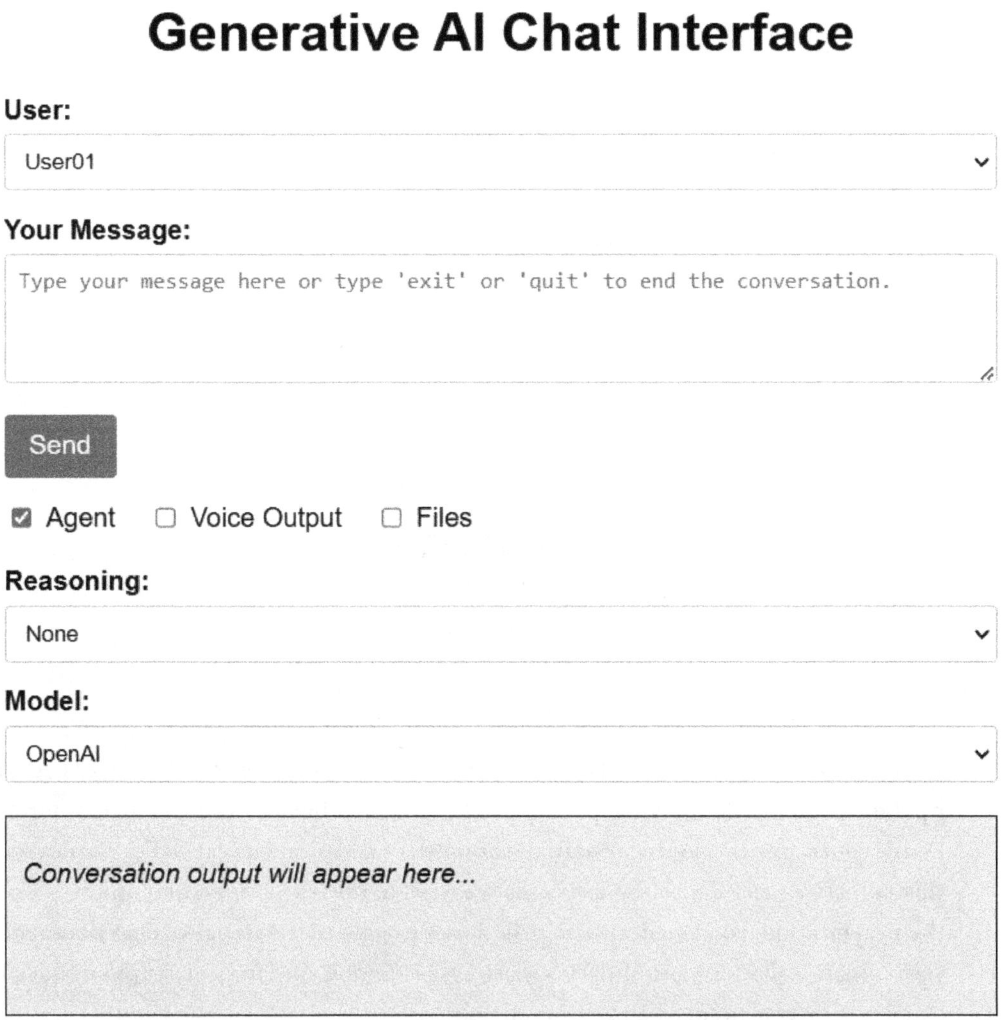

Generative AI Chat Interface

User:

User01

Your Message:

Type your message here or type 'exit' or 'quit' to end the conversation.

Send

Agent Voice Output Files

Reasoning:

None

Model:

OpenAI

Conversation output will appear here...

Figure 10.3: Generative AI Chat Interface

The descriptions provided during your presentation need to be user-friendly and intuitive. You must adapt your level of technical terminology to the audience. Start with clear terms and then expand when necessary to provide more technical or even more straightforward explanations as needed. The flexibility of our explanations must match the flexibility of the GenAISys we've built.

The following are some ways to address the many questions an audience can ask and what we must address. The technical details have been described throughout the book. The idea here is to prepare you to answer tricky questions. Let's go through each component:

- **Generative AI chat interface**: This term is intuitive. The customer can choose any other term that suits the project's needs.
- **User**: Explain that your GenAISys possesses multi-user functionality. We can limit the user to one or expand it to *n* users. Each user can have a specific name, such as "John Doe," a role, such as "Marketing," or any other username depending on the project. In this case, the user is selected from a list:

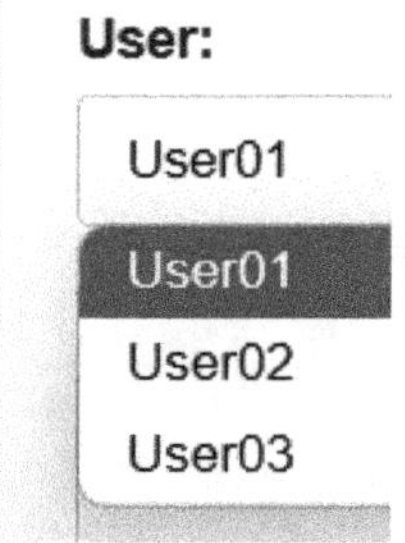

Figure 10.4: Manual or automatic user selection

- **Your message**: We need to explain that the user message can contain keywords decided by the customer. The messages will thus trigger AI behaviors such as text generation, image generation, reasoning (**chain of thought**, or **CoT**), and advanced text analysis. At this point, we can refer to the use cases we wish to present. Depending on the scope of the presentation, you can decide to drill down (toggle to the Google Colab notebook or your local installation) and show your use cases or not. Don't forget to explain that a full conversation can be saved and summed up.
- **Agent, Voice Output, and Files checkboxes**: We can explain these options and decide whether to drill down or not, depending on the customer's interest. You can also expand with creative ideas if they are within the scope of the project the customer is looking for.
- **Reasoning**: The reasoning agents will show how effective your GenAISys can be. Make sure to describe the options in *Figure 10.5* to showcase the power of the GenAISys when it comes to neuroscientific text analysis, CoT generation, or generic mobility features. Briefly explain the CoT approaches discussed in this book. Go into detail if asked using Google Pro or your local installation (personal computer or server).

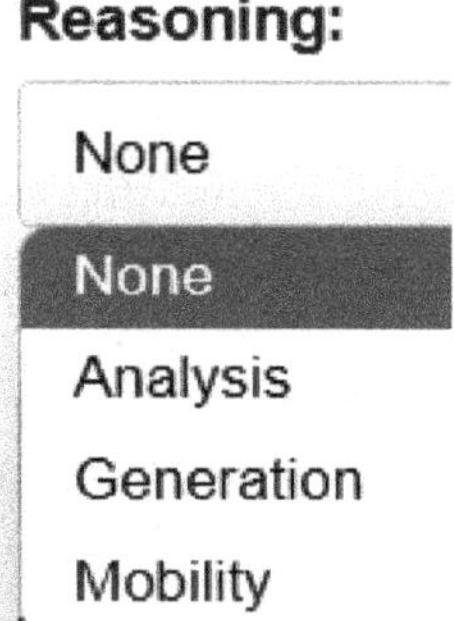

Figure 10.5: Selecting a reasoning function

- **Model**: The choice of model, as shown in *Figure 10.6*, perfectly illustrates the system's flexibility. The customer can decide to use an API or a locally installed open source model. Explain that the GenAISys can support other models if necessary.

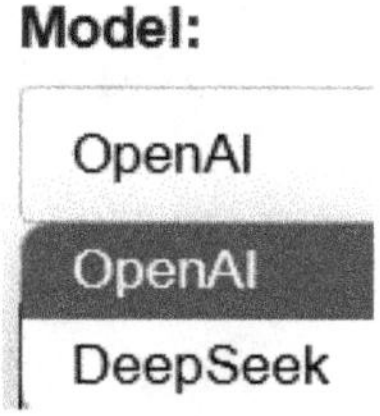

Figure 10.6: Selecting a generative AI model

- **Output area**: We can explain that the output generates text and images. If requested, we can decide whether to show an example.

We can conclude by explaining that the GenAISys interface is just the frontend. The GenAISys proof of concept is a sophisticated, scalable, and secure structure. The interface provides a seamless, multi-user, high-performance system that allows interactions from around the world. Also explain that this proof of concept can be deployed seamlessly onto cloud services such as AWS, Microsoft Azure, IBM Cloud, or Google Cloud. At this point, we can delve into the details of the functionality developed in this book and adapted to the specific use case at hand.

2. Presenting the vector store

A vector store is an important component of the GenAISys you built. First, explain that Pinecone was implemented, but that you can implement another vector store if needed. Then, demonstrate how visual elements can inform the user of the Pinecone index's status on the frontend interface.

Start by duplicating `01.html` to a new file named `02.html` to keep the interface's initial presentation as it was. Why? A fully populated web interface can confuse your audience, but a step-by-step approach assures you don't lose anybody along the way. Also, maybe the customer might not want to see the Pinecone connection. Or the customer might not want to go further and start the project with `01.html`. We must remain flexible and adaptable in case of all these scenarios.

In `02.html`, first add a clear Pinecone connection indicator:

```
<div class="container">
  <h1>Generative AI Chat Interface</h1>
  <!-- Pinecone Connection Status -->
  <div id="pineconeStatus" style="text-align: center; margin-bottom:
10px;">
    <span style="color: green;">&#9679;</span> Connected to Pinecone Index
  </div>
  <!-- Existing form elements -->
  ...
</div>
```

The customer may also want to review the Pinecone retrieval before augmenting the input to the generative AI model. We could add a new section, in this case:

```
<div class="output-area">
  <p><em>Conversation output will appear here...</em></p>
</div>
<div id="pineconeResults" style="background: #fff; padding: 10px; border:
1px solid #ccc; margin-top: 10px;">
  <h3>Context Retrieved from Pinecone</h3>
  <p><em>No results yet.</em></p>
</div>
```

We can then run the cell to display the interface:

```
display_interface("/content/02.html")
```

The interface now contains the possible enhancements we could apply, as shown here:

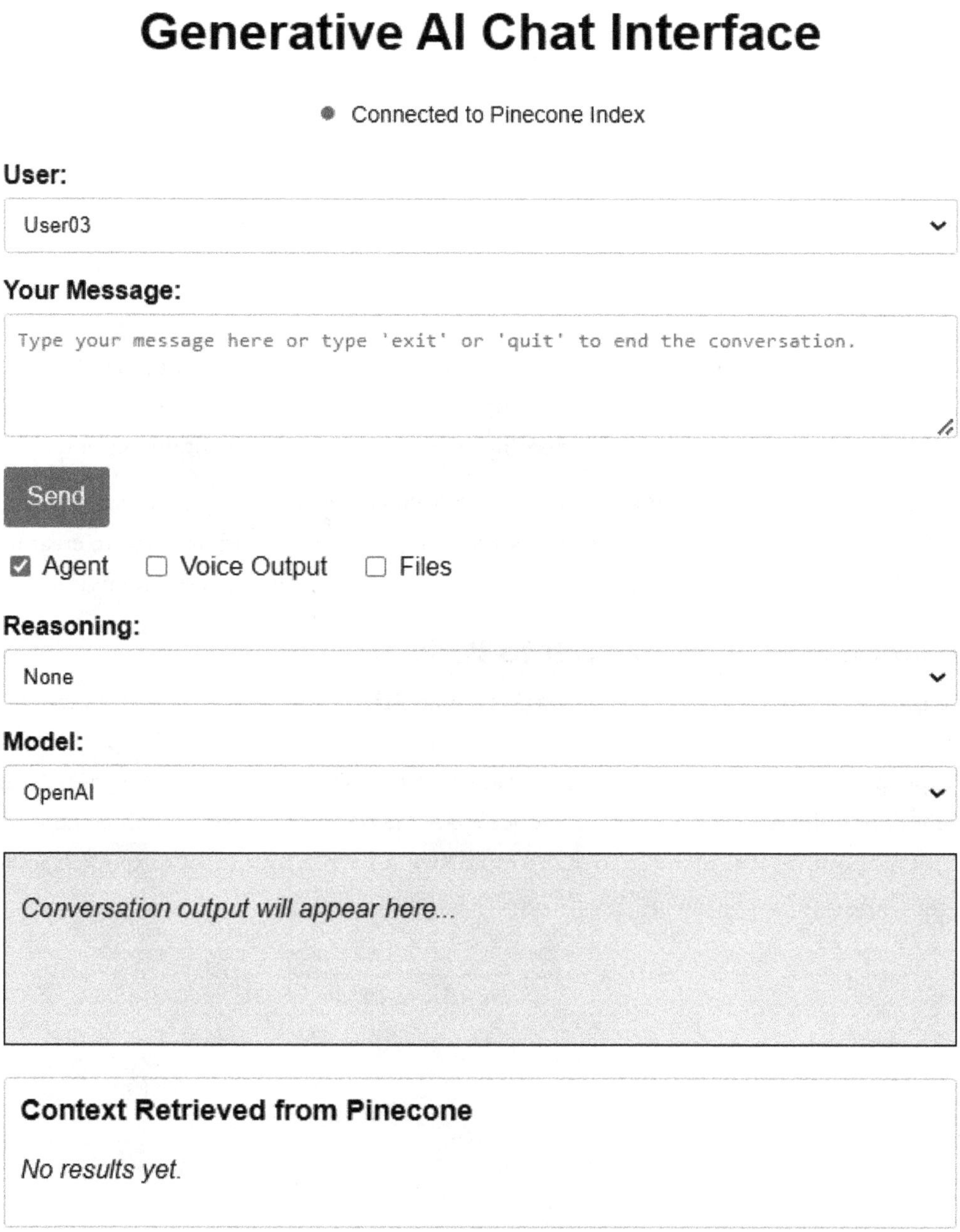

Figure 10.7: Enhancing the interface with Pinecone features

To present Pinecone, we could drill down further and show the structure of a vector store by going to our Pinecone console at `https://www.pinecone.io/`. You can summarize what was covered in *Chapter 3* while showing the `Chapter03/Pinecone_RAG.ipynb` version of the notebook with a checklist such as the following one:

- Pinecone installation
- OpenAI installation
- Chunking
- Embedding
- Upserting
- Querying

Before drilling down into a notebook, however, make sure your audience understands the meaning of vectors and vector stores. Otherwise, simply explain that the vector store provides a highly efficient way of retrieving data and augmenting generative AI inputs to obtain better results.

You can also run any other notebook that queries the Pinecone index or one you built. Note that the notebooks in this book are educational, so you might want to build on them to create your own notebooks for a presentation. Let's now introduce KPIs in a human-centric environment.

3. Human-centric approach to KPIs

A human-centric implementation of a GenAISys remains the best way to deploy AI. A human-centered approach seems counterintuitive at first. Why not just replace employees? Why not lay off all the people who can be replaced? Why not get a lot of cash out of firing employees? After all, why not drastically reduce the workforce since a GenAISys can save so much time and money? Why not simply invest the money and earn a rapid ROI? A simple, counter-intuitive *ROI Scenario 1* through layoffs or "replacement" would look as follows:

$$\text{ROI Scenario1} = \frac{\text{Layoff savings}}{\text{GenAISys project cost}}$$

Indeed, certain corporations have pursued extensive layoffs, choosing to rely heavily on AI. They might even appear successful. However, before following that path, consider my personal experience and perspective, then choose how you wish to proceed. Let's go through my vision of ROI through growth.

ROI through growth

Figure 10.8 illustrates my vision of AI implementations and policy based on real-life AI projects over the past decades. It represents ROI by increasing sales through GenAISys neuroscientific marketing, decreasing time to market production-to-delivery with generative AI, including CoT, and boosting sales with a GenAISys.

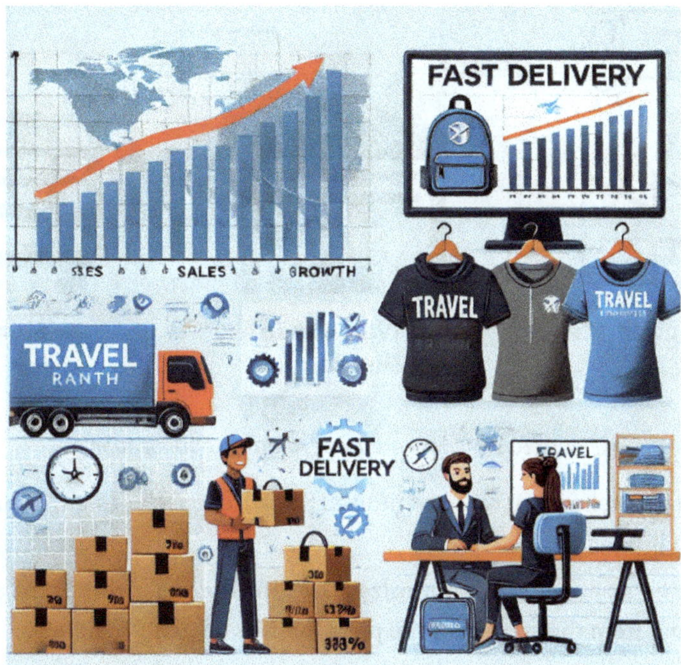

Figure 10.8: Evolution through growth

How is that possible? Am I an idealist? Let me break this down with a real-life AI project I implemented for a distribution center in a mega-warehouse of 100,000+ square meters, which is roughly 1.076+ million square feet. The project was a ground-up hybrid AI and advanced mathematics project:

- The AI location optimizer determined in which location the warehouse should be built by calculating the barycenter of thousands of deliveries from the production sites (A) to the distribution centers (B) and then to the local customers (C), such as supermarkets and large stores, as shown in *Figure 10.9*:

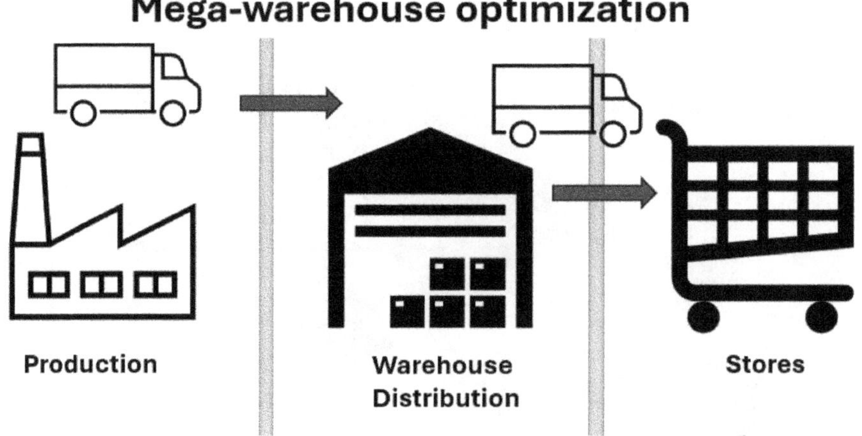

Figure 10.9: Mega-warehouse AI-driven optimization

- The internal warehouse AI agent optimized the warehouse's design, including the number of piers and zones.
- The AI warehouse agent then optimized the flow from the piers to the storage locations (from A) and from the storage to the piers for delivery (C).
- The key issue was the number of messages for large, unplanned events, such as late truck arrivals (due to traffic or weather) and faulty products. A GenAISys, such as the one we built in this book and adapted to this use case, could automate these messages to a certain extent.

So why not replace the many planners who work 24/7 to solve these issues automatically? Everybody first thought this was possible. I found that this could be done to a certain extent. However, I discovered quickly that many parameters the planners considered were not predictable. Generative AI, or even rule-based systems, excel at detecting *patterns* but are at a loss when there are none! They struggle significantly when encountering unprecedented, non-repeating scenarios.

Consider this actual warehouse scenario: stickers are missing from crate #100234, while another crate contains a faulty product (#343445). Simultaneously, an **automated guided vehicle** (**AGV**) is inexplicably operating slower than usual. The truck at Pier 94 must depart immediately, as it's nearly 11:00 p.m. on a Saturday, and trucks cannot use highways on Sundays. Any delay now risks severe penalties and customer dissatisfaction.

I found that many similar problems occurred daily. So, yes, a hybrid GenAISys with powerful mathematical functions could automate many decision-making processes. And yes, this represented a lot of gross margins over a year in large warehouses. But what about all the unsolved problems that an AI cannot solve? Remember, in distribution supply chains, late deliveries incur stiff penalties and possibly customer distrust, leading to the loss of many contracts.

At that point, I worked 24/7 on another approach. Let's keep the personnel, especially the planners, even if increased productivity requires fewer personnel overall in the warehouse. Why and how? I designed an interface that contained a KPI like the one we will add in this section with an AI agent and a human expert. *Figure 10.10* illustrates how an expert can visualize KPIs in real time. Every time the expert ran the AI system and completed the decision, taking complex unplanned events into account, the system displayed how much money was saved, such as not having to use a second truck, switching orders to optimize personnel, and having the extra time to fix errors.

Figure 10.10: Human-centric GenAISys collaboration

The expert planner thus increased the speed of the incoming and outgoing storage events. The warehouse could process more units without having to build additional piers, purchase new AGVs, or hire more hands. The productivity gains represented a 3% gross margin increase for the mega-warehouse. Such a significant surge in gross margin in warehouse management generated substantial profit.

You can design the ROI ratio with the parameters you wish, but the core concept is that collaborative human-AI growth is highly productive. The ROI through growth produced far more margin with a human-centric system than laying off personnel:

$$\text{ROI Scenario 2: Growth} > \text{GenAISys project cost}$$

This successful project had a huge impact on my reputation and sales:

- The word spread that Rothman was ethical and that his AI systems could be trusted.
- Human-centric successes built an image of AI projects that would federate teams, not destroy them.
- AI expertise could work 24/7 to help teams boost their performance, thus generating growth and obtaining bonuses!

Ultimately, the choice between aggressive layoffs and strategic growth remains yours. This example simply reflects my personal approach and experiences. In any case, if the customer wants a real-time KPI, let's add it to the interface.

Adding a real-time KPI to the GenAISys web interface

If your customer is interested, here's how we can implement a real-time KPI into the existing web interface, emphasizing the human-centric approach:

1. Copy the previous interface file (`02.html`) to a new version (`03.html`) to preserve previous interface states. Then, add the KPI panel:

    ```
    <!-- Gross Margin KPI Panel -->
    <div class="kpi-panel" id="grossMarginKPI">
      <h2>Gross Margin</h2>
      <div class="kpi-value" id="grossMarginValue">Loading...</div>
    </div>
    ```

2. Then, we will add the CSS styling:

```
    .kpi-panel {
      background: #fff;
      padding: 15px;
      margin: 20px auto;
      border: 1px solid #ccc;
      border-radius: 5px;
      text-align: center;
      box-shadow: 0 2px 4px rgba(0,0,0,0.1);
      max-width: 300px;
    }
    .kpi-panel h2 {
      margin-bottom: 10px;
      font-size: 20px;
    }
    .kpi-value {
      font-size: 36px;
      font-weight: bold;
      color: #007bff;
    }
```

3. Finally, we will add real-time update simulations in JavaScript to illustrate what the KPI would look like if implemented in real-time situations:

```
    // Example function to update the KPI
    function updateGrossMargin() {
      // Replace this with your real-time data-fetching logic
      const grossMargin = fetchGrossMarginFromBackend(); // your API
call here
      document.getElementById('grossMarginValue').textContent =
`$${grossMargin.toFixed(2)}`;
    }
    // Simulate real-time update every 5 seconds
    setInterval(updateGrossMargin, 5000);
```

Now run the cell that displays `03.html` in section *3. KPI* of the notebook:

```
display_interface("/content/03.html")
```

This simulation displays real-time updates of the gross margin KPI, visually reinforcing the value human planners bring to GenAISys implementations:

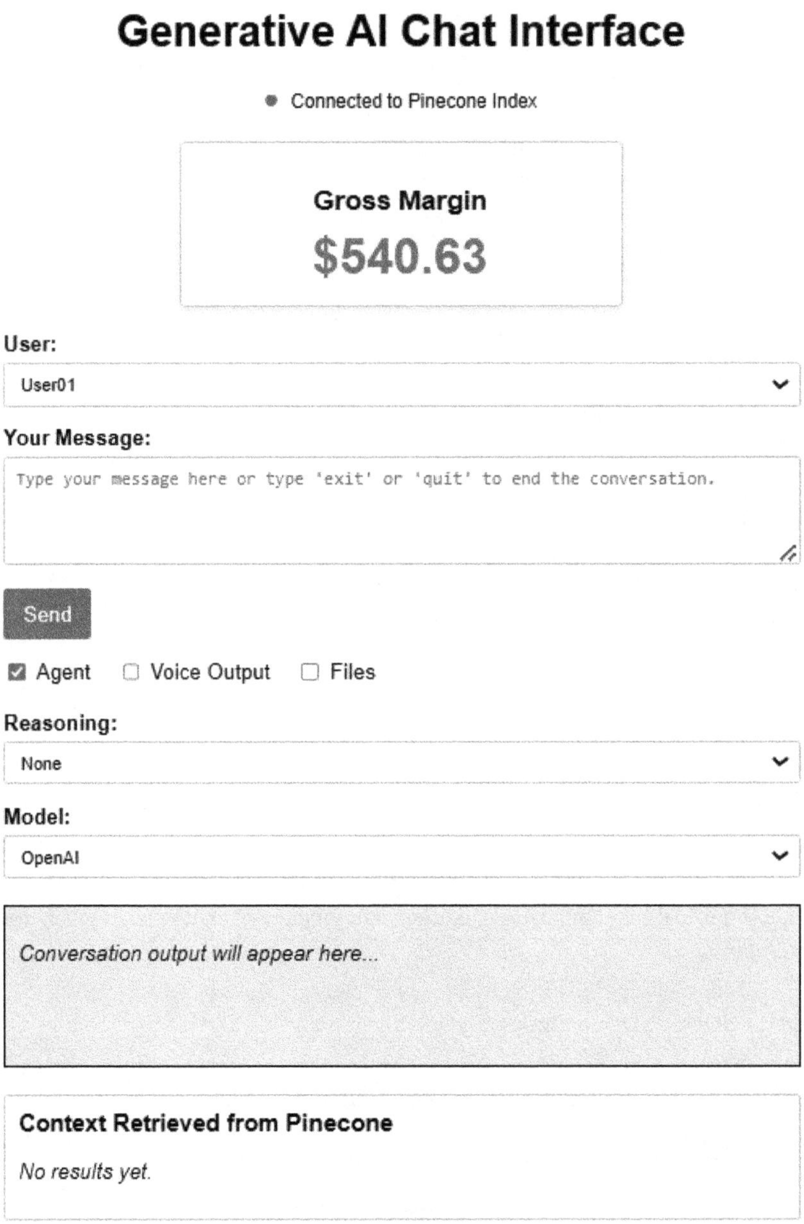

Figure 10.11: A human-KPI relationship

The customer may wish to delete this function or enhance it to adapt it to their project. Our role is to adapt to the project's needs, regardless of the level of automation required. We will now proceed to present the integration of our GenAISys in the customer's environment.

4. Integration: Platforms and frameworks

Integration is often among the most challenging stages of deploying a GenAISys, particularly when it comes to selecting platforms, operating systems, and frameworks. While start-ups or internal projects might initially have the luxury of freely selecting platforms, real-world scenarios often involve stringent constraints from investors, customers, or internal policies. The following examples are drawn from real-world situations that illustrate the potential challenges you may face:

- "Our company only works with Windows Server with ISO with native security. We will not accept your Ubuntu service. It's not open to discussion. It's our policy." (`https://www.microsoft.com/en-us/evalcenter/evaluate-windows-server-2025`)
- "Our company only works with Ubuntu Server. We will not accept any Windows applications. It's not open to discussion. It's our policy." (`https://ubuntu.com/download/server`)
- "As an investor, we need to make sure that your GenAISys meets AWS security and privacy standards that adhere to US-European regulations through the **Data Privacy Framework** (**DPF**). We don't care about the third-party components you installed. Either you use AWS's framework, or we will not invest." (`https://aws.amazon.com/compliance/eu-us-data-privacy-framework/`)
- "Are you joking? Do you really think we are going to let you into our network with a GenAISys? Impossible. So, forget about the security and privacy components you are presenting, including AWS's framework. You're on a highly secure aerospace site. We are going to install your GenAISys, not you, on an isolated VM with absolutely no access to the web. Only certified users will have access to it. The connections are monitored by our security department 24/7 in real time. Any attempt to write unauthorized prompts will be filtered before they reach your system, accompanied by a high-security alert. We will not tell you more. We'll just give you a hint. One of our strategies is our IBM **Intrusion Detection and Prevention Systems** (**IDPSs**)." (`https://www.ibm.com/docs/en/snips/4.6.0?topic=introducing-security-network-intrusion-prevention-system-ips-productn`)
- "We have been working with Google Cloud for many years and will only accept systems that fit into our hosting, security, and privacy framework." (`https://cloud.google.com/`)

- "We only work with Microsoft Azure for everything: hosting, AI, security, and privacy. Let's assume that you have an innovative GenAISys. How will it fit in our framework?" (`https://azure.microsoft.com/en-us/`)

The list of possible hosting platforms, security, and privacy frameworks is limitless! Your GenAISys proof of concept must remain *flexible*. This flexibility is why the frontend HTML page we've built is designed for quick adaptation to specific customer requirements—without necessitating deep modifications at the backend. Then, you can take the customer as deep as requested with the educational components built in this book and the ones you have added to prepare a professional-specific presentation.

But integration doesn't stop here! Creative and unexpected requests can arise. A potential customer might ask you the following two questions:

- "We really love your GenAISys! The multi-user feature is fantastic, especially since it supports activating or deactivating the AI agent! Great! We'll purchase your system immediately if you can get it to work with the Zoom API. That would be tremendous! How long would it take you to integrate your GenAISys with Zoom?" (`https://developers.zoom.us/docs/api/`)
- "We love love love your Zoom integration! It's great! But our company policy is only to use Microsoft Teams. How long would it take you to integrate your GenAISys with Teams?" (`https://learn.microsoft.com/en-us/graph/api/resources/teams-api-overview?view=graph-rest-1.0`)

Once you address these challenges, further integration demands may emerge around enterprise-level ERPs (such as SAP or Oracle), specific database solutions, or even certifications to authenticate your GenAISys's compliance and security measures. For that I say, welcome to the real world! You must hold on to one basic principle: No matter how far you develop your GenAISys, potential customers will often ask for more or a different way to implement your system. Your GenAISys provides legitimacy. Beyond that, remain a flexible expert!

Let's now demonstrate our willingness to adapt to any situation. Copy `03.html` and name it `04.html` to keep the interface's previous step as is. Add this section to the HTML frontend page:

```
    <!-- ERP Integration Dropdown -->
    <div class="form-group">
      <label for="erpIntegrationSelector">ERP, database, platform and
meeting Integration:</label>
      <select id="erpIntegrationSelector">
```

```
            <option value="none" selected>Select ERP or Meeting API</option>
            <option value="SAP">SAP</option>
            <option value="Oracle">Oracle</option>
            <option value="Microsoft Dynamics">Microsoft Dynamics</option>
            <option value="NetSuite">NetSuite</option>
            <option value="AWS">AWS</option>
            <option value="Google Cloud">Google Cloud</option>
            <option value="Azure">Azure</option>
            <option value="Zoom API">Zoom</option>
            <option value="Teams API">Teams</option>
            <option value="Other">Other</option>
        </select>
    </div>
```

Now, run the page to display the updated interface:

```
display_interface("/content/04.html")
```

This newly added integration selection appears just below the **Reasoning** dropdown (as shown in *Figure 10.12*). It's an excellent visual aid to discuss multimodal CoT, ERP integrations, or specific platform compatibilities.

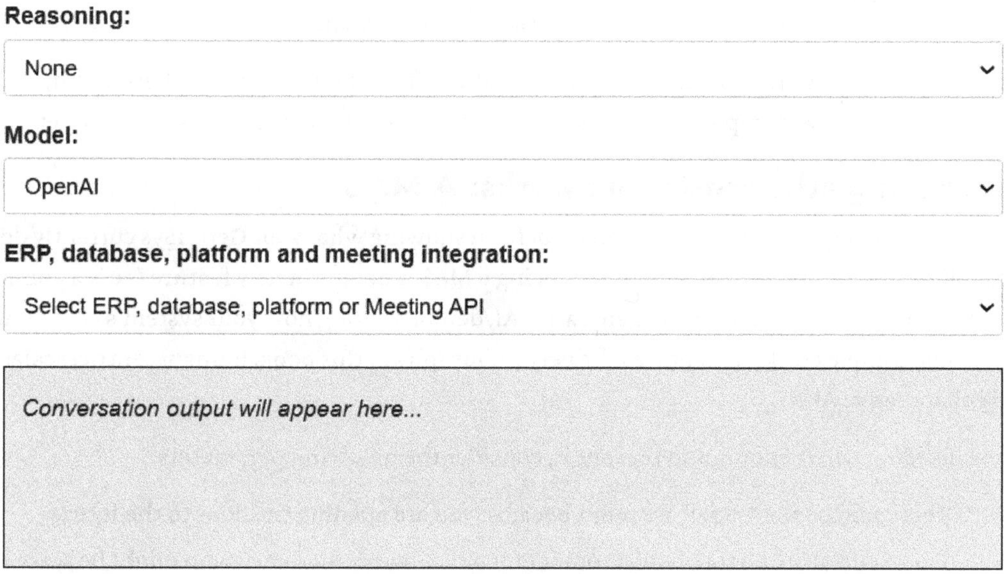

Reasoning:

None

Model:

OpenAI

ERP, database, platform and meeting integration:

Select ERP, database, platform or Meeting API

Conversation output will appear here...

Figure 10.12: ERP, database, platform, and meeting integration list

Before adding this dropdown, ensure you're thoroughly prepared. Familiarize yourself with each listed integration (illustrated in *Figure 10.13*), as your audience may ask specific, detailed questions on each item:

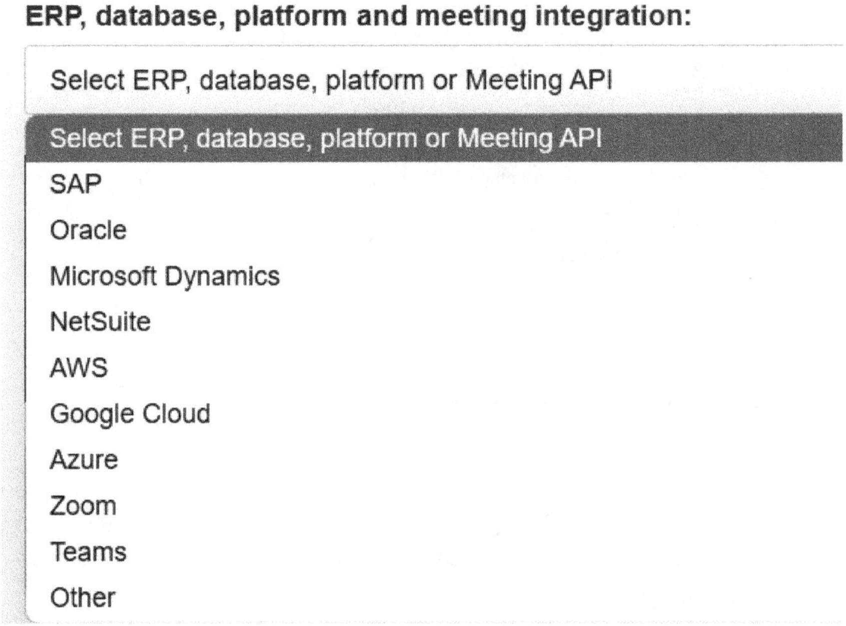

ERP, database, platform and meeting integration:

Select ERP, database, platform or Meeting API

Select ERP, database, platform or Meeting API
SAP
Oracle
Microsoft Dynamics
NetSuite
AWS
Google Cloud
Azure
Zoom
Teams
Other

Figure 10.13: The list of some of the possible integrations

Remember, the integration flexibility your GenAISys offers extends beyond these examples. It might even include MAS implementations—another sophisticated approach we will now explore.

Showcasing advanced frameworks: A MAS

At this point in the presentation, you can pivot from showing what your GenAISys currently does to highlighting the future it enables. Introducing a MAS is not just a new feature—it's a glimpse into the transformative potential of generative AI, demonstrating how your system's core design can handle complex tasks. It captures the very philosophy of this book: human-centric, scalable, and collaborative AI.

When deciding whether or not to present it, consider the following parameters:

- This could be the "wow" moment because you are opening the door to the future.
- This could also be a terrifying moment for a risk-averse audience who might be scared of the implications, no matter what you say: job displacement, replacement, or destruction.

- Even if it is a "wow" moment, the content might be too complex and maybe should wait until the GenAISys is in implementation. If this happens, you can add this MAS to the handler registry of the GenAISys and let a user decide whether to activate it or not.
- Even if it is a "wow" moment and the audience wants to go further, remind them before letting it run on its own that the human-centered GenAISys with a human-machine interface ensures that experts control the output, even within advanced MAS implementations.

If you choose to present the MAS, frame it clearly as an innovative game-changer. It can operate autonomously or partially autonomously—either standalone or through a simplified interface within your existing GenAISys.

To see the full implementation of this MAS, open the `GenAISYS_&_MAS.ipynb` notebook. The program's flow and code are designed to be highly self-explanatory, with each section having a clear explanation of its purpose. We, meaning the human author and Gemini 2.5 Pro/ChatGPT, my co-workers for this project, have taken care to add detailed comments directly in the code to explain not just *what* the functions do but also the design decisions behind them—from the prompt engineering that guides the agents to the asynchronous patterns that enable the swarm to run efficiently.

The notebook has three key features:

- **The MAS**: This notebook demonstrates an educational MAS, as shown in *Figure 10.14*. This MAS orchestrates a swarm of AI agents for concurrent task processing. The program is organized by first setting up the asynchronous environment and defining the core components: independent worker agents that create and solve tasks, a summarizer agent for synthesis, and a central orchestrator to manage the two-stage workflow. The final cells execute the complete simulation, showcasing an efficient, parallel approach to complex problem-solving.
- **Human-AI co-worker collaboration paradox**: Although this program demonstrates AI agents automating tasks, a concept often tied to human replacement, the notebook itself is a testament to human-AI co-worker collaboration. Authored by a human, it was then refined with ChatGPT and evolved into this MAS in partnership with Gemini. This process mirrors the book's central, human-centric theme: the future is not AI versus humans but AI as a co-worker, augmenting our own creativity and productivity.

- **"From-scratch" approach**: A deliberate choice was made to build this system from foundational libraries such as `asyncio` and `aiohttp` rather than using a pre-existing agentic framework. This from-scratch approach serves a dual purpose. First, it demystifies the core mechanics of how multi-agent orchestration works, an educational goal of this project. Second, it ensures the resulting framework is as flexible and modular as possible, free from the constraints of any single platform, and ready to be adapted to any real-world business requirements.

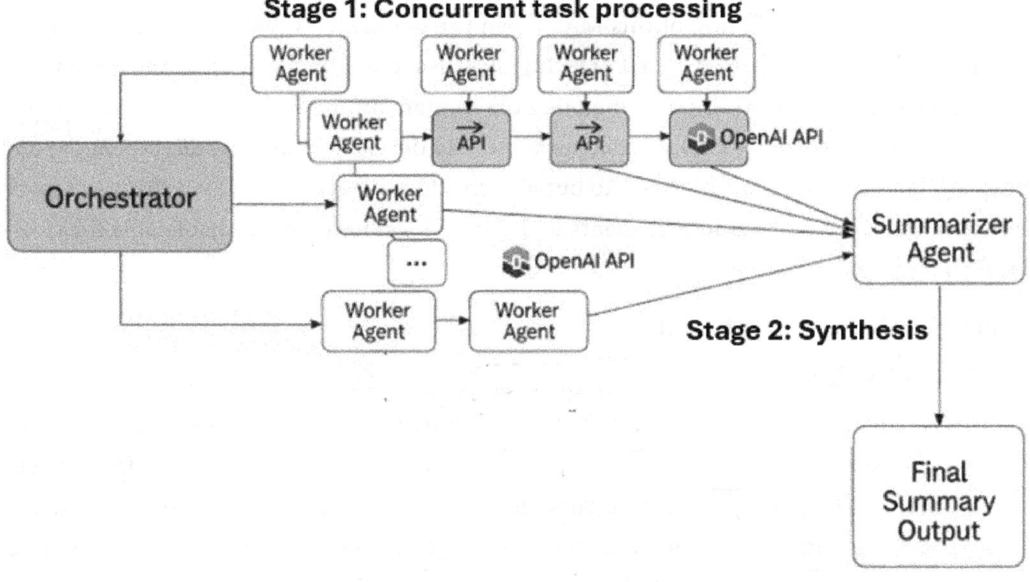

Figure 10.14: MAS workflow overview

The preceding figure shows the main components of the MAS:

- The **central orchestrator** directs the workflow with helpers that display the outputs of the agents in less than a second per task in the chapter notebook
- **Worker agents** process tasks concurrently, interacting with OpenAI
- The **summarizer agent** synthesizes responses, enabling the agents to interact through the outputs to form a summary of the MAS's work

Let's now look into the strategic integration options.

Strategic integration options for the MAS

Now that we've established the *what* and the *why* of this MAS, the crucial business question is *how* it fits with the GenAISys we've presented.

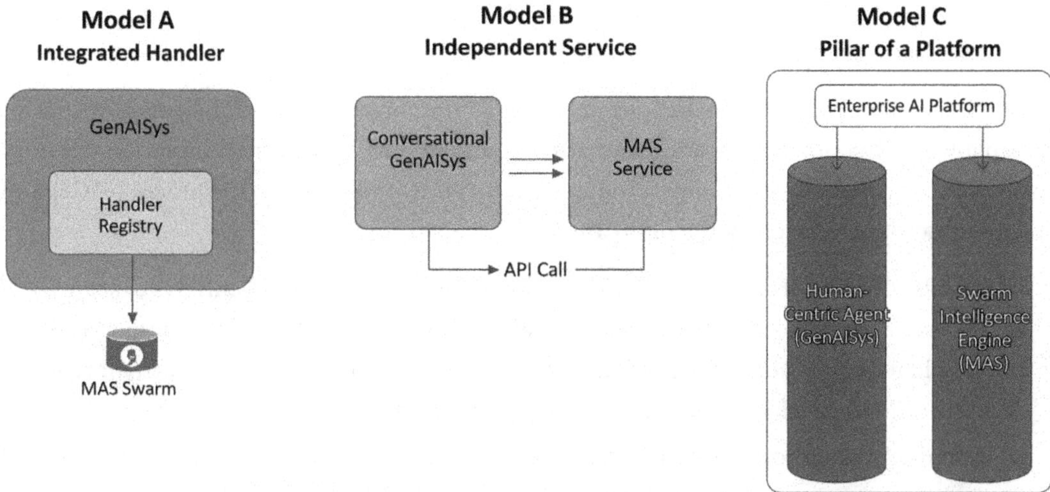

Figure 10.15: The strategic integration models for the MAS

As shown in *Figure 10.15*, we can offer three strategic visions for its deployment:

- **Model A as an integrated handler**: For many use cases, the entire swarm process can be encapsulated as a new function and added to the handler registry of the core GenAISys. A user could trigger this massive parallel task by simply selecting the **Swarm Processing** option from the **Reasoning** dropdown. This demonstrates the modularity and extensibility of the primary system.
- **Model B as an independent, specialized service**: For a massive, dedicated need (e.g., analyzing millions of documents or real-time data streams), the MAS can be deployed as a separate, highly optimized service. The main conversational GenAISys could then call this service when needed. This showcases the ability to build a robust, microservices-style architecture for enterprise-level problems.

- **Model C as the pillar of a larger AI platform:** We can frame the entire GenAISys as a platform with two main pillars. The first is the conversational and reasoning agent for human-centric collaboration. The second is the **Swarm Intelligence Engine (MAS)** for large-scale automation. This presents the most ambitious and powerful vision of the work, not just a tool but a comprehensive, business-ready AI platform.

This integrated vision elevates your GenAISys beyond mere functionality into a fully realized, enterprise-ready AI ecosystem.

Having built your GenAISys from the ground up, you now possess a powerful, flexible AI controller capable of dynamic orchestration and advanced reasoning. This solid foundation uniquely positions you to embrace emerging standards such as the **Model Context Protocol (MCP)** and the **Agent Communication Protocol (ACP)**:

- **Model Context Protocol:** MCP offers a standardized way for your GenAISys to seamlessly access and integrate with diverse external tools and data sources. If a customer wants to implement this, you now have the skills to complete the job. You can learn more about MCP and its specifications here: `https://modelcontextprotocol.io/`.
- **Agent Communication Protocol:** ACP provides the blueprint for your sophisticated AI agents to collaborate effortlessly with other specialized agents. You have built many agents throughout the book preparing you for this protocol. You can find more information about ACP and its course on DeepLearning.AI's platform, often in partnership with IBM Research's BeeAI: `https://www.deeplearning.ai/short-courses/agent-communication-protocol/`.

Your expertise in designing intelligent AI pipelines and agentic workflows means you are well equipped to leverage and even contribute to these powerful interoperability frameworks, taking your business-ready AI solutions to the next level. Now comes a more difficult aspect: security constraints and privacy regulations.

5. Security and privacy

Security and privacy often inherit constraints from the selected hosting platforms or frameworks decided in the initial integration phase. However, even if your system aligns with established standards, you must clearly demonstrate your moderation and data security components within the GenAISys. Always be ready to face rigorous questions on these topics. It's strongly recommended to have a certified security expert alongside you during the presentation. If that's not possible, ensure you demonstrate your willingness to adapt your GenAISys according to the customer's established protocols and frameworks.

Here are essential security layers that your GenAISys might need to integrate or align with—review them carefully to be ready for challenging questions, even if another team or external service manages these security layers:

- **Encrypted communications**: Communicate how data encryption is managed. Refer to robust resources such as IBM's overview of encryption to ensure you're prepared (`https://www.ibm.com/think/topics/encryption`).
- **Audit reports and penetration testing**: Customers often demand proof of regular security testing. Make sure you understand the penetration testing standards of major providers such as AWS (`https://aws.amazon.com/security/penetration-testing/`).
- **Access controls and authentication:** Discuss how your GenAISys integrates with industry-standard access management protocols. Familiarize yourself with frameworks such as Google Cloud's access control methodologies (`https://cloud.google.com/storage/docs/access-control/`).
- **Monitoring and incident response**: Articulate your system's ability to handle security incidents. It's beneficial to reference frameworks such as IBM's incident response guidelines (`https://www.ibm.com/think/topics/incident-response`).

You don't have to be an expert, but at minimum, you should comfortably discuss these areas using accurate terminology. Review the provided links carefully to ensure you are up to date with contemporary security practices.

Let's showcase how your system might handle real-time security alerts effectively. As usual, to avoid overwhelming your audience, copy `04.html` to create `05.html`, ensuring the previous presentation state remains untouched if you decide not to proceed further in the demonstration.

We'll add a simple, visually intuitive red alert banner to our flexible HTML interface:

```
<!-- Place this alert banner near the top of your container -->
<div id="securityAlert" style="display:none; background-color: #ffdddd;
color: #a94442; padding: 15px; border: 1px solid #ebccd1; border-radius:
4px; margin-bottom: 20px;">
  <strong>Security Alert:</strong> Suspicious activity detected.
</div>
```

Then, in your JavaScript, you can easily toggle its visibility based on certain conditions:

```
// Example function to display the alert
function showSecurityAlert() {
  document.getElementById('securityAlert').style.display = 'block';
```

```
}
// Example function to hide the alert
function hideSecurityAlert() {
  document.getElementById('securityAlert').style.display = 'none';
}
// Simulate a security event after 5 seconds (for demonstration purposes)
setTimeout(showSecurityAlert, 5000);

red alert banner / 5 seconds
```

In this example, the alert banner appears automatically after five seconds, visually illustrating how a real-time security notification might look. Read the code to be able to modify it in real time or at least rapidly during a meeting if requested. Some customers might dislike the way it's displayed or its color.

Sometimes, modifying the frontend page shows your willingness to adapt to the customer's needs quickly, as we've reiterated several times now. On the other hand, careful and considered modifications signal caution and reliability. Make a strategic choice depending on your understanding of your audience's needs and expectations.

The alert can come from your system or the hosting environment. Now, let's run the code to display the updated interface:

```
display_interface("/content/05.html")
```

A security alert will be displayed after five seconds, as illustrated here:

Generative AI Chat Interface

Security Alert: Suspicious activity detected.

Figure 10.16: A security alert banner in real time

The security alert conditions must be precisely defined during the project's implementation stage to ensure alignment with the customer's exact security policy. With security and privacy considerations addressed, let's now move on to the nuanced and often challenging area of customization.

6. Customization

As with most software, the GenAISys's path to success relies on our ability to customize an application. No matter how hard we try, the end users will request interface and process evolutions. We can ignore them or accept them. The middle ground is to accept the requests that are feasible and within the scope of the project, and find as many realistic workarounds as possible for evolutions that would require fundamental modifications.

There are many possible customization approaches; here, we focus on a practical three-phase, human-centric customization method:

- **Phase 1:** Customizing the frontend HTML page through brainstorming workshops with as many groups as necessary
- **Phase 2:** Creating an isolated sandbox to add features to the GenAISys you built progressively
- **Phase 3:** Integrating the sandbox evolutions step by step in the HTML page (documentation and stabilization)

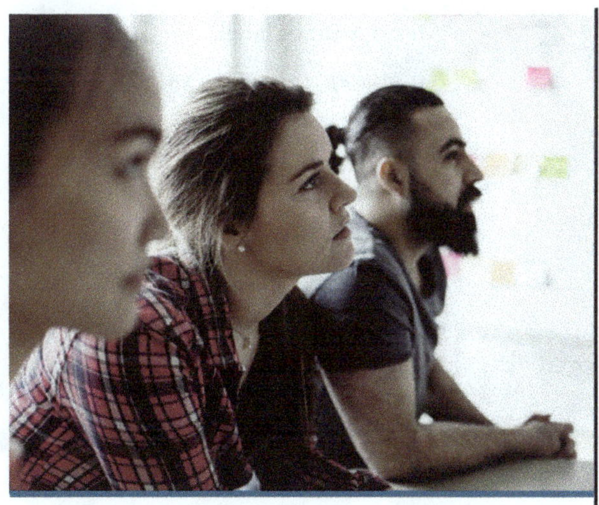

Three-Phase Customization Process
• Phase 1 involves brainstorming workshops for frontend HTML. customization.
• Phase 2 focuses on developing an isolated sandbox for feature additions.
• Phase 3 integrates sandbox evolutions into the HTML page progressively.
• The process is iterative, allowing for continuous improvement.
• All phases work in cycles to enhance collaboration and outcomes.

Figure 10.17: A three-phase human-centric customization process

All three phases will progress continuously in workshop cycles. Reinforce your participants' roles as active co-creators, rather than passive consumers of technology who only express their wishes.

Again, copy `05.html` and name it `06.html`. Add some icons to trigger reactions from the audience so that they can express their creativity, ideas, and needs. If you succeed in involving your audience, you're already on your way to success.

The icons in this code come from Font Awesome. They are made available by including the `Font Awesome` CSS library via a link in `<head>`. Then, the icons are inserted throughout the HTML using `<i>` tags with the appropriate Font Awesome classes. Here are the key snippets that you can customize in real time or near-real time during your workshops:

1. Import the `Font Awesome` library:

```
<link rel="stylesheet" href="https://cdnjs.cloudflare.com/ajax/libs/
font-awesome/6.0.0-beta3/css/all.min.css">
```

2. Integrate the icons in the HTML:

   - Gross margin KPI panel header:

   ```
   <h2><i class="fas fa-chart-line"></i> Gross Margin</h2>
   ```

   - User selector label:

   ```
   <label for="userSelector"><i class="fas fa-user"></i> User:</
   label>
   ```

   - Message input label:

   ```
   <label for="messageInput"><i class="fas fa-comment"></i> Your
   Message:</label>
   ```

   - Send button:

   ```
   <button id="sendButton" title="Send your message"><i
   class="fas fa-paper-plane"></i> Send</button>
   ```

   - Reasoning selector label:

   ```
   <label for="reasoningSelector"><i class="fas fa-brain"></i>
   Reasoning:</label>
   ```

   - Model selector label:

   ```
   <label for="modelSelector"><i class="fas fa-robot"></i>
   Model:</label>
   ```

Each snippet leverages Font Awesome's intuitive icon library (such as `fas fa-chart-line`, `fas fa-user`, etc.), ensuring clear visual communication of each component's purpose. Here, too, you'll have to be prepared to swiftly modify these icons in real time during workshops to demonstrate your agility in adapting to user feedback.

Now, run the web page to visualize your icon-rich interface:

```
display_interface("/content/06.html")
```

The page now contains icons, as illustrated in the following excerpt of the HTML page:

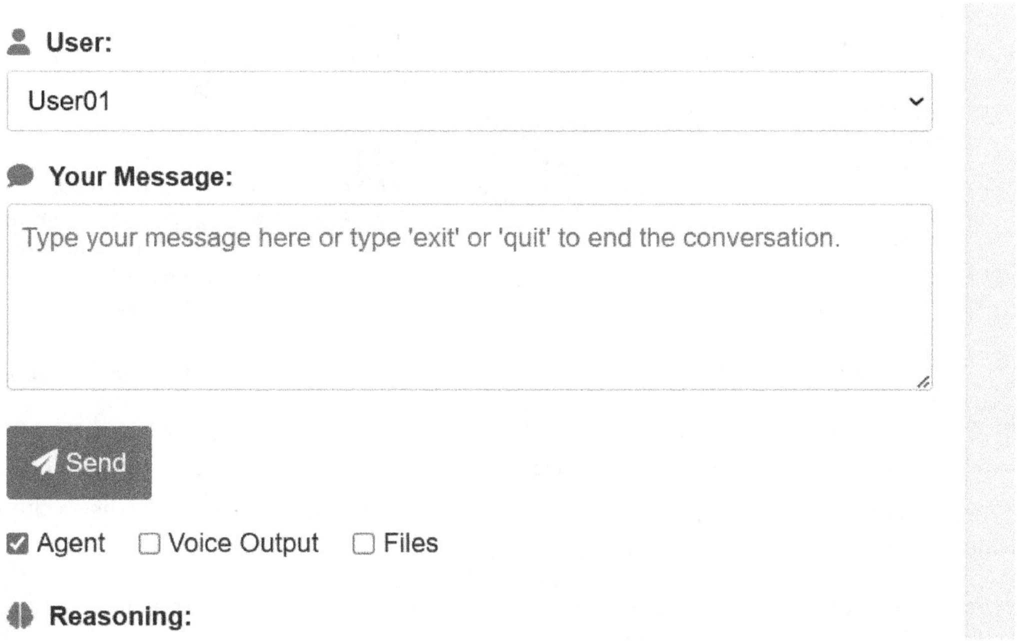

Figure 10.18: Customizing the web page with icons

Do not take this topic lightly! An end user only sees the frontend web page and its outputs (text, images, and popups). Spend all the necessary time listening to the customer (your colleagues or external customers). The more the end users work with you in this area, the more it becomes *their* GenAISys.

The tricky part is fitting in phase 2 (GenAISys) development and phase 3 (GenAISys and web page integration). You can fast-track the project by implementing all three phases in parallel, as illustrated in *Figure 10.19*.

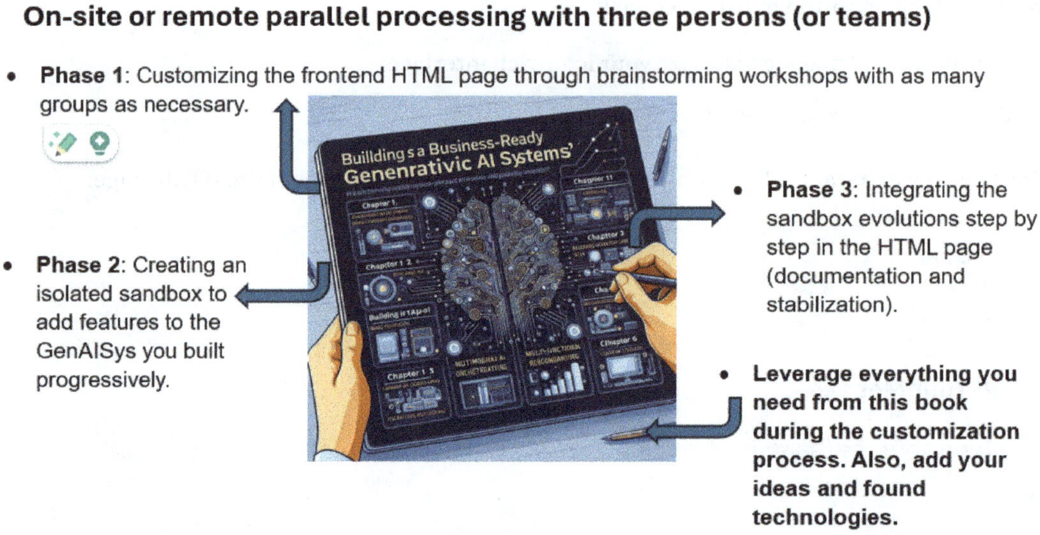

On-site or remote parallel processing with three persons (or teams)

- **Phase 1**: Customizing the frontend HTML page through brainstorming workshops with as many groups as necessary.
- **Phase 2**: Creating an isolated sandbox to add features to the GenAISys you built progressively.
- **Phase 3**: Integrating the sandbox evolutions step by step in the HTML page (documentation and stabilization).
- **Leverage everything you need from this book during the customization process. Also, add your ideas and found technologies.**

Figure 10.19: Working in teams on all three phases in parallel on-site or remotely

The choice depends on the project and your vision. Your deployment approach will create the continual momentum required for a fast-tracked, eventful, optimized time-to-market project. However, each project advances at its own speed, and the best pace is the one the users choose, giving them time to adapt to your cutting-edge GenAISys.

At some point, you will inevitably need to explain the resources required for the project and its associated costs. So, let's move on and discuss that.

7. GenAISys resources (RACI)

Every successful software project clearly specifies the human and machine resources required. Machine resources usually align closely with the platforms identified during the earlier integration discussion (refer to the *Integration: Platforms and frameworks* section in this chapter). Human resources, however, require more careful consideration and explicit detailing using a structured approach such as the **RACI** matrix outlined in *Chapter 1*. Recall that a RACI matrix classifies project responsibilities into four categories:

- **R (Responsible)**: The person(s) who works actively on the task
- **A (Accountable)**: The person(s) who is answerable for the success or failure of a task
- **C (Consulted)**: In a complex project, this is the person(s) providing input, advice, and feedback to help the others in a team, but they are not responsible for executing the work
- **I (Informed)**: These are the people who are kept informed about a task's progress or outcome

Section 7 (RACI) of the notebook provides a practical example of a RACI matrix specifically designed for the GenAISys we've developed in this book. *Figure 10.20* shows an excerpt from this RACI ideation, providing a concrete starting point to build your project-specific RACI:

7.RACI Matrix for GenAISys Implementation

Task
1. Requirements & Planning Define project scope for integrating DeepSeek, OpenAI, Pinecone, and the interactive interface.
2. Environment Setup & File Downloading Implement the initial file download script (e.g., fetching `grequests.py` via curl) and set up the basic Python environment.
3. DeepSeek Activation & Setup Configure DeepSeek as shown in the "DeepSeek Activation Guide" and set the appropriate flags (`deepseek=True` , `hf=True` , etc.).
4. GPU & Cache Configuration Mount Google Drive, check GPU status (via `nvidia-smi`), and set cache paths for Hugging Face as detailed in the notebook.
5. Hugging Face Environment Installation & Model Loading Install transformers, check version, and load the DeepSeek model using `AutoTokenizer` and `AutoModelForCausalLM` .
6. OpenAI Integration Run OpenAI installation scripts, initialize the API key (via `openai_setup.py`), and import the API call functions from `reason.py` .

Figure 10.20: Excerpt of the RACI ideation for the GenAISys built in this book

While presenting, you will need to explain that the **RACI Matrix for GenAISys Implementation** ideation clearly outlines a coordinated, cross-functional approach to deploying an advanced AI system:

- The project manager provides overall strategic oversight and leads planning and documentation
- At the same time, the AI/ML engineer drives the technical aspects—ranging from DeepSeek activation and model integration to API setups, machine learning functionalities, and comprehensive testing
- The DevOps engineer's responsibility is to set up the environment, configure hardware resources, and manage deployment and scalability

- In contrast, the software developer focuses on coding interactive interfaces and handler registries and supporting features such as text to speech
- Meanwhile, the business/CTO role offers strategic guidance, ensures key decisions are aligned with organizational goals, and upholds compliance and security measures, ensuring every stage—from initial setup through ongoing maintenance—is effectively managed

Using the RACI matrix as your reference point will take you directly to the costs you must estimate and negotiate (in your company, for investors, or with a customer). Clearly defined roles will help you put a number to the human hours, allocate budgets effectively, and identify areas requiring specialized expertise. This allows you to transparently estimate project costs, negotiate resource allocation internally or externally, and segment your project into manageable milestones or deliverables.

And with that, you are now business-ready for a GenAISys. The market is yours to conquer!

Summary

In this chapter, we designed a strategic framework for showcasing a business-ready GenAISys amid intense market competition. We saw that technological innovation alone is not enough to secure customers or funding. A compelling business narrative must accompany the technical proof of concept to demonstrate real-world value. One fast-track approach is to begin with a concise introduction of the team and quickly establish our credibility and expertise. Then, pivot the presentation to a clear explanation of the GenAISys we have built, underpinned by real-time KPIs.

We designed a *flexible* frontend web page to demonstrate the system's capabilities. The idea is to have an adaptable HTML page that can be modified in real time or rapidly, depending on the situation. The frontend web page translates the IPython interface we built into an accessible, customizable HTML display. Live data updates and real-time simulations reinforce the system's practical impact. We designed a narrative that stresses the GenAISys's adaptability and ability to evolve with customer needs. At all times, you saw how you could drill down into the main concepts taught in this book and enhance them with your developments. You also have your own GenAISys built from scratch and ready to present, which can be adapted to the specifications of the project you are working on.

This chapter favored a human-centric approach, ensuring AI augments rather than replaces expert decision-making. The system is positioned as a collaborative tool designed to boost operational efficiency. Real-time KPIs can reinforce the GenAISys's profitability and strengthen the bonds between the GenAISys and the teams using it. Innovative features such as vector stores enhance data retrieval for improved outputs. We showed how to present a forward-looking vision by demonstrating a prospective evolution of the system into a swarm-based MAS, proving the framework's scalability and introducing the powerful narrative of human-AI co-worker collaboration. We presented integration with major cloud platforms and ERP systems. We showed the possibility of integrating services such as AWS, Google Cloud, and Microsoft Azure, highlighting the versatility. Security and privacy were addressed rigorously through possible real-time alerts and adaptive measures.

Flexibility remains key, enabling the system to meet diverse enterprise requirements. The approach balances cutting-edge technology with tangible business outcomes. An iterative process encourages customization based on ongoing client feedback. We presented interactive workshops as a means of tailoring the system on the fly. A RACI matrix was introduced to delineate roles and streamline project management clearly. Ultimately, the chapter provided a comprehensive blueprint for deploying a dynamic, market-ready GenAISys presentation. We balanced technical depth with clear business benefits.

You are now business-ready to design, build, and deliver a GenAISys!

Questions

1. Presenting a business-ready **generative AI system (GenAISys)** requires no preparation. (True or False)
2. There is no need for a frontend GenAISys for demonstration purposes. (True or False)
3. There is no need to explain what a vector store is during a presentation. (True or False)
4. Pinecone is the best vector store on the market. (True or False)
5. Only OpenAI and DeepSeek can provide generative AI models. (True or False)
6. Open source generative AI models are superior to those that do not share their code. (True or False)
7. We need to develop our own security and privacy software for a GenAISys. (True or False)
8. There is no need to customize our GenAISys if we think it is good. (True or False)
9. All customization meetings should be held face to face and on-site. (True or False)
10. There is no need to present the resources necessary to implement our GenAISys. (True or False)

References

- IBM, *What is encryption?*: https://www.ibm.com/think/topics/encryption
- AWS, *Penetration Testing*: https://aws.amazon.com/security/penetration-testing/
- Google Cloud, *Overview of access control*: https://cloud.google.com/storage/docs/access-control/
- IBM, *What is incident response?*: https://www.ibm.com/think/topics/incident-response
- Microsoft Teams API: https://learn.microsoft.com/en-us/graph/api/resources/teams-api-overview?view=graph-rest-1.0
- Zoom API: https://developers.zoom.us/docs/api/

Further reading

- Amazon documentation: https://aws.amazon.com/
- Microsoft Azure: https://azure.microsoft.com/en-us/
- Google Cloud: https://cloud.google.com/?hl=en
- IBM Cloud: https://www.ibm.com/consulting/cloud

Subscribe for a Free eBook

New frameworks, evolving architectures, research drops, production breakdowns—*AI_Distilled* filters the noise into a weekly briefing for engineers and researchers working hands-on with LLMs and GenAI systems. Subscribe now and receive a free eBook, along with weekly insights that help you stay focused and informed.

Subscribe at https://packt.link/TRO5B or scan the QR code below.

Answers

Chapter 1

1. No, a GPT model performs tasks only. The AI controller is necessary to orchestrate and manage tasks dynamically.
2. No, memoryless sessions process each request independently and remember nothing.
3. Yes, RAG retrieves specific data in documents.
4. Yes, human expertise is necessary from design to production and maintenance.
5. Yes, the AI controller dynamically adapts task order based on input context.
6. Yes, the goal is to optimize development cost and time.
7. Yes, these systems require computing power, skilled teams, and substantial budgets.
8. Yes, long-term memory retains exchanges beyond a single session.
9. Yes, vector stores can store knowledge and instruction scenarios.
10. No, contextual awareness is fundamental to detect the key points of an LLM request.

Chapter 2

1. False. A ChatGPT-like AI agent is a highly integrated generative AI system with several components.
2. False. Although online platforms have seamless frameworks, the systems are built with intelligent controllers.
3. True. The fundamental rule in GenAISys is: "No humans, no GenAISys."

4. True. In general, the prompt of a GenAISys will contain a precise task tag such as "summarize this text."
5. True. In some cases, an AI controller orchestrator can determine the most probable task to perform based on the user input.
6. False. GPT-4o can natively perform semantic text similarity tasks.
7. True. Each project is different and requires different types of conversations that we must design and code.
8. False. Sometimes, we need to have access to the full history of a conversation with a copilot to make a decision.
9. False. GPT-4o can run in several languages.
10. False. The illusion of sentient AI agents or self-consciousness comes from the fact that the user dialogues are seamless.

Chapter 3

1. False. Generative AI systems requires teams of AI specialists, ML experts, and many more experts. Also, computing and, therefore automating tasks, comes with a cost.
2. False. We must continually monitor the cost-benefits of the features we are building into GenAISys.
3. True. When a vector search is performed, the retrieval system can target specific areas of data.
4. False. Complex data may require more dimensions, but for simpler data, smaller dimensions are sufficient.
5. True. Upserting is the process of uploading vectorized data into a Pinecone database.
6. False. A namespace is only a subset of records in a Pinecone database.
7. True. A namespace can be used to query specific parts of a dataset.
8. True. To perform vector similarity searches, we must embed the input into vectors also.
9. True. A metric such as cosine similarity will enable the system to find a vector with data similar to the user input. However, other metrics are possible depending on the need, such as Euclidean distances.
10. True. A comprehensive GenAISys requires an AI controller, an AI conversational agent, and more.

Chapter 4

1. True. The complexity of generative AI must not be a constraint for the users. The goal is to help them in their workplace.
2. False. Each project requires specific interfaces that can be built with web frameworks or classical software interfaces or integrated seamlessly into existing software.
3. True and False. The GenAISys needs to provide knowledge that a generative AI model such as GPT-4o cannot know, such as specific company data. However, in other cases, a model such as GPT-4o can provide sufficient information and perform tasks quite well.
4. False. Your imagination is the limit! Each project may require custom features that are not available in standard ChatGPT-like copilots.
5. True. We can store embedded instruction scenarios in Pinecone, retrieve them, and augment the input to a generative AI model with those instructions.
6. False. The namespace can be used to distinguish instruction scenarios from data.
7. True. A generative AI model such as GPT-4o learns general information but not specific memories within an organization. Fine-tuning personal memories in a model can be costly and time-consuming. Storing them in Pinecone and managing them there can be effective.
8. True and False. True if the users only want to use the AI conversational agent with a vector store. False in a multi-user conversation in which the AI conversational agent is only a participant like all the other members of a team.
9. True and False. True because we could add a querying functionality to the GenAISys interface. False because we can implement response RAG triggers that the conversational agent will manage.
10. True. The best way to help users is to provide intuitive interfaces.

Chapter 5

1. False. A seamless GenAISys shows that the interface is seamless, but it does not prove that it did not take much work to get the job done.
2. False. Building a GenAISys requires a great amount of work.
3. True and False. It is possible to build an interface for an AI application that is not event-driven, but it might lack flexibility.
4. True. An AI system can mimic human reasoning as in a chain-of-thought process.
5. False. A classical sequence of functions cannot match the creativity of a generative AI chain of tasks in which each task reacts to the output of the previous task.

6. False. A CoT can be multimodal.
7. True. A CoT process can include AI functions with intermediate calls to non-AI functions.
8. True. This is one of the productive features of a reasoning GenAISys.
9. True. The challenges of a growing and accelerating economy require automation. AI is an effective way to automate problem-solving.
10. True. Using AI to boost the productivity of a team can alleviate difficult tasks and leave room for more decision-making time and creativity.

Chapter 6

1. True. Emotional memory creates a bond between a person and a promotional message.
2. True. OpenAI's o3 can reason and perform chain-of-thought tasks.
3. False. Humans can remember emotions many years after an event and as far back as childhood.
4. False. A generative AI model such as o3 can understand complex prompts and memory structures.
5. True. Generative AI models can now process numerical values to some extent, as well as natural language.
6. True. A Pinecone index can contain vectorized instructions that can be retrieved with a similarity query.
7. True. The early generative AI models could only process relatively simple prompts. Now, they can understand complex steps of tasks and perform them well.
8. True. A simple user input can trigger a complex thread-of-reasoning scenario.
9. True. A reasoning model analyzes a prompt and has the ability to go through several steps to perform a complex set of tasks to process reviews.
10. False. A generative AI system can interpret complex prompts and perform the set of tasks requested.

Chapter 7

1. False. DeepSeek was trained as other LLMs' but with efficient techniques.
2. True.
3. True.
4. False. DeepSeek-R1 was the teacher, and the Llama model was the student.

5. True. It was a complex cycle. R1 was derived from V3 to learn reasoning, and V3 learned reasoning from R1.
6. True. A handler registry can contain as many handlers that point to AI functions as we need. The handlers all being in the same format makes the GenAISys highly scalable and expandable.
7. True. A handler selection mechanism can remain unchanged if we design the handlers in a unique format. We can enhance the mechanism if necessary, but the core process remains unchanged.
8. True and False. True because these models were trained on massive amounts of data and contexts. False because in some cases, RAG will be necessary, for example. At one point, fine-tuning may also be necessary, although the scope of generative AI models is continually growing.
9. True. A well-designed GenAISys that fits the specifications of a project will be able to expand with user requests and as the AI market evolves.

Chapter 8

1. False. A trajectory can be any sequence of events.
2. True. Solid, well-designed synthetic data can save many resources and accelerate GenAISys development.
3. False. Generative AI can now solve mathematical problems and time sequences, and many perform a growing number of tasks beyond LLMs.
4. False. The whole purpose of expanding GenAISys is for everybody to be able to use this technology.
5. True. We can use generative AI to design effective, complex prompts.
6. False. Trajectory predictions for missing data can be applied to a wide range of domains, including fire disasters.
7. True. We see GenAISys and agentic innovations released continually.

Chapter 9

1. True. A vector search can find specific words and expressions that are otherwise challenging.
2. False. The GenAISys will not process inappropriate content if detected.
3. False. The OpenAI moderation's API provides a wide range of categories.

4. False. The GenAISys will flag the content and reject the request.
5. True. Real-time services that adapt to ever-changing conditions will give a company a competitive edge.
6. True. Generative AI systems are evolving at full speed and the trend will be multi-user, cross-domain, multimodal, and more as AI is increasingly integrating sensors.
7. False. WhatsApp, Zoom, and Teams, for example, are proven examples.
8. True. By automating many tasks but also providing interactive AI features, teams that use a GenAISys will surpass companies that do not have the technology.
9. True. A customer panel is the best way to get real-time feedback. Naturally, the decision is up to the team developing the GenAISys. But the recommendation stands: listen to the market represented by customer panels if possible.

Chapter 10

1. False. Presenting a GenAISys requires careful preparation, or we will be unprepared for the tough questions.
2. True and False. A flexible frontend interface that can be rapidly modified can be useful for a complex project's presentation. However, if the GenAISys is completely finalized, this step may not be necessary. It's a strategic presentation choice.
3. False. A simple reminder for experts should suffice. However, non-AI specialists might want an educational explanation.
4. False. Although Pinecone is an excellent vector store, the best vector store is the one that best fits the requirements of a given project.
5. False. Many other models can be found on Hugging Face and on various platforms. The best model is the one that fits your project's requirements.
6. False. Whether or not to choose open-source generative AI models depends on the goals of the project we are working on.
7. True and False. We might have to implement our own security and privacy components if it is necessary for a project. We might develop everything or add functions to existing solutions. However, many customers might ask us to rely on existing platform frameworks such as AWS, Microsoft Azure, and Google Cloud.

8. True and False. You might want to control the versions of your product with a development plan and not customize it for each request. However, some projects will require customization, or GenAISys might not cover the scope.
9. True and False. Face-to-face meetings enable more human interactions. However, if the teams are spread out in different locations, remote meetings can do the job.
10. False. In practically every project, human and machine resources are a constraint we must address from the start to establish our credibility.

8. [illegible] You might want to control [illegible] not customize it [illegible] it will require [illegible] documentation [illegible]

9. [illegible] in different locations [illegible]

10. [illegible] any project, [illegible] we [illegible] to establish our [illegible]

packtpub.com

Subscribe to our online digital library for full access to over 7,000 books and videos, as well as industry leading tools to help you plan your personal development and advance your career. For more information, please visit our website.

Why subscribe?

- Spend less time learning and more time coding with practical eBooks and Videos from over 4,000 industry professionals
- Improve your learning with Skill Plans built especially for you
- Get a free eBook or video every month
- Fully searchable for easy access to vital information
- Copy and paste, print, and bookmark content

At www.packt.com, you can also read a collection of free technical articles, sign up for a range of free newsletters, and receive exclusive discounts and offers on Packt books and eBooks.

Other Books You May Enjoy

If you enjoyed this book, you may be interested in these other books by Packt:

In color
Building AI Agents with LLMs, RAG, and Knowledge Graphs
A practical guide to autonomous and modern AI agents
Salvatore Raieli | Gabriele Iuculano
‹packt›

Building AI Agents with LLMs, RAG, and Knowledge Graphs

Salvatore Raieli, Gabriele Iuculano

ISBN: 978-1-83508-038-2

- Learn how LLMs work, their structure, uses, and limits, and design RAG pipelines to link them to external data
- Build and query knowledge graphs for structured context and factual grounding
- Develop AI agents that plan, reason, and use tools to complete tasks
- Integrate LLMs with external APIs and databases to incorporate live data
- Apply techniques to minimize hallucinations and ensure accurate outputs
- Orchestrate multiple agents to solve complex, multi-step problems
- Optimize prompts, memory, and context handling for long-running tasks
- Deploy and monitor AI agents in production environments

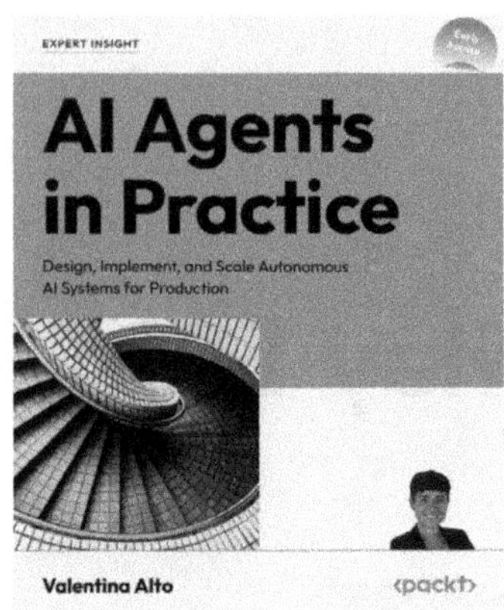

AI Agents in Practice

Valentina Alto

ISBN: 978-1-80580-134-4

- Master core agent components like LLMs, memory systems, tool integration, and context management
- Build production-ready AI agents using frameworks like LangChain with code
- Create effective multi-agent systems using orchestration patterns for problem-solving
- Implement industry-specific agents for e-commerce, customer support, and more
- Design robust memory architectures for agents with short and long-term recall
- Apply responsible AI practices with monitoring, guardrails, and human oversight
- Optimize AI agent performance and cost for production environments

Packt is searching for authors like you

If you're interested in becoming an author for Packt, please visit `authors.packtpub.com` and apply today. We have worked with thousands of developers and tech professionals, just like you, to help them share their insight with the global tech community. You can make a general application, apply for a specific hot topic that we are recruiting an author for, or submit your own idea.

Share your thoughts

Now you've finished *Building Business-Ready Generative AI Systems*, we'd love to hear your thoughts! Scan the QR code below to go straight to the Amazon review page for this book and share your feedback or leave a review on the site that you purchased it from.

`https://packt.link/r/1837020698`

Your review is important to us and the tech community and will help us make sure we're delivering excellent quality content.

Index

D

E

F

G

H

I

R

S

T

Join our Discord and Reddit space

You're not the only one navigating fragmented tools, constant updates, and unclear best practices. Join a growing community of professionals exchanging insights that don't make it into documentation.

Stay informed with updates, discussions, and behind-the-scenes insights from our authors. Join our Discord at `https://packt.link/z8ivB` or scan the QR code below: 	Connect with peers, share ideas, and discuss real-world GenAI challenges. Follow us on Reddit at `https://packt.link/0rExL` or scan the QR code below: 

www.ingramcontent.com/pod-product-compliance
Lightning Source LLC
Chambersburg PA
CBHW081037220326
41598CB00038B/6900

* 9 7 8 1 8 3 7 0 2 0 6 9 0 *